Closing the Achievement Gap from an International Perspective

Julia V. Clark
Editor

Closing the Achievement Gap from an International Perspective

Transforming STEM for Effective Education

 Springer

Editor
Julia V. Clark
Chevy Chase
Maryland
USA

ISBN 978-94-007-4356-4   ISBN 978-94-007-4357-1 (eBook)
DOI 10.1007/978-94-007-4357-1
Springer Dordrecht Heidelberg London New York

Library of Congress Control Number: 2013951533

Printed on acid-free paper

Springer is part of Springer Science+Business Media (www.springer.com)

To the memory of my mother and father,
Bessie M. Clark and Frank Clark, who
taught me the value of a strong educational
background and who by example taught
me the power of principle; to my sister,
Mary C. Lewis (retired school teacher)
and nephew Kelvin C. Lewis for their
encouragement and support throughout
my professional career; and to all teachers
who believe that all children are capable of
learning and succeeding in school.

Preface

In the very interesting essays that are collected here, one learns that the gap in academic achievement that has caused so much consternation among educators in the United States is not unique to North America. From these descriptions of exemplary efforts at making more efficient the teaching and learning of science, technology, engineering and mathematics we see, let alone what can and should be done pedagogically, we are reminded that the underproductivity of teaching and learning with some populations of students is an international problem that is bigger than education alone can solve.

In his brilliantly conceived book, *Paths Toward a Clearing,* Professor Michael Jackson problematizes the task of the anthropologists who seek to discern meaning from their observations of the lived experience of the other. Jackson interrogates the insufficiently addressed issue, "with what degree of certainty can I interpret the meaning to the other person of the experiences she, herself, has lived."

The late distinguished anthropologist, Ogbu (1978), avoided this problem in much of his work by focusing not on interpreting meaning to the other, but on the correlates and apparent consequences of the life lived by the other. In Ogbu's insightful studies of the lives of people in different parts of the world who experienced low caste and caste-like status, he concluded that the status of the experiencing person influenced the degree to which that person became socialized to the standards and values of the hegemonic culture. Low-status, marginalized and underresourced people all over the world tend to fair poorly in the educational institutions to which they are exposed. To the extent that the inferior status is perceived as immutable, as in low-caste or low-caste-like status, this association between status and achievement tends to be even more prominent. Thus, when one's assigned status is identifiable by physiognomic characteristics, such as skin color (even in the absence of caste), the identification as low-class functions as caste. Thus, Ogbu describes Blacks in the United States as a caste-like group, and he describes academic achievement and life-outcome characteristics as consistent with the expectations associated with that caste-like status. Ogbu goes on to report similar relationships in caste or caste-like groups in Australia, India, Japan and the United States.

In the collection of essays assembled here at The Achievement Gap in International Perspective, we see Ogbu's perspective reflected again. Concern with the achievement gap may have first come to recent public notice in the United States in the contexts of Black/White comparisons of academic achievement. Clark (1954), Coleman (1966), Miller (1995), and National Task Force on Minority High Achievement (1999). However, the essayists included in this collection remind us that disparities in academic achievement are worldwide phenomena begging for a solution. Guidance toward a possible solution may be found in an interrogation of Jackson's and Ogbu's perspectives on this issue.

Much of the work underway on this problem in the United States and around the world is directed at helping underdeveloped persons to achieve what I have written about and labeled as intellective competence. I have defined such competence as the universal currency of technologically developed societies. I generally refer to such attributes as the capacity to access and utilize information that is grounded in such disciplines as the humanities, mathematics and the sciences. The use of such information is reflected in the capacity to create and manage engineering and technology projects. In addition, I refer to the intentional command and control of one's affective, cognitive and situative mental capacities as essential in intellective competence. This is the complex of human competencies that we recognize to be the by-products of one's having studied and mastered the so called STEM disciplines. But these indicators of intellective competence are largely identified with the cultures of the social orders that have also used these competencies to subordinate their low-class or low-caste members. To close the achievement gap, these subordinated populations must embrace the standards, skills and values of the very people who have used these factors in the service of their own advancement and the suppression of the low-status members.

Following Jackson, we cannot be certain that we are accurately interpreting the meaning of the lived experiences associated with ones having been forced to learn the ways of the dominating other. But Ogbu's work with Black boys in Compton, California suggests that the low-status persons are very much aware of the absence of a sense of polity for themselves. As a result, according to Ogbu, they constrain their investment of effort in the mastery of the hegemonic cultural forms, information, techniques and values. Should this way of conceptualizing the problem prove to be correct, our domestic and international efforts at improving our strategies for improving the teaching and learning of STEM and other essential academic subjects may be limited until we also find a way to bring a greater sense of polity into the lives of low-class and low-caste members of the societies of the world.

In the collection of essays that follow, we see examples of exceptional pedagogical efforts directed at the more effective teaching and learning of STEM subjects, but our elevation of this work makes the task seem too easy. Solid curriculum, good and creative teaching, and engaged learners are important, but they may not provide an adequate solution to a problem that involves more than schooling. Even with excellent teaching and well-resourced schools, it may be necessary for attention to be given to the personal attributions that are assigned to the lived experiences of the persons who must do the learning. Jackson reminds us that such interpretations are one of the challenges to the anthropologist. I claim that understanding the lived experience of the learner, and appropriately adapting our teaching to it, is one of the continuing challenges to those of us who teach. Yet we know that the personal attributions assigned to the actual experiences of one's life are only a part of the dynamic. Those actual lived experiences do influence the attributions that are assigned by the persons living the experience. Thus pedagogical efforts at closing the achievement gap in the United States and in other countries around the world must be viewed as more than problems of the goodness of the teaching and learning of STEM and other subject matter. To close the academic achievement gap in societies where people live lives of inegalitarian conditions and statuses, education may need to be thought of more comprehensively and thought of as inclusive of both the conditions of life for the learners and the meanings that they assign to their conditions of life. Education may have to begin with serious attention being given to the improvement of the quality of life for the learners. Obviously, education is not synonymous with schooling. These essays remind us that this admonition is true worldwide.

Edmund W. Gordon is the John M. Musser Professor of Psychology, Emeritus at Yale University and the Richard March Hoe Professor of Psychology and Education, Emeritus at Teachers College, Columbia University

References

Clark, K. B. (1954). *Segregated Schools in New York City.* Paper presented at the Child Apart, Northside Center for Child Development, New Lincoln School NY.

Coleman, J. S., Campbell, E. Q. Hobson, C. J., McPartland, J., Mood, A. M., Weinfeld, F. D., et al. (1966). *Equality of educational opportunity.* Washington, DC: U.S. Government Printing Office.

Jackson, M. (1989). *Paths toward a clearing: Radical empiricism and ethnographic inquiry.* Bloomington: Indiana University Press.

Miller, L. S. (1995). *An American imperative: Accelerating minority educational advancement* New Haven: Yale University Press.

National Task Force on Minority High Achievement. (1999). Reaching the top: A Report of the National Task Force on Minority High Achievement. New York: College Board.

Ogbu, J. (1978). *Minority education and caste: The American system in cross-cultural perspective.* New York: Academic Press.

Acknowledgements

I have often felt that much of my current and future anticipated success can be attributed to a broad educational foundation, a plethora of professional and community experiences, networking opportunities, and a commitment to teaching and learning. Many people have contributed very extensively to my learning over the years. Without family, writing would be intolerable. First and foremost, I express my heartfelt thanks to my beloved deceased parents for always giving me their blessings. They have been my inspiration and motivation for continuing to improve my knowledge and move my career forward. They have been my rock, and I dedicate this book to them. For individual and collective contribution to this effort, encouragement, and moral support, I wish to acknowledge with special thanks my sister, Mary C. Lewis (a retired teacher), and my nephew Kelvin C. Lewis.

I wish to thank several colleagues, teachers and friends for directly or indirectly influencing my thinking on the writing of this book and for their encouragement and interest in my professional growth, especially Drs. Willie Pearson, Marvin Druger, and Madelyn "Faye" Neathery. Unreserved thanks to Faye for her many years of friendship and for her valuable advice and support.

Special thanks to Professor Edmund W. Gordon, my mentor and friend, whose scholarly work and interest in the quality education of all children inspired and motivated me to excel to my fullest potential. His professional input in the writing of this book is appreciated immensely.

I also dedicate this book to the teachers, administrators, and policy-makers committed to the improvement of education. I thank them for their willingness to challenge mainstream viewpoints. These people are involved in changing systems to ensure universal education in the future.

My thanks to Drs. Gavin W. Fulmer, Joseph Krajcik, and Lorenzo Taylor for assisting me in selecting some of the contributors to this book. In the end, I believe that this team of authors has provided the perfect blend of knowledge and skills that went into this book.

Thanks to Leonard S. Rosenbaum for his superb editing of the book. He not only noticed the mistakes but also suggested those additions that can seem obvious in hindsight but that would never occur to you.

I express my sincere gratitude to my colleagues affiliated with the National Association for Research in Science Teaching (NARST), American Association for the Advancement of Science (AAAS), Sigma Xi: The Scientific Research Society, and the Kellogg National Fellowship Program (KNFP) for their inspiration and encouragement in carrying out this project.

Most importantly, I would like to thank the authors of this book for devoting their time and effort. They have helped to bring a new dimension to our knowledge of what is happening on an international scale. Many thanks to Drs. Edmund W. Gordon, Linda Darling Hammond, Motoko Akiba, Guodong Lang, Larry D. Yore, Leslie Francis Pelton, Brian W. Neill, Tim W. Pelton, John Anderson, Todd M. Milford, Armando Sanchez Martinez, Marcos A. Rangel, Ricardo A. Madeiria, Geoff Whitty, Jake Anders, Mustafa Sami Topcu, Gaoming Zhang, Yong Zhao, Sonya N. Martin, Seung-Um Choe, Chan-Jong Kim, Youngsun Kwak, Jason TAN, Nick Taylor, Johan Muller, and Debra Panizzon. Working with all of you made this a magnificent experience. The support received from all of you was vital to the success of the project. I appreciate that you believed in me to provide the leadership and knowledge to make this book a reality. You have helped bring a new dimension to our knowledge of what is happening on an international scale. As researchers, you were interested in addressing achievement gap issues as they related to school effectiveness from an international perspective involving academies, practitioners and policy makers that play major roles in reform efforts in different parts of the world. Your chapters will influence global changes in the world. Through this book and my travel, I will continue to spread your message of support to others wishing to improve the teaching and learning of *all* students.

Julia V. Clark

Contents

Part I   Chronicling Educational Challenges and Development

1   Introduction 3
Julia V. Clark

2   Closing the Achievement Gap: A Systemic View 7
Linda Darling-Hammond

3   Teacher Qualification and the Achievement Gap:
A Cross-National Analysis of 50 Countries 21
Motoko Akiba and Guodong Liang

Part II   North America

4   Addressing the Achievement Gap in the United States 43
Julia V. Clark

5   Closing the Science, Mathematics, and Reading Gaps from
a Canadian Perspective: Implications for Stem Mainstream
and Pipeline Literacy 73
Larry D. Yore, Leslee Francis Pelton, Brian W. Neill,
Tim W. Pelton, John O. Anderson and Todd M. Milford

6   Achievement Gap in Mexico: Present and Outlook 105
Armando Sanchez-Martinez

Part III   South America

7   Racial Achievement Gaps in Another America: Discussing
Schooling Outcomes and Affirmative Action in Brazil 127
Ricardo A. Madeira and Marcos A. Rangel

Part IV   Europe

**8   Narrowing the Achievement Gap: Policy and Practice in
England, 1997–2010** 163
Geoff Whitty and Jake Anders

**9   The Achievement Gap in Science and Mathematics:
A Turkish Perspective** 193
Mustafa Sami Topcu

Part V   Asia

10 Achievement Gap in China 217
Gaoming Zhang and Yong Zhao

**11 Employing a Sociohistorical Perspective for Understanding
the Impact of Ideology and Policy on Educational
Achievement in the Republic of Korea** 229
Sonya N. Martin, Seung-Urn Choe, Chan-Jong Kim
and Youngsun Kwak

12 Closing the Achievement Gap in Singapore 251
Jason TAN

Part VI   Africa

**13 Equity Deferred: South African Schooling
Two Decades into Democracy** 265
Nick Taylor and Johan Muller

Part VII   Australia

**14 Securing STEM Pathways for Australian High School
Students from Low-SES Localities: Science and
Mathematics Academy at Flinders (SMAF)** 285
Debra Panizzon, Martin Westwell and Katrina Elliott

**15 The Road to Excellence: Promoting Access and Equity to
Close the Achievement Gap Internationally** 307
Julia V. Clark

The Contributors 317

Contributors

Motoko Akiba Department of Educational Leadership and Policy Studies, Florida State University, Tallahassee, FL, USA

Jake Anders Institute of Education, University of London, London, UK

John O. Anderson University of Victoria, Victoria, BC Canada

Seung-Urn Choe Seoul National University, Seoul, Republic of Korea

Julia V. Clark Chevy Chase, MD, USA

Linda Darling-Hammond Stanford University, Stanford, CA, USA

Katrina Elliott Department of Education and Child Development, Adelaide, SA, Australia

Chan-Jong Kim Seoul National University, Seoul, Republic of Korea

Youngsun Kwak Korean Institute for Curriculum and Evaluation, Seoul, Republic of Korea

Guodong Liang Community Training and Assistance Center (CTAC), Boston, MA, USA

Ricardo A. Madeira University of Sao Paulo (USP), Sao Paulo, Brazil

Sonya N. Martin Seoul National University, Seoul, Republic of Korea

Todd M. Milford Griffith University, Mt. Gravatt, QLD Australia

Johan Muller School of Education, University of Cape Town, Cape Town, South Africa

Brian W. Neill University of Victoria, Victoria, BC Canada

Debra Panizzon Faculty of Education, Monash University, Clayton, Victoria, Australia

Leslee Francis Pelton University of Victoria, Victoria, BC Canada

Tim W. Pelton University of Victoria, Victoria, BC Canada

Marcos A. Rangel University of Sao Paulo (USP), Sao Paulo, Brazil

NORC/University of Chicago, Chicago, IL, USA

Abdul Latif Jameel Poverty Action Lab (MIT), Cambridge, MA, USA

Armando Sanchez-Martinez Mexican Education Consultant, Editorial Manager, Santillana, Mexico, North America

Jason TAN Policy and Leadership Studies, National Institute of Education, Singapore, Singapore

Nick Taylor National Education Evaluation and Development Unit, Ministry of Basic Education, Pretoria, South Africa

Mustafa Sami Topcu Department of Elementary Science Education, Muğla Sıtkı Koçman University, Mugla, Turkey

Martin Westwell Flinders Centre for Science Education in the 21st Century, Flinders University, Bedford Park, Adelaide, Australia

Geoff Whitty Institute of Education, University of London, London, UK

Larry D. Yore University of Victoria, Victoria, BC Canada

Gaoming Zhang Department of Teacher Education, School of Education, University of Indianapolis, Indianapolis, IN, USA

Yong Zhao College of Education, University of Oregon, Eugene, OR, USA

About the Editor

Julia V. Clark   Julia V. Clark is Program Director in the Division of Research on Learning in Formal and Informal Settings (DRL) in the Directorate for Education and Human Resources at the National Science Foundation (NSF). She spent four years on a detail assignment as a Legislative Fellow to Congress in the House of Representatives and Senate. She served as Principal Advisor for Science and Technology Issues to Congressional Members of the Science Committee (Rep. Bob Etheridge [D-NC]) and the Education and Workforce Committee (Rep. Rush Holt [D-NJ]) in the House of Representatives. She also served as an advisor on education and science-related issues and policy to the Committee on Health, Education, Labor and Pensions in the Office of Senator Edward Kennedy (D-MA).

She received a Bachelor of Science in Natural Science from Fort Valley State University, Masters in Science Education (with concentration in the Biological Sciences) from the University of Georgia, and Doctorate in Science Education from Rutgers University. She has completed postgraduate studies in Environmental Science at Yale University, Radiation Biology at the University of California-Berkeley, and Chemistry and Biology at Emory University.

She has a distinguished record as a scholar, educator, and administrator. Prior to working at National Science Foundation (NSF) in 1990, she served as a tenured Associate Professor of Science Education at Texas A&M University, Assistant Professor of Science and Mathematics Education at Howard University, Associate Professor of Biology and Science Education at Clark Atlanta University, and Assistant Professor of Biology and Botany at Albany State University and Morris Brown College. She has also served as a Visiting Professor of Education at Memorial University in Newfoundland Canada and Visiting Professor of Education at the University of Maryland-College Park and George Mason University. At George Mason University, she taught education courses to Islamic teachers. She was a high school teacher of physics, chemistry, biology, and physical science for eight years, mostly in Atlanta, Georgia, prior to becoming a college professor.

Throughout her career, Dr. Clark has published and conducted research in cognitive development, science curriculum, epidemiology, and women and minorities in science and leadership development. She wrote a book chapter on women leadership and a book entitled *Redirecting Science Education: Reform for a Culturally*

Diverse Classroom. She has traveled extensively internationally as a speaker and as a workshop and seminar leader.

Her honors and awards include the 2012 NSF Director's Award for Distinguished Service; 2006 Distinguished Alumni Lifetime Achievement Award (University of Georgia); American Association for the Advancement of Science (AAAS) Fellow in Science Education, W. K. Kellogg National Fellow, Outstanding Achievement in Government Award in Science and Technology; Exceptional Women in Science and Technology Award, Lily Foundation Award, Distinguished Alumni Award (Fort Valley State University) and Outstanding Young Women of America Award.

She has served on several national boards and committees: Chair, AAAS Education Committee (2003–2006); Advisory Committee for Student Science Enrichment Program of the Burroughs Wellcome Fund; Education and Diversity Committees, Sigma Xi: The Scientific Research Society; Editorial Board, *Journal of College Science Teaching*; Advisory Board, Education and Instructional Technology, Public Service Satellite Consortium; National Technical Advisory Board, National Urban Coalition; Women Equity Action League; and Board of Directors, Consortium of Southern Colleges for Teacher Education. Memberships in professional organizations include Sigma Xi, AAAS, National Association for Research in Science Teaching (NARST), AAUW; Phi Delta Kappa, and STEM Leading Ladies.

Part I
Chronicling Educational Challenges and Development

Chapter 1
Introduction

Julia V. Clark

*The views expressed in this book are the views of the editor and
not those of the National Science Foundation.*

One of the most troubling problems facing education in schools today is the *achieve-ment gap*—the observed disparity in a number of educational measures in academic performance between different groups of students, especially groups defined by race/ethnicity, gender, and socioeconomic status. In the USA, *achievement gap* is typically used to describe the disparity in test scores between minorities, usually between Blacks and Hispanics and their White and Asian peers. Similar gaps exist more broadly between high-poverty students and their more wealthy counterparts. At each grade level, racial disparities in an array of achievement variables demonstrate a wide gap in performance, especially in mathematics and science, particularly among disadvantaged minorities from urban and rural communities. These disparities start as early as kindergarten, persist across grades, and in most cases widen over time.

Although standardized tests are the standard measurements used in the USA, a variety of measures, including standardized test scores, grade point average, drop-out rates, and college-enrollment and college-completion rates, are used in other countries.

What causes the achievement gap? The factors are numerous, but some of the strongest include poverty, early-childhood learning, teacher quality, and strength of the curriculum. In the USA, many initiatives have been implemented to narrow the gap. Although some progress has been made, a wide achievement gap remains. There are a few success stories in some of the states, and these will be highlighted in the chapter on the discussion of the achievement gap in America.

Data from the National Assessment of Educational Progress (NAEP, the Nation's Report Card) indicate that Blacks and Hispanics made strides in closing the gap until the mid-1980s, at which point these gains began to level off. For example, in the

J. V. Clark (✉)
5600 Wisconsin Avenue, Suite 1205, Chevy Chase, MD 20815, USA
e-mail: jvclark@starpower.net

J. V. Clark (ed.), *Closing the Achievement Gap from an International Perspective*,
DOI 10.1007/978-94-007-4357-1_1, © Springer Science+Business Media B.V. 2014

2009 NAEP results, the gap between Black and Hispanic fourth-grade students and their White counterparts in mathematics was more than 20 points. In eighth-grade mathematics, the gap was more than 26 points.

The achievement gap in education is not unique to the USA. A wide achievement gap among various groups of students is common in many countries and has become a focal point of education reform efforts in these countries. Like the USA, many countries are faced with the failure of their schools to adequately prepare all students.

Across the globe, education is essential, bettering the lives of individuals and nations from poverty to affluence. Yet educational opportunity and the upward mobility it can bring have not always been equally available to everyone. In rich and poor nations alike, the disadvantaged—defined by gender and geography, race and religion, and class and caste—fall behind, losing the chance to improve their lives and depriving society of the contributions they might have made.

As in some states in the USA, many countries have developed and implemented unique education models to meet the demands of their students. Almost every education system has been involved in restructuring. School administrators, teachers, students, and parents have found themselves responding to structural, administrative, and curriculum changes that governments claim will improve the quality of education, and many of these changes have been documented and discussed. Some school districts have shown that all students—regardless of race, ethnicity, income, and background—can achieve at high levels when provided with the appropriate opportunities. This book highlights these success stories. Around the world, education is the path to progress, for both nations and individual citizens. Ensuring the equity of educational opportunity and the improved life chances that education can bring are important for narrowing the achievement gap.

Nothing is more important in education than ensuring that all children have the tools and opportunities they need to succeed. The USA and other countries around the world face problems of an increasingly global nature that sometimes require collaborative efforts that would be advantageous to all. Closing the achievement gap is one of those problems.

This book was written in response to the growing concern for the improvement of quality education provided for all students. This book can be broadly characterized as providing global change research that offers a wide array of benefits to the nation, in terms of closing the educational achievement gap with particular interest in science, technology, engineering, and mathematics (STEM).

To assist in helping to close the achievement gap, a group of researchers from selected countries from around the world that have had or are having similar problems are sharing how their countries are tackling the problems. These countries were selected because of their uniqueness and the work they are doing in their educational school system to change a practice that will help poor, low-income students and students of color to succeed. The researchers contributing to the publishing of this book provide information on the achievement gap in the following countries: the USA, Brazil, Canada, China, England, Korea, Mexico, Singapore, South Africa, Turkey, and Australia.

For a number of reasons, looking at these countries may assist those interested in improving student achievement. First, the school systems are diverse. The countries bring a broad array of international experience to the subject. Each country offers us something to learn. Many countries, especially Asian countries, have developed and implemented unique models to meet the demands of today's learners. For example, in Singapore, the education system is flexible and caters to every child's abilities, interests, and aptitudes so as to help each develop to their fullest potential. It focuses on the development of human resources to meet Singapore's need for an educated and skilled workforce. It also facilitates the inclusion of social moral values in the curriculum to serve as cultural ballast in the face of rapid progress and change.

China illustrates the difficulties brought about by size. Further, China has a huge span of levels of wealth. Information about this country can help us gain a better understanding of how the dimensions of size and wealth can influence the ability of a country to educate its people. Furthermore, China, like many countries in the Asia-Pacific region, seems to outperform those that we usually hear about. Singapore, Korea, and Australia perform at levels similar to the European countries that are noted for their high quality of education. In places such as the USA and England, Asian students generally outperform those from other ethnic backgrounds. We might discover reasons for this high level of performance that would have a message for countries that do not perform as well.

Reforms are underway in South Africa to deal with the achievement gap between advantaged and disadvantaged students. Poor communities, in particular those of rural Africans, bear the brunt of the country's past inequalities.

Although the achievement gap in Canada is smaller than that in the USA, it is far from trivial. Socioeconomic status in high-poverty communities is an issue that Canadians have been dealing with in closing the gap. Schools in Canada have raised their test scores and graduation rates by providing more resources.

In the USA, different schools have different effects on similar students. Children of color, especially Black and Hispanic students, tend to be concentrated in low-achieving, highly segregated schools. These minority students are more likely to come from low-income households, meaning that minority students are more likely to attend poorly funded schools based on the districting patterns within the school system. Schools in lower-income districts tend to employ less-qualified teachers and tend to have fewer educational resources. Research shows that teacher effectiveness is the most important in-school factor affecting student learning.

Because every culture is different, the contours of the problem vary from place to place; what counts as failure in one country may look enviable in another. Everywhere, however, eliminating educational gaps is a complicated endeavor that demands concerted effort from politicians, bureaucrats, teachers, university administrators, and policymakers.

Education systems around the world have recognized the need for schools to change the way in which they function. Many systems are moving from a quality education system for a few students to a quality education system for most students. The challenge now is to move from having a quality education system for most students to having a quality education system for all students.

Closing the achievement gap between low- and high-achieving students is an important goal of US education. As the USA and other countries work to build their capacity in STEM education, they will need to interact with each other in order to enhance their efforts in international scientific engagement and capacity building to provide quality education to all of their students.

This book will link all of the countries together in solving a global problem in society. Closing the achievement gap is a global problem. This book is about understanding the factors that will promote progress and the factors that contributed to the progress in closing the achievement gap that has occurred in some countries. This book will establish some of the commonalities and differences that exist between countries. By sharing these stories, hopefully we will gain a better understanding of what might have contributed to the progress and will probe the reasons why progress was halted in the hope of finding clues and directions for moving forward in narrowing the achievement gap.

The time is right to address this global problem. Decades of research by the various countries and others reinforce the need to improve and validate the needs of our educational systems. It is time to take definitive action to begin closing the achievement gap between different groups of children. This book will help to do just that, by providing essential information and guidance produced by decades of research.

Through collaboration with other countries, we are sharing and promoting the best practices for closing the achievement gap. We looked at how students' data correlated with classroom practices, teacher instruction, and academic programming, as part of our efforts toward measuring student growth. Qualitative and quantitative data have been produced to provide evidence not only for the problem but also for the solution.

Sharing the experiences from countries with diverse backgrounds helps us to solve the problem of a global society. The stories that researchers share from their countries will help us identify and learn from other places that are involved in providing all young children with the education they need and deserve. Each country is an important player in the orchestra of education.

Chapter 2
Closing the Achievement Gap: A Systemic View

Linda Darling-Hammond

"Closing the achievement gap" has become an American mantra over the last decade, as federal and state policies have sought to reduce unequal educational outcomes largely by setting targets and sanctions based on student test scores. And while some progress has been made since 1990, gaps in achievement between affluent and low-income students in the USA have remained large and persistent, while a number of other countries around the world have made stunning strides over the last 30 years in both raising overall achievement and reducing differentials across students and schools, including those from low-income communities and historical minority groups.

What did these nations do? In *The Flat World and Education* (Darling-Hammond 2010), the practices of many nations that have become high achieving and substantially equitable in their education outcomes are reviewed. Among their commonalitiesare a number of societal and educational factors, including:

- Secure housing, food, and health care, so that children can come to school ready to learn;
- Supportive early learning environments;
- Equitably funded schools that provide equitable access to high-quality teaching;
- Well-prepared and well-supported teachers;
- Standards, curriculum, and assessments focused on twenty-first-century learning goals; and
- Schools organized for in-depth student and teacher learning.

Efforts both outside and inside of the educational system have been key to their success: Outside of school, they have created a functional social safety net and a set of early learning supports for children which allow them to come to school ready to learn. Within the educational domain, they have created a *teaching and learning system* that enables a coherent approach to providing high-quality education in an equitable way. Such a system not only prepares all teachers and school leaders

L. Darling-Hammond (✉)
Stanford University, Stanford, CA, USA
e-mail: ldh@stanford.edu

J. V. Clark (ed.), *Closing the Achievement Gap from an International Perspective*,
DOI 10.1007/978-94-007-4357-1_2, © Springer Science+Business Media B.V. 2014

well for the challenging work they are asked to do, but it ensures that schools are organized to support both student and teacher learning and that the standards, curriculum, and assessments that guide their work encourage the kind of knowledge and abilities needed in the twenty-first century.

This chapter reviews the sources of the achievement gap in the USA and then discusses the policies and practices of three nations that have made particularly noteworthy strides toward high and more equitable achievement over the last 30 years and that now top the international rankings on assessments like the Program for International Student Assessment (PISA) : Finland, Singapore, and South Korea. This chapter also draws on other international data to describe how some of these practices appear in other jurisdictions around the world. In the course of this discussion, the chapter emphasizes that what occurs inside education systems is reinforced or undermined by the contexts within which they operate and that the challenges for the USA are to pursue equity both within schools and within the society as a whole.

The Achievement Gap in the USA

US policymakers have been trumpeting the need for educational reform for nearly three decades, during which there has been no shortage of handwringing or high-blown rhetoric. In 1983, *A Nation at Risk* decried a "rising tide of mediocrity" in education and called for sweeping reforms. In 1989, then-President George H. W. Bush and the 50 governors announced a set of national goals that included ranking first in the world in mathematics and science by the year 2000. No Child Left Behind set targets and created sanctions for schools to drive achievement and to close the gaping gaps in performance between groups of students.

However, by 2006, on the PISA, a test conducted by the Organisation for Economic Co-operation and Development (OECD), the USA ranked 21st of 30 OECD countries in science and 25th of 30 in mathematics—a drop in both raw scores and rankings from 3 years earlier (OECD 2007). When non-OECD members from Eastern Europe and Asia are added to the list, the US rankings dropped to 29th out of 40 developed countries in science, sandwiched between Latvia and Lithuania, and 35th out of 40 in mathematics, between Azerbaijan and Croatia. Although the USA made small gains over the next 3 years, ranking 14th in reading, 20th in science, and 27th in mathematics in 2009 (OECD 2010), it still remained far from those heady aspirations of two decades earlier.

The hidden story about US achievement rankings are the large disparities that are a function of growing inequality—specifically the very different performance of high- and low-income children, Whites and Asians in comparison to African Americans and Latinos, and those in low-poverty schools vs. those in high-poverty schools. In fact, Whites and Asians in the USA score above the OECD average in mathematics, reading, science, and problem solving on the PISA (OECD 2007), and US students in low-poverty schools actually score at the very top of the internation-

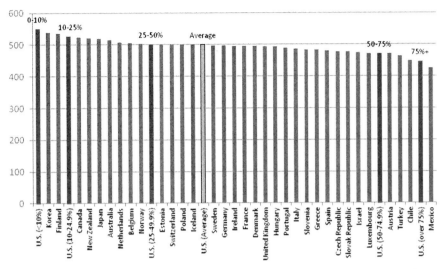

Fig. 2.1 Scores on PISA reading assessments, 2009, by country and poverty rates in U.S. schools

al rankings in reading, while those in schools of concentrated poverty are near the bottom (see Fig. 2.1). Similar patterns are also found in mathematics and science, although overall achievement in mathematics is lower in the USA—a function of teacher recruitment, training, and curriculum policies which will be addressed later.

Five factors create the major building blocks of unequal and inadequate educational outcomes in the USA:

- The high level of poverty and the low levels of social supports for low-income children's health and welfare, including their early learning opportunities;
- The unequal allocation of school resources, which is made politically easier by the increasing re-segregation of schools;
- Inadequate systems for providing high-quality teachers and teaching to all children in all communities;
- Rationing of high-quality curriculum through tracking and inter-school disparities; and
- Factory model school designs that have created dysfunctional learning environments for students and unsupportive settings for strong teaching.

Poverty and Unequal Resources

The root of inequity in educational outcomes in the USA is growing poverty and re-segregation. US childhood poverty rates have grown by more than 60% since the 1970s and are now by far the highest among OECD nations, reaching 22% in the last published measures and rising since then due to the economic recession

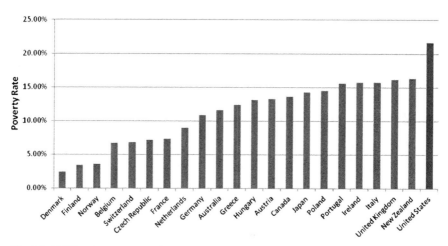

Fig. 2.2 Childhood poverty rates in PISA countries (before government transfers)

(UNICEF 2007) (see Fig. 2.2). The disparity is even greater when poverty rates are calculated after government transfers that support housing, health care, food, child-care assistance, and other essentials: These transfers bring most OECD nations' childhood poverty rates down to well under 10% but, because our safety net for families is so tattered, the recalculation hardly changes the US rate (Bell et al. 2008). American children living in poverty have a much weaker safety net than their peers in other industrialized countries, where universal health care, housing subsidies, and high-quality universally available childcare are the norm.

In addition to the direct effects of poverty on children's home resources, low-income children are much less likely to have access to early learning opportunities in the USA than their more affluent peers. As a result, an estimated 30–40% of children enter kindergarten without the social and emotional skills and language experiences needed to be initially successful in school (Zigler et al. 2006, p. 23). Studies have found that the size of the working vocabulary of 4-year-old children from low-income families is approximately one-third that of children from middle-income families, which makes it much more difficult for them to read with comprehension or to engage in academic learning relying on that vocabulary, even when they can decode text. By first grade, only half as many first graders from poor families are proficient in understanding words in context and engaging in basic mathematics as first graders from nonpoor families (Denton and West 2002).

Although there is significant evidence that high-quality preschool programs improve achievement and attainment, with estimated returns of about $ 4–$ 10 for every dollar invested (Reynolds and Temple 2006, p. 50), only a few states have committed to high-quality universally available preschool for all students. Thus, the achievement gap that is already present at the start of kindergarten has not been addressed in most communities.

Beyond the large and growing inequalities that exist among families and communities, profound inequalities in resource allocations to schools have been reinforced by the increasing re-segregation of schools over the decades of the 1980s and 1990s. During that 20-year span, desegregation policies and funding assistance were largely abandoned by the federal government and the courts, and state governments generally followed suit (Rumberger and Palardy 2005). As a consequence, the gains in desegregation made in the 1960s and 1970s were substantially rolled back. By 2000, 72% of the nation's black students attended predominantly minority schools, up significantly from the low point of 63% in 1980. The proportion of students of color in intensely segregated schools also increased. Nearly 40% of African American and Latino students attend schools with a minority enrollment of 90–100% (NCES 2001).

These intensely segregated schools serving concentrations of children in poverty are also located in districts that less well resourced than those serving more advantaged students. Recent analyses of data prepared for school equity cases in more than 20 states have found that on every tangible measure—from qualified teachers and reasonable class sizes, to adequate textbooks, computers, facilities, and curriculum offerings—schools serving large numbers of students of color have significantly fewer resources than schools serving more affluent, White students (Darling-Hammond 2010). Many such schools are so severely overcrowded that they run a multitrack schedule with a shortened school day and school year, lack basic textbooks and materials, do not offer the courses students would need to be eligible for college, and are staffed by a parade of untrained, inexperienced, and temporary teachers (Oakes 2004).

These inequities are in part a function of how public education is funded in the USA. In most cases, education costs are supported primarily by local property taxes, along with state grants-in-aid that are somewhat equalizing, but typically not sufficient to close the gaps caused by differences in local property values. In most states, the wealthiest districts spend at least three times what the poorest districts can spend per pupil, differentials that translate into dramatically different salaries for educators, as well as different learning conditions for students (Adamson and Darling-Hammond 2011). Furthermore, the wealthiest states spend about three times what the poorer states spend (Baker et al. 2010; Darling-Hammond 2010). Therefore, the advantages available to children in the wealthiest communities of high-spending and high-achieving states like Massachusetts, Connecticut, Vermont, and New Jersey are dramatically different from the schooling experiences of those in the poorest communities of low-spending states like California, Mississippi, Alabama, and Louisiana, where buildings are often crumbling, classes are overcrowded, instructional materials are often absent, and staff are often transient.

Although many US educators and civil rights advocates have fought for higher quality and more equitable education over many years—in battles for desegregation, school finance reform, and equitable treatment of students within schools—progress has been in many states over the last two decades as segregation has worsened and disparities have grown. While students in the highest-achieving states and districts in the USA do as well as their peers in high-achieving nations, our continu-ing comfort with profound inequality is the Achilles' heel of American education.

Unequal Distribution of Curriculum and Teachers

These inequalities translate into disparities in the number and quality of teachers and other educators available to students, and to unequal access to high-quality curriculum.

In a case brought to challenge school desegregation efforts in Jefferson County, Kentucky, and Seattle, WA, more than 550 scholars signed onto a social science report filed as an amicus brief, which summarized an extensive body of research showing the persisting inequalities of segregated minority schools. The scholars concluded that:

> …(M)ore often than not, segregated minority schools offer profoundly unequal educational opportunities. This inequality is manifested in many ways, including fewer qualified, expe-rienced teachers, greater instability caused by rapid turnover of faculty, fewer educational resources, and limited exposure to peers who can positively influence academic learning. No doubt as a result of these disparities, measures of educational outcomes, such as scores on standardized achievement tests and high school graduation rates, are lower in schools with high percentages of non-White students (American Educational Research Association 2006).

As segregation and school funding disparities grew worse throughout the 1980s and 1990s, the practice of lowering or waiving credentialing standards to fill classrooms in high-minority, low-income schools—a practice that is unheard of in high-achiev-ing nations and in other professions—became commonplace in many US states, especially those with large minority and immigrant populations, like California, Texas, Florida, and New York.

In many states where school funding litigation has been brought, plaintiffs have documented the fact that teachers in high-need schools have, on average, lower levels of experience and education, are less likely to be credentialed for the field they teach, and have lower scores on both certification tests and other measures of academic achievement. Furthermore, a growing body of research has shown that these kinds of qualifications matter for student achievement. Studies at the state, district, school, and individual student level have found that teachers' academic background, preparation for teaching, certification status, and experience signifi-cantly affect their students' learning gains (Betts et al. 2000; Boyd et al. 2006; Clot-felter et al. 2007; Darling-Hammond 2000; Darling-Hammond et al. 2005; Fergu-son 1991; Fetler 1999; Goe 2002; Goldhaber and Brewer 2000; Monk 1994).

In combination, teachers' qualifications can have substantial effects. For exam-ple, a large-scale study of high-school student achievement in North Carolina found that students' achievement growth was significantly higher if they were taught by a teacher who was certified in his or her teaching field, fully prepared upon en-try (rather than entering through the state's alternative "lateral entry" route), had higher scores on the teacher licensing test, graduated from a competitive college, had taught for more than 2 years, or was National Board Certified (Clotfelter et al. 2007). Taken individually, each of these qualifications was associated with greater teacher effectiveness. Moreover, the researchers found that the combined influence on achievement growth of having a teacher with most of these qualifications as compared to one with few of them was larger than the effects of race and parent education combined, or the average difference in achievement between a typical White student with college-educated parents and a typical black student with high-school educated parents. While achievement from 1 year to the next is still largely dependent on prior achievement, this finding suggests that the achievement gap might be reduced over time if minority students were more routinely assigned highly qualified teachers, rather than the poorly qualified teachers they most often encounter.

These findings appear to extend around the world. Akiba and Scriber (2007), for example, found that the most significant predictors of mathematics achievement across 46 nations included teacher's certification, a college major in mathematics or mathematics education, and at least 3 years of teaching experience. These same variables—reflecting what teachers have learned about content and how to teach it to a range of learners—show up in study after study as predictors of teachers' effectiveness. This study also found that although the national level of teacher quality in the USA is similar to the international average, the opportunity gap in students' access to qualified teachers between students of high and low socioeconomic status (SES) is among the largest in the world.

These disparities, which have come to appear inevitable in the USA, are *not* the norm in developed nations around the world, which typically fund their education systems centrally and equally, with additional resources often going to the schools where students' needs are greater. These more equitable investments made by high-achieving nations are also steadier and more focused on critical elements of the system: the quality of teachers and teaching, the development of curriculum and assessments that encourage ambitious learning by both students and teachers, and the design of schools as learning organizations that support continuous reflection and improvement. With the exception of a few states with enlightened long-term leadership, the USA, by contrast, has failed to maintain focused investments on these essential elements.

Learning from Others

One wonders what we might accomplish as a nation if we could finally set aside what appears to be our de facto commitment to inequality, so profoundly at odds with our rhetoric of equity, and put the millions of dollars spent continually arguing and litigating into building a high-quality education system for all children. To imagine how that might be done, one can look at nations that started with very little and purposefully built highly productive and equitable systems, sometimes almost from scratch, in the space of only two to three decades.

Consider three very different nations—Finland, Singapore, and South Korea—
that built strong education systems, nearly from the ground up. None of these nations succeeded educationally in the 1970s, when the USA was the unquestioned
education leader in the world. All created productive *teaching and learning systems*
by expanding access while investing purposefully in ambitious educational goals
using strategic approaches to build teaching capacity.

Equitable Access to High-Quality Schools and Teaching

In this chapter, the term "teaching and learning system" is used advisedly to describe a set of elements that, when well designed and connected, reliably support all
students in their learning. These elements ensure that students routinely encounter
well-prepared teachers who work in concert around thoughtful, high-quality curriculum, supported by appropriate materials and assessments. These elements also
help students, teachers, leaders, and the system as a whole continue to learn and
improve. While none of these countries lack problems and challenges, each has created a much more consistently high-quality education system for all of its students
than has the USA. While no system from afar can be transported wholesale into
another context, there is much to learn from the experiences of those who have addressed problems we encounter. A sage person once noted that, although it is useful
to learn from one's own mistakes and experiences, it is even wiser to learn from
those of others.

Although Finland, Singapore, and South Korea are very different from one another culturally and historically, all three have made startling improvements in their
education systems over the last 30 years. Their investments have catapulted them to
the top of international rankings in student achievement and attainment, graduating
more than 90 % of their young people from high school and sending large majorities through college, far more than in the much wealthier USA. Their strategies also
have much in common which are as follows:

- All three nations fund schools adequately and equitably, and add incentives for
  teaching in high-need schools. All three nations have built their education systems on a strong egalitarian ethos, explicitly confronting and addressing potential
  sources of inequality. In South Korea, for example, a wide range of incentives
  is available to induce teachers to serve in rural areas or in urban schools with
  disadvantaged students. In addition to earning bonus points toward promotion,
  incentives for equitable distribution of teachers include smaller class sizes, less
  in-class teaching time, additional stipends, and opportunities to choose later teaching appointments (Kang and Hong 2008). The end result is a highly qualified,
  experienced, and stable teaching force in all schools, providing a foundation for
  strong student learning.
- All three nations organize teaching around national standards and a core curriculum that focus on higher-order thinking, inquiry, and problem solving through

rigorous academic content. Working from lean national curriculum guides that have recommended assessment criteria, teachers collaborate to develop curriculum units and lessons at the school level, and develop school-based

performance assessments—which include research projects, science investigations, and technology applications—to evaluate student learning. In Singapore, these are increasingly part of the examination system. In Finland, the assessments are classroom based, but are guided by the national curriculum, which emphasizes students' abilities to reflect on, evaluate, and manage their own learning.

Unlike in the USA, narrowing the curriculum has not been an issue. Take South Korea: it devotes the large majority of instructional time in every grade to a liberal arts curriculum that includes social studies, science, physical education, music, fine arts, moral education, foreign language (English), practical arts, and a range of extracurricular activities and electives (Huh 2007). Curriculum offerings are similarly comprehensive in Singapore and Finland.

- All three nations eliminated examination systems that had once tracked students into different elementary and middle schools and restricted access to high school. Since adopting national curriculum guidelines, these nations have been committed to helping all students master the same essential skills and content until the beginning of high school—not to devising watered-down versions for some students.
- All three nations use assessments that require in-depth knowledge of content and higher-order skills. All three countries have matriculation exams for admission to college. These are the only external examinations in Finland and South Korea. In Singapore, examinations are given in the sixth and ninth grades as well as at the end of high school. These exams have open-ended questions that require in-depth content knowledge, critical analysis, and writing. Although the matriculation exams are not used to determine high-school graduation, they are taken by nearly all students and they set a high bar for high-school coursework.

In Finland, where there are no external standardized tests used to rank students or schools, most teacher feedback to students is in a narrative form, emphasizing descriptions of their learning progress and areas for growth (Sahlberg 2009). Like the National Assessment of Educational Progress in the USA, Finland uses a centrally developed assessment given to samples of students at the end of the second and ninth grades to inform curriculum and school investments. The focus of these open-ended assessments is to provide information to support learning and problem solving, not to allocate sanctions and punishments.

- All three nations invest in strong teacher education programs that recruit top students, completely subsidize their extensive training programs, and pay them a stipend while they learn to teach. In all three nations, teacher education programs were overhauled to increase teachers' pedagogical knowledge and skills, on top of a deep mastery of the content areas they will teach. Finnish teachers' preparation includes at least a full year of clinical experience in a model school associated with a university. Within these model schools, student teachers participate in problem-solving groups, a common feature in Finnish schools. All teachers are trained in research methods so that they can "contribute to an increase of the problemsolving capacity of the education system" (Buchberger and Buchberger 2004).

Their problem-solving groups engage in a cycle of planning, action, and reflection and evaluation that is reinforced throughout teacher education and is a model for what teachers will plan for their own students, who are expected to engage in similar kinds of research and inquiry in their own studies.

- All three nations pay salaries that are equitable across schools and competitive with other careers, generally comparable to those of engineers. Teachers are viewed as professionally prepared and are well respected. Working conditions are supportive, including substantial participation in decision making about curriculum, instruction, assessment, and professional development.
- All three nations support ongoing teacher learning by ensuring mentoring for beginning teachers and providing 1,525 h a week for all teachers to plan collaboratively and engage in analyses of student learning, lesson study, action research, and observations of one another's classrooms, which help them continually improve their practice. All three nations have incentives for teachers to engage in research on practice, and all three fund ongoing professional development opportunities in collaboration with universities and other schools.
- All three nations pursue consistent, long-term reforms by setting goals for expanding, equalizing, and improving the education system and by steadily implementing these goals, making thoughtful investments in a high-quality educator workforce and in school curriculum and teaching resources that build the underpinnings for success. This has been made possible in part by the fact that these systems are managed by professional ministries of education, which are substantially buffered from shifting political winds. Frequent evaluations of schools and the system as a whole have helped guide reforms. In each nation, persistence and commitment to core values have paid off handsomely, as all three are ranked in the very top tier of countries on international assessments and have among the most equitable outcomes in the world.

All three nations have undertaken these elements in a systemic fashion, rather than pouring energy into a potpourri of innovations and then changing course every few years, as has often been the case in many communities in the USA, especially in large cities. While these three small nations—each comparable in size to a midsize US state—have conducted this work from a national level, similar strategies have been successfully employed

at the state or provincial level in high-scoring Australia, Canada, and New Zealand, and regions such as Hong Kong and Macao in China. They demonstrate how it is possible to build a *system* in which students are routinely taught by well-prepared teachers who are given time to collaboratively reflect on and refine the curriculum, supported by appropriate materials and assessments that foster learning for students, teachers, and schools alike.

Equitable Access to a Strong Curriculum

In the USA, enormous energy is devoted to discussions of the achievement gap. Much less attention, however, is paid to the opportunity gap—the accumulated differences in access to key educational resources that support learning at home and at school. These key resources include high-quality curriculum, good educational materials, expert teachers, personalized attention, and plentiful information resources.

In contrast, nations around the world are transforming their school systems to eliminate opportunity gaps; they are expanding educational access to more and more of their people, and they are revising curriculum, instruction, and assessment to meet the demands of the knowledge economy. Today, there is very little curriculum differentiation until high school in the education offerings for students in high-achieving jurisdictions, such as Finland, Hong Kong, Singapore, and South Korea, which have sought, as part of their reforms, to equalize access to a common, intellectually ambitious curriculum (Korea Institute of Curriculum & Evaluation no date). In the last 2 years of high school, there is often differentiation of program and courses by interest, aptitude, and aspirations, but all courses of study offer high-quality options for later education and careers. By comparison, countries like France and Germany that have continued their tradition of sorting students much earlier are, like the USA, lagging in international assessments.

This is not surprising, as a substantial body of research over the last 40 years has found that (1) the combination of teacher and curriculum quality explains most of a school's contribution to achievement and (2) access to a rich curriculum is a more powerful determinant of achievement than initial achievement levels. That is, when students of similar backgrounds and initial achievement levels are exposed to more or less challenging curriculum material, those given the richer curriculum ultimately outperform those given the less challenging curriculum (Gamoran and Berends 1987; Oakes 2005).

These efforts to reduce tracking have been supported by social policies that reduce childhood poverty and allow students to start school on a level playing field, and that give their teachers much better training and much more noninstructional time to plan and collaborate. In addition, over time, as all children are exposed to similar high-quality lessons, the variance in their knowledge and skills decreases. Ensuring access to a more common curriculum supports greater equity and ultimately makes teaching all students easier.

Finland provides an excellent example. Although there was a sizable achievement gap among students in the 1970s, strongly correlated to socioeconomic status, this gap has been progressively reduced as a result of curriculum reforms starting in the 1980s—and it has continued to grow smaller and smaller in the 2000, 2003, and 2006 PISA assessments. By 2006, Finland's between-school variance on the PISA science scale was only 5%, whereas the average between-school variance in other OECD nations was about 33% (Sahlberg 2007). This small variability is true even in schools in Helsinki that receive large numbers of previously less well-educated immigrants from Africa and the Middle East. (Large between-school variation is generally related to social inequality, including both the differences in achievement across neighborhoods differentiated by wealthand the extent to which schools are funded and organized to reduce or expand inequalities.)

Today's expectation that schools will enable all students, rather than a small minority, to learn challenging skills to high levels creates an entirely new mission for schools. Instead of merely "covering the curriculum" or "getting through the book," this new mission requires that schools substantially enrich the intellectual opportunities they offer while meeting the diverse needs of students. This demands not only more skillful teaching but also a coherent curriculum that engages students in learning essential concepts in ways that develop strong thinking skills.

It is imperative that America closes the achievement gap among its children by addressing the yawning opportunity gap. Given the critical importance of education for individual and societal success in the flat world we now inhabit, inequality in the provision of education is an antiquated tradition the USA can no longer afford. If "no child left behind" is to be anything more than empty rhetoric, we will need a policy strategy that creates a rich and challenging curriculum for all students and supports it with thoughtful assessments, access to knowledgeable, well-supported teachers, and equal access to school resources.

Smart, equitable investments are not only the right thing to do, they will, in the long run, save far more than they cost. The savings will include more than $ 200 billion we now lose in wages, taxes, and social costs annually due to dropouts; the $ 50 billion we pay for lost wages and for incarceration tied to illiteracy and school failure; and the many tens of billions wasted each year on reforms that fail, fads that don't stick, unnecessary teacher turnover, avoidable special education placements, remedial education, grade retention, summer school, lost productivity, and jobs that move overseas (Levin 2007; Western et al. 2003).

The path to our mutual well-being is built on educational opportunity. Central to our collective future is the recognition that our capacity to survive and thrive ultimately depends on ensuring for all of our people what should be an unquestioned entitlement—an inalienable right to learn.

References

Adamson, F., & Darling-Hammond, L. (2011). *Speaking of salaries: What It will take to get qualified, effective teachers in all communities.* Washington, DC: Center for American Progress.

Akiba, G. L., & Scriber, J. (2007). Teacher quality, opportunity gap, and national achievement in 46 countries. *Educational Researcher, 36,* 369–387.

American Educational Research Association. (2006). Brief amicus curiae filed in parents involved in community schools v. seattle school district no. 1. 127 S. Ct. 2738 (2007).

Baker, B., Sciarra, D., & Farrie, D. (2010). *Is school funding fair? A national report card.* Newark, NJ: Education Law Center.

Bell, K., Bernstein, J., & Greenberg, M. (2008). *Lessons for the United States from other advanced economies in tackling child poverty. In big ideas for children: Investing in our nation's future* (pp. 81–92). Washington, DC: First Focus.

Betts, J. R., Rueben, K. S., Danenberg, A. (2000). Equal resources, equal outcomes? The distribution of school resources and student achievement in California. San Francisco: Public Policy Institute of California.

Boyd, D., Grossman, P., Lankford, H., Loeb, S., & Wyckoff, J. (2006). How changes in entry requirements alter the teacher workforce and affect student achievement. *Education Finance & Policy, 1*(2), 176–216.

Buchberger, F., & Buchberger, I. (2004). Problem solving capacity of a teacher education system as a condition of success? An analysis of the "finnish case,". In F. Buchberger & S. Berghammer (Eds.), *Education policy analysis in a comparative perspective* (pp. 222–237). Linz: Trauner.

Clotfelter, C. T., Ladd, H. F., & Vigdor, J. L. (2007). *How and why do teacher credentials matter for student achievement? (NBER Working Paper 12828)*. Cambridge, MA: National Bureau of Economic Research.

Darling-Hammond, L. (2000, January). Teacher quality and student achievement: A review of state policy evidence. *Educational Policy Analysis Archives, 8*(1). http://epaa.asu.edu/epaa/v8n1.

Darling-Hammond, L. (2010). *The flat world and education: How America's commitment to equity will determine our future.* NY: Teachers College Press.

Darling-Hammond, L., Holtzman, D., Gatlin, S. J., & Heilig, J. V. (2005). Does teacher preparation matter? Evidence about teacher certification, Teach for America, and teacher effectiveness. *Education Policy Analysis Archives, 13*(42). http://epaa.asu.edu/v13n42/.

Denton, K., & West, J. (2002). Children's reading and mathematics achievement in Kindergarten and first grade. (NCES 2002–125). *Education Statistics Quarterly, 44*(1). Retrieved on 3/14/09 at http://nces.ed.gov/programs/quarterly/Vol_4/4_1/q3-1.asp.

Ferguson, R. F. (1991, Summer). Paying for public education: New evidence on how and why money matters. *Harvard Journal on Legislation, 28*(2), 465–498.

Fetler, M. (1999). High school staff characteristics and mathematics test results. *Education Policy Analysis Archives, 7*(March 24). http://epaa.asu.edu

Gamoran, A., & Berends, M. (1987). The effects of stratification in secondary schools: Synthesis of survey and ethnographic research. *Review of Educational Research, 57*(4), 415–436.

Goe, L. (2002). Legislating equity: The distribution of emergency permit teachers in California. *Educational Policy Analysis Archives, 10*(42). http://epaa.asu.edu/epaa/v10n42/.

Goldhaber, D. D., & Brewer, D. J. (2000). Does teacher certification matter? High school certification status and student achievement. *Educational Evaluation and Policy Analysis, 22,* 129–145.

Huh, K. (2007). Understanding Korean education: Volume 1, school curriculum in Korea Seoul: Korean Educational Development Institute.

Kang, N., & Hong, M. (2008). Achieving excellence in teacher workforce and equity in learning opportunities in South Korea. *Educational Researcher, 37*(4), 203.

Korea Institute of Curriculum & Evaluation (no date). National Curriculum and Evaluation. Retrieved November 16th, 2006, from http://www.kice.re.kr/kice/eng/info/info.2.jsp.

Levin, H. M. (2007). *The Costs and benefits of an excellent education.* New York, NY: Teachers College, Columbia University.

Monk, D. H. (1994). Subject matter preparation of secondary mathematics and science teachers and student achievement. *Economics of Education Review, 13*(2), 125–145.

NCES (2001). Common core of data, 2000–01.

Oakes, J. (2004). Investigating the claims in Williams v. State of California: An unconstitutional denial of education's basic tools? *Teachers College Record, 106*(10), 1889–1906.

Oakes, J. (2005). *Keeping track: How schools structure inequality.* New Haven, CT: Yale University Press.

Organisation for Economic Co-operation and Development (OECD). (2007). *PISA 2006: Science competencies for tomorrow's world.* Paris: OECD.

Organisation for Economic Co-operation and Development (OECD). (2010). PISA 2009 results: What students know and can do—Student performance in reading, mathematics and science (Vol. 1). http://dx.doi.org/10.1787/9789264091450-en

Reynolds, A., & Temple, J. (2006). Economic returns of investments in preschool education. In E. Zigler, W. S. Gilliam & S. M. Jones, S. M. (Eds.), *A vision for universal preschool education* (pp. 37–68). New York: Cambridge University Press.

Rumberger, R. W., & Palardy, G. J. (2005). Does resegregation matter? The impact of social composition on academic achievement in Southern high schools. In J. C. Boger & G. Orfield (Eds.), *School resegregation: Must the south turn back?* (pp. 127–147). Chapel Hill: The University of North Carolina Press.

Sahlberg, P. (2007). Education policies for raising student learning: The finnish approach. *Journal of Education Policy, 22*(2), 147–171.

Sahlberg, P. (2009). Educational change in Finland. In A. Hargreaves, A. Lieberman, M. Fullan, & D. Hopkins (Eds.), *Second international handbook of educational change* (pp. 323–348). Dordrecht, Netherlands: Springer.

UNICEF. (2007). *Child well-being in perspective: An overview of child well-being in rich countries. Innocenti report card 7.* Florence: UNICEF Innocenti Research Centre.

Western, B., Schiraldi, V., & Ziedenberg, J. (2003). *Education and incarceration.* Washington, DC: Justice Policy Institute.

Zigler, E., Gilliam, W. S., & Jones, S. M. (2006). *A vision for universal preschool education.* New York: Cambridge University Press.

Chapter 3
Teacher Qualification and the Achievement Gap: A Cross-National Analysis of 50 Countries

Motoko Akiba and Guodong Liang

Introduction

Previous studies have shown a major gap between wealthy and high-poverty students' and between white and ethnic-minority students' access to qualified teachers. High-poverty students and ethnic minority students are twice as likely as are wealthy and white students to be assigned novice teachers (Ascher and Fruchter 2001; National Center for Education Statistics 2000; Peske and Haycock 2006). Further, they are more likely to be taught by uncertified teachers (Ascher and Fruchter 2001; Darling-Hammond 2004; Shen et al. 2004), out-of-field teachers (those without a major in the subject they teach; Ingersoll 1999, 2002; Jerald and Ingersoll 2002; Akiba and LeTendre 2009), or teachers with low American College Testing (ACT) or Scholastic Assessment Test (SAT) scores (Shen et al. 2004). Teachers in high-poverty and ethnically diverse schools are also more likely to leave schools or leave the teaching profession altogether, creating a major instability in students' opportunity to learn (Ingersoll 2002). Such inequality, however, is not a problem unique to the USA. Many countries around the world are struggling with how to equalize students' access to qualified teachers (Akiba et al. 2007; UNESCO Institute for Statistics 2006).

How does the major gap in students' access to qualified teachers affect the achievement gap between students? To examine whether the level of achievement gap in a country is driven by the educational system that allows unequal distributions of qualified teachers, it is necessary to conduct a cross-national analysis using data from a large number of countries. The 2007 Trends in International Mathematics and Science Study (TIMSS) data set is the most comprehensive and recent data

M. Akiba (✉)
Department of Educational Leadership and Policy Studies,
Florida State University, Tallahassee, FL, USA
e-mail: makiba@fsu.edu

G. Liang
Community Training and Assistance Center (CTAC), Boston, MA, USA
e-mail: gliang@ctacusa.com

J. V. Clark (ed.), *Closing the Achievement Gap from an International Perspective*,
DOI 10.1007/978-94-007-4357-1_3, © Springer Science+Business Media B.V. 2014

set that includes survey data from students, teachers, and principals. This data set allows us to link students' poverty level with their mathematics teachers' qualifications in order to examine the gap in students' access to qualified teachers in 50 countries.[1]

In this cross-national study, we focus on the measurable characteristics of qualifications of eighth-grade mathematics teachers that share a relatively common meaning across various cultural contexts: (a) certification, (b) mathematics major, (c) mathematics education major, and (d) teaching experience of 3 or more years. These characteristics also align with the requirements for teacher quality in the No Child Left Behind (NCLB) Act; thus, an examination of these teacher-qualification indicators in international contexts will greatly inform US policy makers.

This study is guided by the following research questions:

1. How does the percentage of eighth graders taught by qualified mathematics teachers in the USA differ from that in other countries?
2. How does the size of the gap between high-socioeconomic status (high-SES) and low-SES students in their access to qualified teachers in the USA differ from that in other countries?
3. How are the level of students' access to qualified teachers and the gap in their access to qualified teachers associated with national mathematics achievement and the SES-based achievement gap?

Background

The NCLB Act of 2001 defined *highly qualified teachers* as those who are fully certified, possess a bachelor's degree, and have demonstrated competence in subject knowledge and teaching and required that all teachers be highly qualified by the 2005–2006 academic year. Birman et al. (2009) reported that the percentage of highly qualified teachers increased from 74% in the 2004–2005 academic year to 84% in the 2006–2007 academic year. However, they also reported that the teachers who are not highly qualified are more likely to be teaching in high-poverty schools than in low-poverty schools (5 vs. 1%), in ethnically diverse schools than in white-dominant schools (4 vs. 1%), and in schools with improvement status (as a result of failing to meet Adequate Year Progress targets) than in schools without such status (6 vs. 2%).

Despite the gap in students' access to qualified teachers, a document that supplements the 2011 Quality Counts Report showed that only a small number of states implement a state policy for attracting teachers to high-poverty schools (13 states) or low-performing schools (13 states) (Editorial Projects in Education 2011). Many empirical studies have reported that students achieve better when they are taught by certified teachers, teachers with subject majors, and teachers with at least 3 years of

---

[1] Although Taiwan and Hong Kong are not independent countries, they are considered so in this chapter.

teaching experience (Darling-Hammond and Youngs 2002; Rice 2003; Wayne and Youngs 2003; Wilson et al. 2001, 2002). It is likely that the lack of policy focus on narrowing the gap in students' access to qualified teachers is contributing to long-lasting achievement gap in the USA.

Ensuring students' access to qualified teachers is an important goal of educational policy and reform in other countries. Policy makers in many countries are struggling with the same problems as US policy makers, such as a lack of highly qualified teachers, especially in science- and math-related subjects; low social status and salary of and poor working conditions for teachers; a lack of systemic induction programs; and inequitable distribution of qualified teachers between high-poverty and low-poverty schools (OCED 2005). The United Nations Educational, Scientific and Cultural Organization (UNESCO) also reported a severe teacher shortage in sub-Saharan African countries, the Arab states, and South Asian countries (UNESCO Institute for Statistics 2006).

Several studies have also identified variation in students' access to qualified teachers in other countries. The UNESCO Institute for Statistics (2006) examined the gap in teacher quality among isolated/rural areas, small towns, and large cities in 13 southern and eastern African countries, including South Africa, Botswana, Kenya, and Uganda. A higher percentage of students in isolated/rural areas were taught by teachers with less than 3 years of experience than were students in small towns or large cities. In addition, in Namibia, Tanzania, and Uganda, teachers in isolated/rural schools scored lower when they took a sixth-grade mathematics test than did teachers in large city schools, showing the gap in teachers' mathematics content knowledge (UNESCO Institute for Statistics 2006).

Cross-national comparative studies of teacher quality and policies and contexts influencing teacher quality revealed that the USA differs from other countries in many conditions for promoting teacher quality. A comparative study of teacher qualification using the 2003 TIMSS data showed that, whereas teachers' qualification level in the USA is about the international average, the opportunity gap in students' access to be taught by qualified teachers was the fourth largest among the 39 countries (Akiba et al. 2007). A study conducted by the Educational Testing Service compared the USA with high-achieving countries—Australia, England, Hong Kong, Japan, Korea, the Netherlands, and Singapore—in eighth-grade mathematics and science teacher education and development policies and found that all the countries except the USA and Australia had centralized systems of teacher education and certification with tighter regulatory control by the central government (Wang et al. 2003). All the above countries had screening criteria at multiple time points—entry to the teacher education program, evaluation of field experience, exit from the teacher education program, or certification—whereas in the USA, teacher licensure testing was the only major high-stakes' criterion for determining who could become a teacher.

Teacher salary also influences the quality of teacher candidates. A comparative study of teacher salary level and national achievement in 30 countries showed that US investment in the salary of experienced teachers was lower than the international average, although new teachers in the USA were paid higher than the international average (Akiba et al. 2012). The study also found that the countries with higher average salary for experienced teachers are more likely to have higher national achievement, but the national average salary for new teachers was not significantly associated with national achievement. The low rate of growth in teacher salary in the USA may lead to a high attrition rate and instability in instructional quality and students' opportunity to learn.

A comparative study of teachers' work further revealed that US mathematics teachers are assigned to teach multiple subjects and multiple grade levels more often than are Japanese mathematics teachers, who usually teach only mathematics to only one grade level (LeTendre et al. 2001). US teachers also have a heavier instructional workload than Japanese or Australian teachers do, and they spend less time preparing for instruction (Akiba and LeTendre 2009).

Only a few cross-national comparative studies examined the gap in students' access to qualified teachers and policy and organizational contexts influencing such inequality. Akiba et al. (2007) reported a 14.4% gap (67.6 vs. 53.2%) in eighth-grade students' access to qualified mathematics teachers between high-SES students and low-SES students, compared with the international mean of 2.5% based on 39 countries. *Qualified teachers* were defined as those who are fully certified, majored in mathematics or mathematics education, and having 3 or more years of teaching experience. Akiba and LeTendre (2009) examined teacher hiring and distribution policies in Japan, Australia, and the USA and found that teacher rotation policy in Japan (in which teachers are reassigned to different schools every 4–5 years) and strong teacher incentive policy in Australia (which provides major financial incentives to those who work in remote rural schools with the greatest teacher shortage) contribute to a smaller gap in students' access to qualified teachers than in the USA.

Akiba et al. (2007) further examined the relationship between the size of the opportunity gap in students' access to qualified teachers and the achievement gap based on data from 39 countries, but the relationship was not statistically significant. They suggested that it might be due to other mediating factors in other countries, such as equal professional development opportunities and school resources, which may equalize instructional quality and ameliorate the impact of teacher-qualification gap on the achievement gap.

This study builds on the TIMSS 2003 findings by Akiba et al. (2007) and uses the 2007 TIMSS data set to examine how the level of students' access to qualified teachers and the gap in such an access between high-SES and low-SES students changed from 2003 to 2007. It is important to examine how students' opportunity to be taught by qualified teachers changed after the NCLB target year of 2005–2006 to achieve the goal of all teachers being highly qualified. The data from 50 countries allow us to see (1) where the USA stands with regard to students' access to qualified teachers and the gap in such access in comparison to 49 other countries and (2) how the USA's rank changed from 2003 to 2007. Furthermore, a cross-national analysis of the relationships (1) between students' access to qualified teachers and national achievement and (2) between the size of the gap in students' access to qualified teachers between high-SES students and low-SES students and the national achievement gap using data from a larger number of countries (39 in 2003 vs. 47 in 2007) allows us to reexamine the potential importance of teacher qualification in influencing student learning. By examining these relationships, we attempt to provide empirical findings to inform US federal and state policy making for improving teacher quality and equalizing students' access to qualified teachers.

Method

Data

The TIMSS was developed by the International Association for the Evaluation of Educational Achievement (IEA) to measure trends in students' mathematics and science achievement in more than 50 nations around the world. This study focused on data from eighth graders and their mathematics teachers. A two-stage stratified sampling method was used to sample secondary schools first and then eighth-grade classrooms from the sampled schools. The schools were first stratified by type of school, region of the country, type of location, and percentage of minority students. A probability-proportional-to-size technique was used in the process of selecting schools to give a higher probability of selection to larger schools (Olson et al. 2009). One or two mathematics classrooms were chosen randomly from each sampled school based on the list of eighth-grade classrooms. The mathematics teachers of these classrooms were selected, and they filled out a teacher questionnaire. This study analyzed the 2007 data collected from eighth graders and their mathematics teachers in 50 countries with at least one measure of teacher qualification. The sample sizes of eighth graders and eighth-grade teachers from which the national variables were developed ranged from 3,060 students in Morocco to 7,377 students in the USA and from 116 teachers in Malta to 463 teachers in Sweden.

Measures and Analysis

We measured the national level of students' access to qualified teachers (research question 1) by the percentages of students taught by: (a) teachers with certification; (b) teachers with a mathematics major; (c) teachers with a mathematics education major; (d) teachers with 3 or more years of teaching experience; and (e) teachers with certification, a mathematics or mathematics education major, and 3 or more years of teaching experience (overall measure of teacher qualification).

Teacher-qualification data came from teachers' "Yes" (1) or "No" (0) responses regarding whether or not teachers have (1) a full certification or license, (2) a major in mathematics, and (3) a major in mathematics education. For teaching experience, mathematics teachers were asked, "By the end of this school year, how many years will you have been teaching altogether?" and the teachers reported the number of years, which were recorded as: $0 = none$ to 2 years, $1 = 3$ or more years.

To measure the national-level gap in students' access to qualified teachers (research question 2), we developed five variables based on the difference between the percentage of high-SES students (standard deviation of 1 or higher) and the percentage of low-SES students (standard deviation of −1 or lower) who were taught by qualified teachers based on the five teacher-qualification variables listed above. The measure of the SES of students was created based on the education level of their parents, the existence of educational resources at home (calculator, computer, study desk or table, dictionary, and Internet connection), and the number of books at home. It was standardized around the mean in each nation.

For our last research question, we conducted multiple regression analysis to examine the relationships between (a) students' access to qualified teachers and national achievement and (b) the gap in students' access to qualified teachers and the achievement gap, controlling for educational expenditure as percentage of gross domestic product (GDP) and GDP per capita. For student achievement measures, we developed two national-level variables: (a) the national mean mathematics achievement of eighth graders and (b) the achievement gap measured by the difference in the mean mathematics score between high-SES students (standard deviation of 1 or higher) and low-SES students (standard deviation of −1 or lower). Educational expenditure as a percentage of GDP and GDP per capita was collected from the (UNESCO Institute for Statistics, n.d.). The data from 2007 were collected to match the TIMSS 2007 data. For the countries without 2007 data, the data from the closest year were used. The educational expenditure as a percentage of GDP varied from 2.1% in Qatar to 8.0% in Botswana, with a mean of 4.6% and a standard deviation of 1.3. The GDP per capita in US$ 1,000 ranged from 1.4 (US$ 1,400) in Ghana to 77.4 (US$ 77,400) in Qatar, with a mean of 21.1 and a standard deviation of 16.9. Due to the complex sample design in TIMSS, this study used the International Database Analyzer software (version 2.0), developed by the IEA Data Processing and Research Center, and used appropriate sampling weights and replicate weights for the Jackknife Repeated Replication method in all the data analyses.

Results

National Achievement and Achievement Gap in Eighth-Grade Mathematics

We first examined the levels of national achievement and achievement gap based on eighth-graders' mathematics scores in the TIMSS 2007. Figure 3.1 presents the national mean mathematics achievement of eighth graders in 50 countries, with the size of the achievement gap represented in the vertical lines attached to the

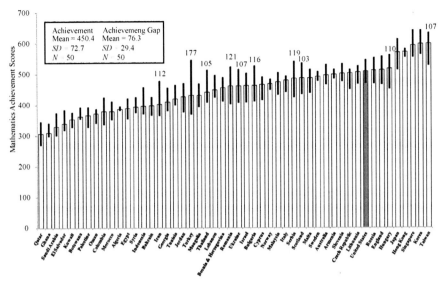

Fig. 3.1 Comparison of mathematics achievement scores and the achievement gap in 50 countries in 2007. (Note. The line attached to each bar represents the size of the achievement gap measured by the difference in mean achievement between students of high and low socioeconomic status. The ten countries with the largest achievement gaps have numbers above their bars showing the size of the gaps. Data are from the 2007 Trends in International Mathematics and Science Study [TIMSS] mathematics assessment).

bar graphs. The USA is highlighted in the graph, and 10 countries with the highest achievement gap are indicated with the size of the gap attached to the vertical lines. Among the 50 countries, the USA is ranked ninth in national achievement and 21st in the size of the achievement gap. National achievement scores varied from 307 in Qatar to 598 in Taiwan. Eighth graders in USA scored 509 on average, higher than the international average of 450. Although the US eighth graders' achievement was higher than the international average, the achievement gap between high-SES and low-SES students was similar to the international average (78 vs. 76). Algeria showed the smallest achievement gap (14), and Turkey showed the largest (177).

We can see from the figure that both high-achieving countries (e.g., Taiwan and Hungary) and low-achieving countries (e.g., Iran) produce large achievement gaps between high-SES and low-SES students. The Pearson correlation coefficient (r) for the relationship between national achievement and the achievement gap was 0.20 and not statistically significant ($p=0.20$). This means that high-achieving countries do not necessarily produce a smaller achievement gap between high- and low-SES students.

National Level of Students' Access to Qualified Teachers

How does the percentage of eighth graders taught by qualified mathematics teachers in the USA differ from that in other countries? Table 3.1 shows the percentage ranking from the highest to the lowest in the percentage of students taught by teachers with each of the four qualifications (full certification, mathematics major, mathematics education major, and teachers with 3 or more years of teaching experience), as well as the percentage of students taught by teachers with full certification, mathematics major or mathematics education major, and 3 or more years of teaching experience. In the USA, 96.6% of eighth graders are taught by fully certified teachers, which is higher than the international average of 91.4%. In Iran, Japan, Korea, Malaysia, and Scotland, all students were taught by fully certified mathematics teachers. In these five countries, it is likely that strict government regulations prevent teachers from entering the teaching profession without certification, although the requirements for certification may differ across countries. In contrast, only 62.5% of students in Algeria and 55.8% of students in Morocco were taught by fully certified teachers.

When we look at mathematics major, only 42.4% of US students are taught mathematics by teachers with a major in mathematics, a smaller percentage than the international mean of 70.1%. The USA ranked 46th among 50 countries in this indicator of teacher qualification. The data suggest that, in a majority of countries, unlike in the USA, possession of a mathematics degree is a common characteristic of teachers teaching mathematics to eighth graders. Cross-nationally, the percentage varies from only 8.8% in Slovenia to 98.5% in Russia.

In the USA, a higher percentage of eighth graders are taught by teachers with a mathematics education major who received both subject content and pedagogical preparation. The data show that 48.5% of US eighth graders are taught by teachers with a mathematics education major, and the USA is ranked the 29th among 48 countries. The cross-national average is 53.8%, which indicates that having majored in *mathematics education* is a less common characteristic among teachers teaching mathematics than is having majored in *mathematics* in many countries. Here, we see a major cross-national variation ranging from 4.5% in Thailand to 95.7% in Hungary. These percentages in the USA indicate that a significant proportion of US eighth graders were taught mathematics by teachers without a subject-specific major.

Teaching experience is another indicator of teacher qualification associated with higher student achievement in the USA; 88.6% of US eighth graders were taught by teachers with 3 or more years of teaching experience, a figure similar to the international average of 90.7%. Cross-nationally, over 70% of eighth graders are taught by experienced teachers with 3 or more years of experience; this percentage ranged from 71.4% in Singapore to 99.5% in Georgia.

We also created an overall measure of students' access to *qualified teachers*, defined by those who have a full certification, mathematics major or mathematics education major, and 3 or more years of teaching experience. On average, in the

Table 3.1 Students' access to qualified teachers: percentage of students taught by qualified teachers

	Certified teachers[a]	%		Teachers with a math major	%		Teachers with a math education major[b]	%		Teachers with 3+ years' experience	%		Overall teacher quality	%
1	Iran	100	1	Russia	98.5	1	Hungary	95.7	1	Georgia	99.5	1	Russia	96.7
1	Japan	100	2	Armenia	98.0	2	Slovenia	94.1	2	Russia	99.0	2	Armenia	94.0
1	Korea	100	3	Mongolia	97.1	3	Armenia	90.8	3	Bulgaria	98.9	3	Lithuania	92.6
1	Malaysia	100	4	Romania	96.7	4	Czech Republic	85.3	4	Egypt	98.5	4	Romania	89.8
1	Scotland	100	5	Hungary	96.6	5	Georgia	78.1	5	Kuwait	98.2	5	Ukraine	89.6
6	Botswana	99.9	6	Cyprus	96.3	6	Malta	75.5	6	Armenia	97.9	6	Bosnia & Herzegovina	88.7
7	Russia	99.7	7	Bulgaria	96.0	7	Indonesia	74.8	7	Algeria	97.8	7	Bulgaria	87.8
8	Cyprus	99.4	8	Bosnia & Herzegovina	95.9	8	Bulgaria	73.6	8	Iran	97.8	8	Czech Republic	87.2
9	Norway	99.4	9	Serbia	94.8	9	Bahrain	72.9	9	Lithuania	97.7	9	Iran	84.9
10	Turkey	99.3	10	Lithuania	93.5	10	Oman	72.6	10	Colombia	97.6	10	Georgia	83.4
10	Bulgaria	99.3	11	Tunisia	92.2	11	Egypt	71.2	11	Hungary	97.5	11	Israel	82.2
12	Bosnia & Herzegovina	98.6	12	Syria	91.9	12	Russia	70.4	12	El Salvador	96.5	11	Slovenia	82.2
13	Taiwan	98.4	13	Scotland	87.2	13	Korea	69.9	13	Ukraine	96.4	13	Cyprus	82.1
13	Ukraine	98.4	14	Morocco	86.6	14	Turkey	68.8	14	Bosnia & Herzegovina	96.3	14	Bahrain	81.9
15	Romania	98.3	15	Jordan	86.3	15	Qatar	65.3	14	Israel	96.3	15	Serbia	81.2
16	Singapore	98.2	16	Algeria	85.3	15	Romania	65.3	16	Czech Republic	95.9	16	Korea	78.9
17	Thailand	98.1	17	Taiwan	80.7	17	Sweden	63.3	17	Romania	95.5	17	Scotland	77.8
18	England	97.9	18	Colombia	79.1	18	Botswana	61.4	18	Italy	95.0	18	Malta	75.7
18	Israel	97.9	19	Palestine	77.9	19	Kuwait	61.0	19	Bahrain	94.4	19	Indonesia	75.6
20	Lithuania	97.9	20	Lebanon	77.0	20	Ukraine	58.7	20	Norway	94.0	20	Botswana	75.1
21	Bahrain	97.8	21	Israel	76.4	21	Colombia	57.9	21	Qatar	93.5	21	Japan	73.0
22	Hong Kong	97.4	22	Japan	76.3	22	Hong Kong	57.6	21	Indonesia	93.5	22	Turkey	71.4
23	USA	96.6	23	Botswana	73.3	23	Iran	57.3	23	Sweden	92.7	23	Qatar	70.7
24	Czech Republic	96.5	24	England	72.3	23	Israel	57.3	24	Slovenia	92.5	24	Taiwan	70.1
24	Australia	96.5	25	Thailand	72.1	25	Ghana	56.8	25	Morocco	92.3	25	Kuwait	67.5
26	Armenia	96.0	26	Oman	71.2	26	Japan	51.6	26	Malaysia	92.2	26	Mongolia	67.3

Table 3.1 (continued)

#	Certified teachers[a]	%	Teachers with a math major	%	Teachers with a math education major[b]	%	Teachers with 3+ years' experience	%	Overall teacher quality	%
27	El Salvador	95.3	Malta	70.7	Taiwan	50.3	Serbia	91.8	Syria	66.3
28	Sweden	94.9	Qatar	70.1	Singapore	48.7	Tunisia	91.1	El Salvador	63.2
29	Malta	94.2	Saudi Arabia	69.9	*USA*	48.5	Mongolia	90.7	England	63.2
30	Georgia	93.9	Ghana	68.8	Palestine	47.1	Lebanon	90.4	Sweden	61.6
31	Oman	92.7	Singapore	68.8	Scotland	46.5	Cyprus	89.0	Jordan	61.6
32	Indonesia	91.1	El Salvador	66.2	Australia	45.7	Taiwan	88.7	Oman	61.5
33	Slovenia	90.9	Egypt	63.0	El Salvador	45.6	*USA*	88.6	Egypt	61.4
34	Qatar	89.3	Bahrain	62.4	Malaysia	44.7	Korea	88.3	Thailand	60.7
35	Mongolia	86.6	Czech Republic	62.3	Jordan	44.1	Malta	87.7	Tunisia	60.4
36	Tunisia	85.6	Hong Kong	61.9	Saudi Arabia	42.8	Australia	87.6	Malaysia	60.4
37	Serbia	84.6	Kuwait	61.6	Serbia	41.4	Scotland	87.6	*USA*	60.2
38	Kuwait	83.6	Iran	57.5	Lebanon	40.4	Japan	87.3	Hong Kong	57.2
39	Italy	83.3	Ukraine	52.6	England	39.0	Syria	87.0	Singapore	57.1
40	Syria	82.6	Sweden	49.5	Bosnia & Herzegovina	36.5	Saudi Arabia	85.7	Australia	52.2
41	Egypt	80.1	Turkey	49.5	Lithuania	32.7	Jordan	84.5	Ghana	48.4
42	Jordan	78.9	Australia	49.2	Cyprus	27.5	Thailand	83.8	Lebanon	45.0
43	Ghana	75.7	Malaysia	46.2	Morocco	26.6	England	83.1	Algeria	44.5
44	Lebanon	71.1	Georgia	46.1	Algeria	25.9	Botswana	82.2	Palestine	44.4
45	Palestine	63.7	Indonesia	44.1	Syria	17.4	Palestine	82.2	Morocco	36.2
46	Algeria	62.5	*USA*	42.4	Tunisia	12.6	Turkey	81.5	Norway	35.8
47	Morocco	55.8	Norway	40.6	Norway	7.8	Hong Kong	81.3	Italy	14.6
48			Korea	27.7	Thailand	4.5	Ghana	73.6		
49			Italy	15.8			Oman	71.5		
50			Slovenia	8.8			Singapore	71.4		
	Mean	*91.4*	*Mean*	*70.1*	*Mean*	*53.8*	*Mean*	*90.7*	*Mean*	*68.6*
	SD	*11.0*	*SD*	*21.8*	*SD*	*21.5*	*SD*	*7.1*	*SD*	*17.4*

[a] Data on certification were not available from Colombia, Hungary, and Saudi Arabia

[b] Data on mathematics education major were not available from Italy and Mongolia

47 countries, 68.6% of eighth graders were taught by teachers with these qualifications, and it ranged from only 14.6% in Italy to 96.7% in Russia. In the USA,
60.2% of eighth graders are taught by qualified teachers, and this figure is lower
than the international average. The USA is ranked 37th in the level of students' access to qualified teachers.

National Level of Gap in Students' Access to Qualified Teachers

How does students' access to qualified teachers vary by their SES? Table 3.2 presents the difference in the percentage of high-SES students and low-SES students
who were taught by qualified teachers. For the countries with a positive value of the
percentage difference, high-SES students have a greater opportunity to be taught
by qualified teachers than do low-SES students, indicating the existence of unequal
access to qualified teachers and a greater gap. For countries with a negative value
for the percentage difference, low-SES students were more likely than high-SES
students to be taught by qualified teachers, indicating the existence of needs-based
access to qualified teachers and a smaller inequality.[2]

When we look at the international average of 50 countries across all the indicators of teacher qualification, the gap is no more than 4%. On average, many countries are successful in equalizing access to qualified teachers along the line of SES.
However, we can also observe a major variation across the countries in the size of
gap in students' access to qualified teachers.

For students' access to fully certified teachers, the percentage gap varied from
− 13.7 in El Salvador to 11.1 in Algeria. In El Salvador, low-SES students had greater access to certified teachers than did high-SES students, whereas in Algeria, high-
SES students had greater access to certified teachers than did low-SES students.
In the USA, the difference was −2.0, showing that there is no major difference
between high-SES and low-SES students in their access to certified teachers.

When we look at the difference in students' access to teachers with a mathematics major and mathematics education major between high-SES and low-SES
students, the data showed larger variations across countries. The difference varied
from − 14.5 in Tunisia to 20.8 in Malaysia for mathematics major and from − 18.3
in Algeria to 23.4 in Israel for mathematics education major. In the USA, the gap
was 0% (41.0 vs. 41.0%) in mathematics major and 10.3% (53.3 vs. 43.0%) in
mathematics education major, compared with the international average of 2.1 and
0.9%. This shows that US eighth graders have equal access to teachers with a major
in mathematics, but high-SES students are more likely than low-SES students to

---

[2] Readers could argue that when low-SES students have greater access to qualified teachers than
do high-SES students, high-SES students receive an unequal opportunity to be taught by qualified
teachers. However, such a gap is likely the result of a government policy or system that attempts to
promote greater equality in students' opportunity to learn, based on the preexisting disadvantage
of low-SES students as compared with high-SES students. Therefore, we consider the inequality to
be smaller in a national context where low-SES students have greater access to qualified teachers.

Table 3.2 Gap in students' access to qualified teachers: difference between high- and low-SES students in percentage of students taught by qualified teachers

	Certified teachers[a]	%		Teachers with a math major	%		Teachers with a math education major[b]	%		Teachers with >= 3 years of experience	%		Overall teacher quality	%
1	Algeria	11.1	1	Malaysia	20.8	1	Israel	23.4	1	Turkey	26.8	1	Turkey	26.7
2	Serbia	10.5	2	El Salvador	11.9	2	El Salvador	17.9	2	Tunisia	14.5	2	Lebanon	17.0
3	Jordan	9.0	3	Oman	10.7	3	Cyprus	12.7	3	Malaysia	14.4	3	Mongolia	12.7
4	Indonesia	7.1	4	Colombia	10.6	4	Jordan	11.3	4	Mongolia	13.0	4	Algeria	12.1
5	Lebanon	6.7	5	Turkey	10.5	5	USA	10.3	5	Serbia	11.7	5	Malaysia	12.0
6	Syria	6.6	6	Serbia	9.7	6	Ghana	9.9	6	Ghana	11.4	6	Serbia	10.9
7	Mongolia	5.7	7	Italy	9.6	7	Singapore	9.8	7	Jordan	9.8	7	Indonesia	10.7
8	Qatar	5.4	8	Norway	9.1	8	Hong Kong	8.0	8	Saudi Arabia	9.6	8	Syria	10.6
9	Romania	5.2	9	Thailand	8.0	9	Romania	7.3	9	Romania	9.0	9	Bulgaria	9.7
10	Malta	4.4	10	Syria	7.8	10	Botswana	6.7	10	USA	7.7	10	Italy	9.6
11	Bosnia & Herzegovina	3.4	11	Indonesia	7.7	11	Japan	6.3	11	Syria	6.6	11	Taiwan	8.8
11	Lithuania	3.4	12	Qatar	7.3	12	Indonesia	6.1	12	Cyprus	6.3	12	Bosnia & Herzegovina	8.3
13	Singapore	3.1	13	Bulgaria	6.5	13	Taiwan	5.4	13	Malta	6.2	13	Romania	7.9
14	Italy	2.7	14	Egypt	6.2	14	Czech Republic	5.3	13	Lebanon	5.2	14	USA	7.8
15	Czech Republic	2.6	14	Taiwan	6.2	15	Scotland	3.7	15	Indonesia	4.9	15	Jordan	7.7
16	Hong Kong	2.5	16	Lebanon	5.9	15	Palestine	3.7	15	Palestine	4.9	15	Bahrain	7.7
17	Bahrain	1.8	17	Algeria	5.4	17	Colombia	3.6	17	Iran	4.3	17	Norway	7.6
18	Ukraine	0.8	18	Romania	4.9	18	Australia	2.9	18	Bosnia & Herzegovina	4.2	18	Thailand	7.1
19	Norway	0.7	19	Singapore	4.3	19	Oman	1.7	19	Singapore	3.0	19	England	6.9
19	Turkey	0.7	19	Hong Kong	4.3	20	Saudi Arabia	1.6	20	Slovenia	2.9	19	Singapore	6.9
19	Kuwait	0.7	21	Ukraine	4.0	21	Sweden	1.5	21	Korea	2.6	21	Iran	5.4
22	Cyprus	0.6	22	Bosnia & Herzegovina	3.7	22	Russia	0.7	22	Bulgaria	2.2	22	El Salvador	4.8

Table 3.2 (continued)

	Certified teachers[a]	%		Teachers with a math major	%		Teachers with a math education major[b]	%		Teachers with >= 3 years of experience	%		Overall teacher quality	%
23	Palestine	0.5	23	Georgia	3.7	23	Armenia	0.5	23	Hungary	1.9	23	Egypt	3.4
24	Bulgaria	0.3	24	Slovenia	3.7	24	Bahrain	0.4	24	Sweden	1.9	24	Qatar	2.7
25	Taiwan	0.2	25	Australia	3.0	25	Iran	0.0	25	Algeria	1.6	25	Sweden	2.6
26	Oman	0.0	26	Mongolia	2.4	26	Hungary	−0.3	26	Botswana	1.4	26	Cyprus	2.2
26	Armenia	0.0	27	Ghana	1.7	27	England	−0.8	27	Ukraine	1.1	27	Hong Kong	1.6
26	Iran	0.0	28	Russia	1.1	28	Kuwait	−1.1	28	Bahrain	1.0	28	Ghana	1.5
26	Japan	0.0	29	Sweden	1.0	29	Thailand	−1.4	28	Israel	1.0	29	Scotland	0.9
26	Korea	0.0	30	Korea	0.9	30	Norway	−1.7	30	Thailand	0.9	30	Slovenia	0.9
26	Malaysia	0.0	31	Hungary	0.7	31	Syria	−1.8	31	Scotland	0.8	31	Israel	0.7
26	Scotland	0.0	32	Iran	0.1	32	Slovenia	−2.2	32	Egypt	0.7	32	Russia	0.3
26	Slovenia	0.0	33	*USA*	*0.0*	32	Lithuania	−2.2	33	Czech Republic	0.5	33	Malta	0.2
34	Russia	−0.3	33	Cyprus	0.0	34	Korea	−3.0	34	Taiwan	−0.3	34	Czech Republic	0.1
35	Georgia	−0.6	35	Palestine	−0.3	34	Bulgaria	−3.0	35	Russia	−0.8	35	Australia	−0.3
36	Botswana	−0.7	36	Armenia	−0.5	36	Tunisia	−3.3	36	Lithuania	−1.2	36	Lithuania	−0.4
37	England	−1.0	37	Scotland	−0.8	37	Turkey	−4.2	37	Qatar	−1.6	37	Ukraine	−0.7
38	Sweden	−1.5	38	England	−1.2	37	Egypt	−4.4	37	Kuwait	−1.6	38	Korea	−2.4
38	Egypt	−1.5	39	Botswana	−2.1	39	Ukraine	−4.7	39	El Salvador	−2.1	39	Botswana	−2.5
40	*USA*	*−2.0*	39	Lithuania	−2.1	40	Lebanon	−4.8	40	Georgia	−2.2	40	Armenia	−2.8
41	Israel	−2.2	41	Japan	−3.4	41	Serbia	−5.4	41	Armenia	−2.8	41	Oman	−3.1
42	Thailand	−2.9	42	Bahrain	−4.0	42	Georgia	−6.2	42	Colombia	−2.9	42	Palestine	−3.5
43	Australia	−4.2	43	Saudi Arabia	−4.7	43	Bosnia & Herzegovina	−7.3	43	England	−3.0	43	Japan	−4.2
44	Ghana	−7.0	44	Czech Republic	−4.9	43	Morocco	−7.3	44	Australia	−3.1	44	Morocco	−5.5
45	Morocco	−9.8	45	Israel	−7.3	45	Malaysia	−10.9	45	Norway	−4.1	45	Georgia	−6.3
46	Tunisia	−12.8	46	Kuwait	−9.1	46	Qatar	−11.5	46	Italy	−4.1	46	Kuwait	−6.9
47	El Salvador	−13.7	47	Morocco	−9.7	47	Malta	−13.8	47	Oman	−4.8	47	Tunisia	−9.6
			48	Jordan	−10.6	48	Algeria	−18.3	48	Hong Kong	−6.0			
			49	Malta	−11.2				49	Japan	−6.1			
			50	Tunisia	−14.5				50	Morocco	−6.7			
	Mean	*0.7*		*M*	*2.1*		*M*	*0.9*		*M*	*2.8*		*M*	*4.0*
	SD	*5.0*		*SD*	*6.8*		*SD*	*7.8*		*SD*	*6.4*		*SD*	*6.9*

[a] Data on certification were not available from Colombia, Hungary, and Saudi Arabia

[b] Data on mathematics education major were not available from Italy and Mongolia

Table 3.3 Comparison of students' access to qualified teachers and gap in access in 2003 and 2007 in the USA

		2003[a]	2007
Percentage of students taught by qualified teachers	Certified teachers	95.4	96.6
	Teachers with math major	47.3	42.4
	Teachers with math education major	55.3	48.5
	Teachers with 3+years experience	90.8	88.6
	Overall teacher qualification	*60.3*	*60.2*
Gap in percentage of students taught by qualified teachers	Certified teachers	1.8 (96.0 vs. 94.2)	−2.0 (95.4 vs 97.4)
	Teachers with math major	10.0 (54.1 vs. 44.1)	0.0 (41.1 vs. 41.1)
	Teachers with math education major	13.8 (59.9 vs. 46.1)	10.3 (53.3 vs. 43.0)
	Teachers with 3+years experience	3.6 (93.8 vs. 90.2)	7.7 (92.2 vs. 84.5)
	Overall teacher qualification	*14.4 (67.6 vs. 53.2)*	*7.8 (63.5 vs. 55.7)*

[a] Figures are from Akiba et al. (2007)

Overall teacher qualification was measured by having a full certification, having majored in mathematics or mathematics education, and having 3 or more years of teaching experience

be taught by teachers with a mathematics education major. The gap in students' access to teachers with at least 3 years of teaching experience varied from −6.7% in Morocco to 26.8% in Turkey. In the USA, the gap was 7.7%, with 63.5% of high-SES students and 55.7% of low-SES students taught by experienced teachers. This figure was larger than the international average of 2.8%.

In the overall measure of teacher qualification, 63.5% of high-SES students in the USA were taught by teachers with certification, mathematics or mathematics education major, and at least 3 years of teaching experience compared with 55.7% of low-SES students, with a gap of 7.8%. This is larger than the international average of 4.0%. The size of gap varied from −9.6 in Tunisia to 26.7 in Turkey. In 13 countries, including Tunisia, Kuwait, Japan, and Australia, low-SES students have a greater opportunity to be taught by qualified teachers than do high-SES students.

Improvement from 2003 to 2007 in Students' Access to Qualified Teachers and Gap in the Access

The NCLB Act of 2001 required states to ensure that all students are taught by highly qualified teachers by the 2005–2006 academic year. If the NCLB influenced state policy, we are likely to see improvement in students' access to qualified teachers, as well as equalization of such access between low-SES and high-SES students from 2003 to 2007. Table 3.3 compares the figures in 2003, obtained by Akiba et al. (2007) using the same measures of teacher qualification, and the figures in 2007.

We can see that the percentages of students taught by certified teachers and teachers with 3 or more years of teaching experiences did not change much from 2003 to 2007, but the percentages of students who were taught by teachers with a mathematics major or mathematics education major decreased from 47.3 to 42.4% and from 55.3 to 48.5%, respectively. This is a surprising finding considering the focus on the requirement of subject-matter knowledge in highly qualified teachers in the NCLB. It may be due to the fact that most states required teachers to pass a subject-specific test (Praxis II assessment) to meet the subject knowledge requirement rather than requiring a subject area major. Evaluation studies have indeed found that although all states had administered tests of teacher content knowledge (Birman et al. 2009), only 26 states required teachers to have a major in the subject area they teach as of the 2005–2006 academic year (Loeb et al. 2009). The percentage of students taught by teachers who are fully certified, who have majored in mathematics or mathematics education, and who have had 3 or more years of teaching experience remained the same: 60.3% in 2003 and 60.2% in 2007.

The gap in access to qualified teachers between high-SES and low-SES students, however, showed a major improvement. The difference in students' access to qualified teachers between high-SES and low-SES students narrowed from 14.4% (67.6 vs. 53.2%) in 2003 to 7.8% (63.5 vs. 55.7%) in 2007. Although the gap in students' access to experienced teachers became larger (from 3.6 to 7.7%), the gap in access to teachers with a mathematics major disappeared (from 10 to 0%), and the gap in access to teachers with fully certified teachers and teachers with a mathematics education major was narrowed (from 1.8% to −2.0% and from 13.8 to 10.3%, respectively). The major reduction of the gap in students' access to teachers with a major in mathematics may be due to the increased number of alternatively certified teachers with a major in mathematics in low-SES schools. In order to meet the requirement of highly qualified teachers, many states allowed the establishment of alternative certification programs, which recruit those with strong subject content knowledge (e.g., working professionals in mathematics and science fields and those with mathematics majors) to become mathematics teachers (Loeb and Miller 2006). Many federal programs required teacher candidates to work in high-needs schools (low-SES, low-achieving schools) in exchange for fully supporting the cost for pursing alternative certification (e.g., National Science Foundation Noyce Grant). The number of teachers certified through alternative routes dramatically increased from 38,519 in 2003 to 62,000 in 2007 (Feistritzer 2010). It is likely that distribution of alternatively certified teachers to low-SES schools has contributed to narrowing the gap in students' access to teachers with a major in mathematics.

Students' Access to Qualified Teachers, Access Gap, and National Achievement Outcomes

For our last research question, we conducted multiple regression analyses to examine the relationships between (a) students' access to qualified teachers and na-

Table 3.4 Multivariate relationship between students' access to qualified teachers and national achievement

National predictors	Model 1	Model 2	Model 3	Model 4	Model 5
	B (SE)	B (SE)	B (SE)	B (SE)	B (SE)
Teacher qualification					
Teacher certification	2.94 (1.08)*				
Math major		−0.07 (0.50)			
Math education major			0.35 (0.54)		
Teaching experience				−0.06 (1.50)	
Overall teacher qualification					0.81 (0.63)
National variables					
Educational expenditure as % of GDP	0.53 (7.29)	−0.24 (7.91)	1.62 (8.53)	−0.33 (7.96)	2.57 (7.82)
GDP per capita (US$ 1,000)	0.57 (0.59)	1.04 (0.65)	1.07(0.64)	1.07 (0.63)	1.18 (0.61)
R^2	0.22	0.07	0.07	0.07	0.11
N	43	46	46	46	43

B unstandardized regression coefficient, GDP gross domestic product, R^2 percentage of variance in the dependent variable explained by the independent variables
*$p < 0.05$; **$p < 0.01$

tional achievement and (b) the gap in students' access to qualified teachers and the achievement gap controlling for educational expenditure as a percentage of GDP and GDP per capita.

Tables 3.4 and 3.5 present five multiple regression models with each of the teacher qualification indicators. The sample size varied from 43 to 46 because of the unavailability of data on educational expenditure as percentage of GDP and/or GDP per capita in some countries. Table 3.4 shows that the percentage of students taught by certified teachers was associated with national achievement. Countries where a higher percentage of eighth graders was taught by certified teachers achieved higher mathematics scores than did other countries. However, no other teacher qualification indicators including the overall measure of teacher qualification showed a statistically significant relationship with national achievement in mathematics.

In contrast, Table 3.5 shows that the level of the gap measured by the difference in the percentages of high-SES students and low-SES students taught by teachers with multiple qualifications (full certification, mathematics major or mathematics education major, and 3 or more years of teaching experience) was associated with the national level of achievement gap between high-SES and low-SES students in mathematics. In countries where there is a larger gap in students' access to qualified mathematics teachers, the size of the achievement gap in mathematics tends to be larger. When these qualification indicators were examined individually, however, only the gap in students' access to experienced teachers was associated with the achievement gap. This means that inequality in students' access to qualified teachers can be more detrimental when we consider multiple qualifications than when we consider individual qualifications separately. This hypothesis makes sense because teachers who have multiple qualifications are more likely to practice effective instruction than are teachers with only a certification or a major in mathematics.

Table 3.5 Multivariate relationship between gap in students' access to qualified teachers and achievement gap (high-SES vs. low-SES students)

National predictors	Model 1	Model 2	Model 3	Model 4	Model 5
	B (SE)	B (SE)	B (SE)	B (SE)	B (SE)
Opportunity gap					
Teacher certification	1.06 (0.94)				
Math major		0.19 (0.66)			
Math education major			−0.09 (0.60)		
Teaching experience				2.12 (0.60)**	
Overall teacher qualification					1.83 (0.61)**
National variables					
Educational expenditure as % of GDP	−3.69 (3.38)	−3.82 (3.50)	−4.25 (3.28)	−5.63 (2.84)	−2.08 (3.12)
GDP per capita (US$ 1,000)	−0.35 (0.25)	−0.32 (0.25)	−0.32 (0.26)	−0.06 (0.23)	−0.23 (0.23)
R^2	0.11	0.08	0.08	0.28	0.25
N	43	46	44	46	43

B unstandardized regression coefficient, GDP gross domestic product, R^2 percentage of variance in the dependent variable explained by the independent variables
*$p<0.05$; **$p<0.01$

Discussion

This cross-national study of 50 countries investigated an important focus of educational reforms around the world: students' access to qualified teachers and inequality in such access based on student SES. Based on the TIMSS mathematics assessment for eighth graders, the study found that the US students scored more than the international average (509 vs. 450), but the size of the achievement gap was similar to the international average (78 vs. 76). Akiba et al. (2007) reported that US eighth-graders' national achievement level was 504, and their achievement gap was 109 in 2003. This means that whereas the national achievement level has remained stable, the level of the achievement gap based on the TIMSS mathematics assessment for eighth graders has narrowed significantly.

This pattern coincided with the national level of students' access to qualified teachers and the gap in such access between high-SES and low-SES students. The percentage of students who were taught by qualified teachers did not change much from 2003 (60.3) to 2007 (60.2), but the difference in the percentage of students taught by qualified teachers between high-SES and low-SES students narrowed from 14.4 to 7.8 %. Reduced levels of achievement gap and inequality in access to qualified teachers between high-SES and low-SES students are great news, showing the progress toward equalizing students' opportunity to learn in the USA. However, it is also important to keep in mind that about 40 % of the students do not have access to qualified mathematics teachers with a full certification, a mathematics major or mathematics education major, and three or more years of teaching experience. This is larger than the international average of 31.4 % (100−68.6 %) among 47 countries. Even though the gap in students' access to qualified teachers was narrowed from 2003 to 2007, in 2007 only 55.7% of low-SES students were taught by qualified teachers compared with 63.5% of high-SES students. This gap of 7.8% is still larger than the international mean of 4.0%. There is a need to continue with the efforts to increase students' access to qualified teachers and to continue narrowing the gap in such access among students.

Our cross-national analysis of the relationships between students' access to qualified teachers and national achievement showed that the countries with a higher percentage of students taught by qualified teachers are not necessarily producing high national achievement. This is different from the findings based on the 2003 TIMSS data (Akiba et al. 2007) that showed a relationship between students' access to qualified teachers and national achievement. It may be because of the differences in the countries that participated in 2003 and 2007. A total of 15 new countries participated in the 2007 TIMSS, and nine of these are developing countries with the GDP per capita of less than US$ 10,000 (compared with the mean of US$ 21,100 among 50 countries). Several of these countries, such as Bosnia, Herzegovina, Georgia, and Ukraine, have over an 80% national level of student access to qualified teachers; yet, their national achievement level is not among the highest. Future studies may examine the factors that mediate the relationship between teacher qualifications and student achievement in these countries.

The gap in students' access to qualified teachers between high-SES and low-SES students, however, was associated with the size of the achievement gap. Many countries with a large gap in students' access to qualified teachers, including Turkey, Serbia, Bulgaria, Taiwan, and Romania, also have a large achievement gap in mathematics assessment. In contrast, many countries where a larger percentage of low-SES students than high-SES students are taught by qualified teachers (e.g., Tunisia, Kuwait, Armenia, and Lithuania) produced a small achievement gap between these groups of students. It may be that in many of the countries that participated in the 2007 TIMSS, less qualified teachers receive less school resources and professional development opportunities than do more qualified teachers, which contributes to the gap in their instructional quality and to the achievement gap.

It is also important to note that there is no statistically significant relationship between students' access to qualified teachers and the gap in students' access to qualified teachers (Pearson $r = -0.14$, $p = 0.34$), meaning that the countries where a larger percentage of students are taught by qualified teachers do not necessarily ensure equal access to qualified teachers between high-SES and low-SES students. This shows the difficulty in increasing the number of qualified teachers while making sure that students have equal access to these teachers.

A decentralized hiring system at the school or at the district level in the USA makes it challenging to ensure students' access to qualified teachers. Because of the different level of resources and teacher salary level across districts and schools, qualified teachers tend to concentrate in wealthier schools. However, federal involvement in alternative certification programs through providing funding to subsidize the cost of teacher education in mathematics and science areas in exchange for working in high-needs schools seems to have contributed to narrowing the gap in students' access to qualified teachers between high-SES and low-SES students from

2003 to 2007. This shows the promise of federal or state government's involvement in promoting students' equal access to qualified teachers.

Many countries around the world have centralized teacher hiring and distribution policies. For example, Australia has a state-level system to hire and distribute teachers using strong financial incentives (Akiba and LeTendre 2009). Teachers receive a higher salary, an extra bonus, and multiple benefits (e.g., housing subsidy, additional paid leave, and additional professional development leaves) for working in remote rural schools where teacher shortage is most severe. When the hiring system is centralized at the state level, it is possible to offer strong incentives to distribute qualified teachers to work in the schools where such teachers are most needed.

The USA faces a major challenge of increasing qualified teachers while ensuring students' equal access to qualified teachers in a highly decentralized system with a major variation in financial capacity across districts and schools. This financial disparity not only affects districts and schools' capacity to hire qualified teachers but also affects teachers' working conditions and professional development opportunities, which are critical for improving instructional quality. This study provides evidence that the countries that do not ensure students' equal opportunity to be taught by qualified teachers produce a larger achievement gap. The fact that there are many countries that succeeded in equalizing students' access to qualified teachers shows that such success depends on the political will to ensure students' right to be taught by qualified teachers regardless of their individual or home background. Further investigation of these countries with regard to how they achieved equity in students' access to qualified teachers is a fruitful area of study that can produce important policy-related information useful for the many countries that are struggling to achieve such equality.

References

Akiba, M., & LeTendre, G. K. (2009). *Improving teacher quality: The U.S. teaching force in global context*. New York: Teachers College Press.

Akiba, M., LeTendre, G. K., & Scribner, J. P. (2007). Teacher quality, opportunity gap, and achievement gap in 47 countries. *Educational Researcher, 36*(7), 369–387.

Akiba, M., Chiu, Y.-L., Shimizu, K., & Liang, G. (2012). Teacher salary and national achievement: A cross-national analysis of 30 countries. International Journal of Educational Research, *53*, 171–181.

Ascher, C., & Fruchter, N. (2001). Teacher quality and student performance in New York City's low-performing schools. *Journal of Education for Students Placed at Risk, 6*(3), 199–214.

Birman, B. F., Boyle, A., Le Floch, K. C., Elledge, A., Holtzman, D., Song, M., et al (2009). *State and local implementation of the No Child Left Behind Act: Volume VIII—Teacher quality under NCLB: Final report*. Washington, DC: U.S. Department of Education.

Darling-Hammond, L. (2004). Inequality and the right to learn: Access to qualified teachers in California's public schools. *Teachers College Record, 106*(10), 1936–1966.

Darling-Hammond, L., & Youngs, P. (2002). Defining "highly qualified teachers": What does "scientifically-based research" actually tell us? *Educational Researcher, 31*(9), 13–25.

Editorial Projects in Education. (2011). *National Highlights Report 2011: Uncertain Forecast: Education Adjusts to a New Economic Reality*. Bethesda: Editorial Projects in Education.

Feistritzer, E. (2010). *Alternative teacher certification: A state-by-state analysis 2010*. Washington, DC: National Center for Education Information.

Ingersoll, R. M. (1999). The problem of under-qualified teachers in American secondary schools. *Educational Researcher, 28*(2), 26–37.

Ingersoll, R. M. (2002). The teacher shortage: A case of wrong diagnosis and wrong prescription. *NASSP Bulletin, 86*(631), 16–31.

Jerald, C., & Ingersoll, R. M. (2002). *All talk, no action: Putting an end to out-of-field teaching*. Washington, DC: Education Trust. www.edtrust.org/main/documents/AllTalk.pdf. Accessed 15 Nov 2006.

LeTendre, G., Baker, D., Akiba, M., Goesling, B., & Wiseman, A. (2001). Teachers' work: Institutional isomorphism and cultural variation in the U.S., Germany, and Japan. *Educational Researcher, 30*(6), 3–15.

Loeb, S., & Miller, L. C. (2006). *A review of state teacher policies: What are they, what are their effects, and what are their implicatiosn for school finance?* Palo Alto: Institute for Research on Education Policy & Practice (IREPP), Stanford University.

Loeb, S., Miller, L. C., & Strunk, K. O. (2009). The state role in teacher professional development and education throughout teachers' careers. *Education Finance and Policy, 4*(2), 212–228.

National Center for Education Statistics. (2000). *Monitoring quality: An indicators report*. Washington, DC: Author.

Olson, J. F., Martin, M. O., & Mullis, I. V. S. (Eds.). (2009). *TIMSS 2007 technical report*. Boston: TIMSS and PIRLS International Study Center.

Organisation for Economic Co-operation and Development. (2005). *Teachers matter: Attracting, developing and retaining effective teachers*. Paris: Author.

Peske, H. G., & Haycock, K. (2006). *Teaching inequality: How poor and minority students are shortchanged on teacher quality*. Washington, DC: Education Trust.

Rice, J. K. (2003). *Teacher quality: Understanding the effectiveness of teacher attributes*. Washington, DC: Economic Policy Institute.

Shen, J., Mansberger, N. B., & Yang, H. (2004). Teacher quality and students placed at risk: Results from the Baccalaureate and Beyond Longitudinal Study, 1993–1997. *Educational Horizons, 82*(3), 226–235.

UNESCO Institute for Statistics. (n.d.). *Custom Tables*. http://stats.uis.unesco.org/unesco/TableViewer/document.aspx?ReportId=136&IF_Language=eng&BR_Topic=0. Accessed 26 Jan 2012.

UNESCO Institute for Statistics. (2006). *Teachers and educational quality: Monitoring global needs for 2015*. Montréal. Canada: Author.

Wang, A. H., Coleman, A. B., Coley, R. J., & Phelps, R. P. (2003). *Preparing teachers around the world*. Princeton: Educational Testing Service.

Wayne, A. J., & Youngs, P. (2003). Teacher characteristics and student achievement gains: A review. *Review of Educational Research, 73*(1), 89–122.

Wilson, S. M., Floden, R. E., & Ferrini-Mundy, J. (2001). *Teacher preparation research: Current knowledge, gaps, and recommendations*. Seattle: Center for the Study of Teaching and Policy.

Wilson, S. M., Floden, R. E., & Ferrini-Mundy, J. (2002). Teacher preparation research: An insider's view from the outside. *Journal of Teacher Education, 53*(3), 190–204.

Part II
North America

Chapter 4
Addressing the Achievement Gap in the United States

Julia V. Clark

The Achievement Gap that exists in American Education is not a gap in ability, but a gap in resources and a gap in expectations. We know that students from all backgrounds can succeed at the highest levels of education, when they are given the support they need to succeed–the support that is regularly given to students from the top income brackets.

Lee Bollinger, President, Columbia University

Our progress as a nation can be no swifter than our progress in education, our requirements for world leadership, our hopes for economic growth, and the demands of citizenship itself in an Era such as this all require the maximum development of every young American's capacity. The human mind is our fundamental resource.

John Fitzgerald Kennedy

Introduction

At this incredible moment in history in an era of unprecedented American hope and expectation, there has never been a time more fitting with an opportunity to include all children in the National Education Agenda. National and international studies indicate that too many children are being left behind in education, especially in mathematics and science, areas critical to success in a technological world. Numerous studies indicate that schools in the United States are failing to adequately prepare all students, especially minority students (Blacks, Hispanics, and Native Americans), to (1) participate fully in a technological society as informed citizens,

J. V. Clark (✉)
5600 Wisconsin Avenue, Suite 1205, Chevy Chase, MD 20815, USA
e-mail: jvclark@starpower.net

J. V. Clark (ed.), *Closing the Achievement Gap from an International Perspective,*
DOI 10.1007/978-94-007-4357-1_4, © Springer Science+Business Media B.V. 2014

(2) pursue further studies in science and technology, and (3) enter the workforce. America's educational system is not educating the masses. Too many minority students are being left behind.

Blacks, Hispanics, and Native Americans make up 24% of the population but only 7% of the science and engineering workforce. Blacks and Hispanics account for only 4% of the scientists and engineers in the United States. However, this group represents the greatest source of future workers. If present trends continue, 68% of workers entering the labor force between 2005 and 2015 will be minorities.

Minorities need to become an integral part of the technical workforce. A larger number of scientists and engineers must come from the talent pool of minorities, and the United States needs to provide a way to expand its capacity to innovate within a framework of inclusiveness and opportunity for all. The current inadequate preparation of many Americans, particularly minority employees and women, for scientific and technical jobs threatens the nation's ability to compete in the world economy, as well as our security and quality of life. As the generation educated in the 1950s and 1960s prepares to retire, America's colleges and universities are not graduating enough scientific and technical talent to step into research laboratories. This gap represents a shortfall in America's national scientific and technical capabilities. The gap is ignored at our peril. Closing it will require a national commitment to develop more of the talent of all America's citizens, especially the minorities, who comprise a disproportionately small part of the nation's science, technology, engineering, and mathematics (STEM) workforce.

Demographic projections add to the need to increase the number of minorities in STEM fields. The majority of the children who will be born in the United States in the twenty-first century will belong to groups that are underrepresented in careers involving STEM.

Minority children represent the most rapidly growing part of the school-age population. According to the U.S. Census Bureau (2007), the nation will be more racially and ethnically diverse by mid-century. Minorities, now roughly one-third of the U.S. population, are expected to be the majority in 2042, with the nation's projection to be 54% minority in 2050. By 2023, minorities will comprise more than half of all children in the United States (U.S. Department of Commerce 2008).

Another concern in America's education is the student *achievement gap*. Student achievement in mathematics and science is also a national educational concern. Concerns about America's science education performance have come from a series of national commissions and studies over the last decade. Despite the exhortations in the national reports on educational reform issued over the last several years, such as *A Nation at Risk* and *Educating Americans for the 21st Century*, science remains an area for great improvement in America's schools. *A Nation at Risk*, published on April 26, 1983, warned that American schools were being eroded by a "rising tide of mediocrity." *A Nation at Risk*, one of the first comprehensive assessments of the American education system, compared America's educational system to other nations. The results indicated that America's quality of life and competitiveness as a nation depended on reforming the educational system. At the same time, a report to the National Science Board, *Educating Americans for the 21st Century*, emphasized that keeping pace with the technological world we live in means the nation's mathematics and science education will have to improve. Furthermore, ensuring quality education for all students was a prime concern. Prior to these reports, reform attempts had been initiated, but increased awareness raised by these publications established a new movement to improve mathematics and science education and to target minorities who are at risk in the educational system, especially in mathematics and science.

Defining the Gap

The achievement gap in America refers to the disparity in academic performance, as shown by standardized test scores, between groups of students, mainly minorities: Blacks (African Americans), Hispanics (Latinos), Native Americans (American Indians), and their White (and Asian) peers. The gap is usually defined based on students' performance in elementary and secondary school in the subject areas of mathematics, science, and reading. At each grade level, racial disparities on an array of achievement variables demonstrate a wide gap in performance, especially in mathematics and science, particularly among disadvantaged minorities from urban and rural communities. These disparities start as early as kindergarten, persisting across the secondary grades, and in most cases widen over time.

The achievement performance also differs by family income. At each grade level, in both mathematics and science, students from low-income families have lower average scores and are less likely than students from wealthy families to reach the proficient level. These gaps related to family income are substantial. For example, students from low-income families are at least three times less likely to score at or above the proficient level for their grade in both mathematics and science (National Science Board [NSB] 2006). Low income is measured by whether or not a student is eligible for the free or reduced-priced school lunch program.

Raising academic achievement levels for all students is an important issue for education reform at all levels across the United States. Although improvements have been made, gaps among students of different demographic backgrounds and among schools with different student populations have been a persistent challenge in K–12 education in the United States. These gaps are reflected in this chapter, including teacher qualifications, school environment, and, ultimately, learning outcomes.

Data from the National Assessment of Education Progress (NAEP) indicate that Blacks and Hispanics have shown improvement since 1990, but the 2011 NAEP data show that White and Asian/Pacific Islander students continue to outperform students at every grade level (NAEP 2011). In mathematics and science, most 4th-, 8th-, and 12th-grade students did not demonstrate proficiency in the knowledge and skills taught at their grade level. Racial/ethnic minority students and students from poor families and disadvantaged backgrounds lagged behind their more advantaged peers, with these disparities starting as early as kindergarten, persisting across grades, and, for some kinds of skills, widening over time (NSB 2006). Despite the improved performance overall, achievement gaps between these various groups persist and have shown no signs of narrowing since 1990. Black, Hispanic, and Native American students in mathematics and science are performing at lower levels than are White and Asian students. In 2011, White students scored higher on average than all other racial/ethnic groups in science. Asian/Pacific Islander and Native American/ Alaska Native students scored higher on average than Black and Hispanic students, and Hispanic students scored higher than Black students (U.S. Department of Education National Center for Education Statistics, National Assessment of Educational Progress [NCES] 2011). Boys performed slightly better than girls in both subjects.

Overall, large majorities of 4th-, 8th-, and 12th-grade students did not demonstrate proficiency in the knowledge and skills taught at their grade level. Though a majority of 9th-grade students reached proficiency in low-level algebra skills, few mastered higher-level skills. Results of international mathematics and science literacy tests show that 15-year-olds continue to lag behind their peers in many countries, even though their scores have improved in recent years (NSB 2012).

Efforts to improve student achievement include raising high school graduation requirements, strengthening the rigor of curriculum standards, increasing advanced course-taking, promoting early participation in gatekeeper courses such as Algebra I, and improving teacher quality (NSB 2012).

The NAEP, a congressionally mandated program, referred to as the Nation's Report Card, monitors changes in students' academic performance. It assesses the performance of students in grades 4, 8, and 12. It ranks student performance according to three achievement levels: (1) *basic*—student has partial mastery of prerequisite knowledge and skills that are fundamental for proficient work at each grade; (2) *proficient*—student demonstrates solid academic performance for each grade level assessed; students reaching this level have demonstrated competency over challenging subject matter, including subject-matter knowledge, application of such knowledge to real-world situations, and analytical skills appropriate to the subject matter; and (3) *advanced*—student demonstrates superior performance. The levels are set by the National Assessment Governing Board (NAGB) based on recommendations from panels of educators and members of the public of what students should know and should be able to do in the subject assessed.

Research on the Achievement Gap

Much of the research on the minority achievement gap has focused on identifying the factors that drive it. An overview of some factors associated with the achievement gap is presented in the next few pages. This overview is not meant to be exhaustive but is provided to show the complexity of the achievement gap problem and the challenges that must be overcome in order to close it.

Researchers have not reached a consensus about the causes of the academic achievement gap, and they have a lag in minority student performance. Studies cite an array of factors, both cultural and structural, that influence student performance in school. These factors include poverty, resources, academic coursework, tracking and ability groups, teacher quality, and instructional practice. Schools that serve underrepresented minority and low-income students provide them with differing access to educational resources. Lareau (1987) suggested that students who lack middle-class cultural capital and have limited parental involvement are likely to have lower academic achievement than their better-resourced peers. Other researchers suggest that academic achievement is more closely tied to race and socioeconomic status (Hallinan 1994). For example, being raised in a low-income family often means having fewer educational resources, in addition to poor nutrition and limited health care, which can contribute to lower academic performance. Researchers concerned with the achievement gap between the genders cite biological differences, such as brain structure and development, as a possible reason why one gender outperforms the other in certain subjects. The differing maturation speed of boys versus girls' brains affects how each gender processes information, and it could impact their school performance (Sax 2005). *The Bell Curve* (1994) by Hernstein and Murray proposed that genetic variation in average levels of intelligence (IQ) is at the root of racial disparities in achievement; this created much controversy. Other researchers have argued that there is no significant difference in inherent cognitive ability between different races that could explain the achievement gap and that the environment is the root issue (Dickens 2005; Flynn 1980; Jencks and Phillips 1998).

One explanation for racial and ethnic differences in standardized test performance is that some minority children may not be motivated to do their best on these assessments. Claude M. Steele suggested that minority children and adolescents may also experience stereotype threat—the fear that they will be judged as having traits associated with negative appraisals and/or stereotypes of their racial/ethnic group, which produces test anxiety and hampers their test performance. According to Steele, minority test takers experience anxiety, believing that if they do poorly on a test, they will confirm the stereotypes about the inferior intellectual performance of their minority group. As a result, a self-fulfilling prophecy begins, and the child performs at a level beneath his or her inherent abilities. Steele and Johnson (1998) hypothesize that, in some cases, some minority students, especially African Americans, stop trying in school because they do not want to be accused of "acting white" by their peers (Ogbu 1986). It has also been suggested that some minority students simply stop trying because they do not believe they will ever see the true benefits of their hard work. As Ogbu (1986) points out, minority students may feel little motivation to do well in school because they do not believe it will pay off in the form of a better job or upward social mobility. For Ogbu, students will perform better and will be more engaged in school if they are helped to modify parts of their collective identity that reject school success, through caring individual and institutional practices. According to Ogbu, the cultural–ecological theory of minority schooling considers two sets of factors that shape minority students' school adjustment and academic performance: (1) the way society and its institutions treat and have treated minorities (i.e., the system) and (2) the way minorities interpret and respond to their treatment, which depends on their unique history and minority status in America. He refers to the second set of factors as *community forces*. Based on his research in

2003, Ogbu made the following recommendation, among others, to communities and schools for closing the achievement gap: Teachers need to recognize that their expectations affect students' self-concept as learners and achievers and the internalization of negative or positive beliefs about their intelligence.

Different schools have different effects on similar students. Minority students tend to be concentrated in low-achieving, highly segregated schools. In general, minority students are more likely to come from low-income households, meaning that they are more likely to attend poorly funded schools based on the districting patterns within the school system. Schools in lower-income districts tend to employ less-qualified teachers and tend to have fewer educational resources (Roscigno 2006). Research shows that teacher effectiveness is the most important in-school factor affecting student learning. Good teachers can close or eliminate the gaps in achievement on standardized tests that separate White and minority students (Gordon et al. 2006).

Some researchers (e.g., Haycock 2006) believe that (1) minority children are taught differently—many Hispanic and Black children get a lower-level, less rigorous curriculum; (2) the least-qualified teachers are assigned to teach minority students; and (3) less is expected of minority children, which becomes a self-fulfilling prophecy (McRobbie 1998). "An unfortunate reality that characterizes the problem of many minority students in science is that the burden of understaffed and underequipped schools usually falls on minority communities," said Clark (1996). This phenomenon can be especially harmful to a science curriculum because well-trained teachers and laboratory experiences are essential. Minority students also get less-experienced teachers. Out-of-field teachers teach more classes in high-minority schools. Inequities in school funding can highlight the social context of schooling.

Perhaps the most significant resource deficit for achievement in science is access to science courses (Lynch 2000). There are wide differences in the availability and quality of courses offered, particularly at the high school level. As colleges become more selective, lack of access to science courses puts students in schools with limited resources at a serious disadvantage when competing for postsecondary opportunities.

Many teachers have low expectations of minority students and do not hold them to high rigorous standards or encourage them to take more advanced courses (Jencks and Phillips 1998). Though more research is needed in this area, experts contend that teachers' perceptions, expectations, and behaviors probably help sustain and even expand the achievement gap.

Research conducted by SciMath and the Minnesota Department of Children, Families, and Learning (1998) found that teacher behaviors affect minority student achievement in mathematics and that minority students benefit from teachers who expect students of all racial, ethnic, and cultural backgrounds to achieve. These teachers consider students' cultural backgrounds as assets rather than liabilities and recognize that all racial, ethnic, and cultural groups have contributed to the mathematics knowledge base (Holloway 2004). These teachers increase the cognitive level of interactions with minority students using diverse and flexible assessments to determine students' strengths. They vary the instructional styles in the classroom.

Hand in hand with teacher expectations, researchers are also noting that teacher quality can contribute to the achievement gap. Research indicates that children in schools with high concentrations of minority and poor students are more likely to be taught by unqualified teachers (Clark 1996; Darling-Hammond 2000). These findings are emerging in conjunction with other studies, quantifying the damage caused by ineffective teachers (Jencks and Phillips 1998). Consequently, teacher professional development has become one of the major elements of the school reform efforts.

Some schools tend to place students in tracking groups as a means of tailoring lesson plans for different types of learners. However, as a result of schools placing emphasis on socioeconomic status and cultural capital, minority students are vastly overrepresented in lower educational tracks (VanderHart 2006). Similarly, Black and Hispanic students are often wrongly placed into lower tracks based on teachers' and administrators' expectations for minority students. Studies show that tracking groups within schools are detrimental to minority students (Hyunsook Song 2006). Once students are in these lower tracks, they tend to have less-qualified teachers, a less-challenging curriculum, and few opportunities to advance into higher tracks. Research suggests that students in lower tracks suffer from social and psychological consequences of being labeled as slower learners, which leads children to stop trying in school (Hochschild 2003; Lareau 1987). Many sociologists argue that tracking in schools does not provide lasting benefits to any group of students (Gamoran 1992).

In researching high school mathematics education, Davenport (1993) found that homogeneous ability/achievement grouping impacts high school mathematics education. Within schools using tracking, lower-track students (who are usually the underrepresented minority students) have less access to (1) strong mathematics programs, (2) well-qualified mathematics teachers, and (3) classroom opportunities. Davenport found research support for the position that tracking, especially in high school, widens the achievement gap and "generally fails to increase learning." Research conducted by Oakes (1990) supports these findings.

Research also shows that poor and minority students have disproportionately less access to high-quality, early childhood education, which has been shown to have a strong impact on early learning and development. Magnuson and Waldfogel (2005) found that, although Black children are more likely than White children to attend preschool, they often experience lower-quality care. The same study also found that Hispanic children are much less likely than White children to attend preschool. According to the National Institute for Early Education Research (NIEER), families with modest incomes (less than $60,000) have the least access to preschool education (Barnett and Yarosz 2007). Research suggests that dramatic increases in both enrollment and quality of prekindergarten programs would help alleviate the school-readiness gap and ensure that low-income and minority children begin school on even footing with their peers (Magnuson and Waldfogel 2005).

Education Week Quality Counts (1998) finds that there are no "quick fixes" for the achievement gap in high schools. Major differences in both the opportunity to learn and achievement itself appear in the early grades, so that by the time minority and poor students reach the ninth grade, the deficit is difficult to remedy.

According to Tim Simmons (1999), "race, not poverty drives a wedge between the test scores of Black and White children." Simmons's conclusion is based on the results of a 5-month study conducted by *The News and Observer*. The classroom observations; test data; academic research; and parent, teacher, and student interviews showed a link between racism and the Black–White test score gap (Minority Achievement Report, Trends in Subgroup Performance 2001).

Simmons further stated that "skin color determines what adults expect from thousands of children—and what those children ultimately expect of themselves." Similarly, Greg Malhoit observed that "the statistics portray a tragic picture of minority educational achievement. Despite the end of segregation, the quality of a child's education still depends in large part on skin color." This statement suggests that the Black–White test score gap might be a manifestation of a greater societal ill: a racial divide (Minority Achievement Report, Trends in Subgroup Performance 2001).

The National Task Force on Minority Achievement (College Board 1999) concluded in its Reaching the Top report that "while it is difficult to quantify the overall negative impact of prejudice and discrimination on the educational fortunes of underrepresented minority students, we have strong reason to believe that it is large."

Barton (2003) found links between student achievement and core factors related to students' racial, ethnic, and socioeconomic status (Educational Testing Service [ETS] 2006).

In addition to the cultural, environmental, structural, and instructional arguments for closing the achievement gap, there are strong economic arguments for doing so. Ladson-Billings (2006) argues that a focus on the achievement gap is misleading. Instead, we need to look at the "education debt" that has accumulated over time. This debt comprises historical, economic, sociopolitical, and moral components. Ladson-Billings draws an analogy with the concept of national debt, which she contrasts with that of a national budget deficit, to argue the significance of the education debt. A 2009 report by the consulting firm McKinsey & Company asserts that the persistence of the achievement gap in the United States has the economic effect of a permanent national recession. The report claims that, if the gap between Black and Latino performance and White student performance had been narrowed, the gross domestic product (GDP) in 2008 would have been $310–525 billion (2–4%) higher. If the gap between low-income students and their peers had been narrowed, the GDP in the same year would have been $400–670 billion higher (3–5%). McKinsey & Company (2009) has provided strong evidence that narrowing the gap would have a positive economic and social impact. Jeneks and Phillips (1998) have argued that narrowing the Black–White test score gap "would do more to move the United States toward racial equality than any politically plausible alternative."

Factors Contributing to the Achievement Gap

The first step toward closing the achievement gap and attracting more minority students to STEM fields is to understand the dynamics that suppress their achievement. The achievement gap is a matter of race and class. Researchers have tried to pinpoint why race and class are such strong predictors of students' educational attainment. In the 1990s, the controversial book, *The Bell Curve,* claimed that gaps in student achievement were the result of variation in students' genetic makeup and natural ability—an assertion that has since been widely discredited. Many experts have since asserted that achievement gaps are the result of more subtle environmental factors and "opportunity gaps" in the resources available to poor versus wealthy children. Being raised in a low-income family, for example, often means having fewer educational resources at home, in addition to poor health care and nutrition. At the same time, studies have also found that children in poverty whose parents provide engaging learning environments at home do not start school with the same academic readiness gaps seen among poor children generally (U.S. Department of Education 2000; Viadero 2000; Sparks 2011).

Researchers have provided several factors contributing to the achievement gap and preventing Blacks and other minority students from achieving success in education, including mathematics and science. They can be summarized in the following grouping:

- Teachers lack skills to deliver instruction to low-performing students.
- Schools that have a history of low performance lack rigor in mathematics and science programs. The curriculum is often watered down, and instruction is not designed to challenge students to perform at high levels.
- Inadequate resources to deliver challenging instruction in STEM programs.
- Tracking students into classrooms where both teachers and students perform at low levels.
- Racial and linguistic minority students and low-income students historically have not been provided equitable access to resources, instruction, and opportunities to achieve at high levels.
- Placement of teachers with minimal teaching skills and experiences with high needs students.

The author (Clark 2013) believes that the key factors contributing to the achievement gap can be summed up in two words: equity and access. Overall, minority students have less access to: (1) well-qualified mathematics and science teachers, (2) strong mathematics and science curriculum, (3) resources, (4) classroom opportunities, and (5) information.

Minority Students Have Less Access to Well-Qualified Mathematics and Science Teachers

Teacher quality can contribute to the achievement gap. Good teaching matters more than anything else, but Blacks and other minority students get less than their fair share of qualified teachers. Minority students get more inexperienced teachers—teachers with 3 or fewer years of experience. As shown on the table below, inexperienced teachers are twice as likely to be in schools with a high level of minority enrollment than in schools with a low level.

Minority students get more inexperienced* teachers

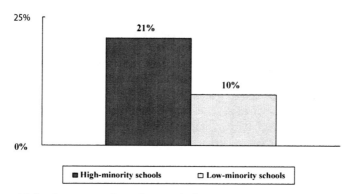

*Teacher with 3 or fewer years of experience. "High" and "low" refer to top and bottom quantities.
Source: National Center for Education Statistics, "An Indicators Report," December 2000.

More classes in high-poverty schools are taught by the least-qualified teachers*

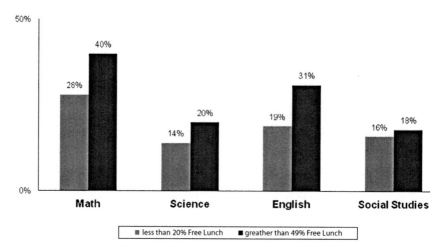

*Teachers who lack a major or minor in the field.
Source: National Commission on Teaching & America's Future, What Matters Most: Teaching for America's Future (p.16) 1996.

The least-qualified teachers are often assigned to teach minority students. More classes in high-minority schools than in low-minority schools are taught by out-of-field teachers—teachers lacking a college major or minor in the field. High-minority schools contain 50% more minority students. Low-minority schools contain 15% or fewer minority students.

More classes in high-minority schools are taught by out-of-field teachers*

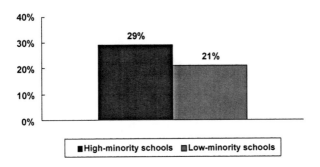

*Teachers lacking a college major or minor in the field.
Source: Education Trust (2003).

Further, teachers and principals in low-income, high-minority, inner-city schools all report problems with teacher interest, motivation, preparation, and competence in mathematics and science instruction. These problems are more evident at the secondary level, where "nearly all types of secondary schools tend to place their least qualified teachers with low-ability classes and their most qualified teachers with high ability classes."

Teacher Experience and Attendance

Percentage of Teachers with Three or Fewer Years of Experience, 1998

Percentage of Fourth-Grade Students in Schools Where Same Teachers Started and Ended the Year, 2000

Percentage of Twelfth-Grade Students Where 6 to 10 Percent of Teachers Are Absent on Average Day, 2000

Level of Minority Enrollment

Low — 10

Medium — 13

High — 21

Race/Ethnicity

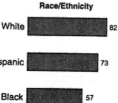

White — 82

Hispanic — 73

Black — 57

Race/Ethnicity

White — 11

Hispanic — 25

Black — 23

Income

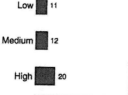

Low — 11

Medium — 12

High — 20

School Lunch Program

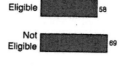

Eligible — 58

Not Eligible — 69

School Lunch Program

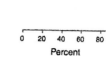

Eligible — 22

Not Eligible — 15

Percent

Note: Low, medium, and high are defined as the schools in the bottom quartile, the middle two quartiles, and the top quartile, respectively. Low income is defined as the percent of students eligible for free or reduced-price lunch.

Source: From Mayer, et al., 2000, which cites the Fast Response Survey System's Teacher Survey on Professional Development and Training, NCES, 1998.

Source: http://nces.ed.gov/ nationsreportcard/ naepdata/getdata.asp, 1/12/03. Data are for public schools.

Source: http://nces.ed.gov/ nationsreportcard/ naepdata/getdata.asp, 1/12/03. Data are for public schools.

Minority Students Have Less Access to a Rigorous High-Level Curriculum

Research shows that students' academic achievement is closely related to the rigor of the curriculum. Poor and minority students have less access to high-level curriculum.

Rigor of Curriculum

Percentage of High School Graduates with Substantial Credits in Academic Courses, 1982 and 1998

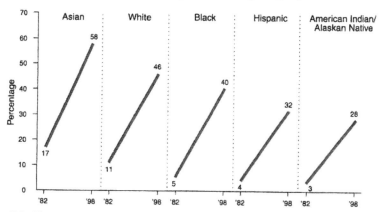

Percentage with four years of English, three years each of social studies and mathematics, and two years of a foreign language

Source: National Center for Education Statistics, *Digest of Education Statistics 2001*, Table 143. Original data from National Center for Education Statistics, *High School Transcript Study.*

Distribution of Advanced Placement Examinations Compared with the Distribution of the High School Population, by Race/Ethnicity, 1999/2002*

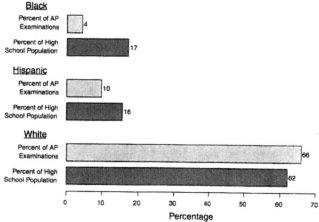

*AP examinations are for 2002; high school population data are for 1999.
Sources: AP data are from the College Board; high school population data are from National Center for Education Statistics, *Digest of Education Statistics 2001*, Table 42.

Minority students consistently achieve and participate less in mathematics and science and have less access to mathematics and science and high-level curriculum. They experience less extensive and less demanding courses and programs. They are less likely to have completed advanced mathematics and science courses.

Differences exist in mathematics and science being taken across racial groups. For example, fewer African-American students are enrolled in Algebra II. Whereas 62% of White and 70% of Asian students had taken Algebra II in 1998, only 52% of African Americans, 48% of Hispanics, and 47% of American Indians had taken this course.

The percentages of African-American, Hispanic, and American-Indian graduates taking chemistry and physics are well below those of White and Asian graduates. In 1998, 63% of White and 72% of Asian high school graduates had taken chemistry, and 31% of White and 46% of Asian students had taken physics. In 1998, 53% of African Americans, 46% Hispanics, and 47% American Indians had taken chemistry; only 21% of African Americans, 19% of Hispanics, and 16% of American Indians had taken physics (NSB 2002).

Minority high school graduates are also less likely to have completed advanced mathematics and science courses, and they are less likely to be enrolled in a full college-prep track.

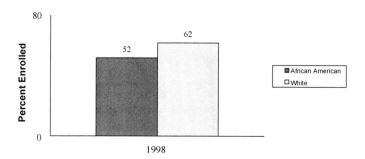

Source: CCSSO, State Indicators of Science and Mathematics Education, 2001.

Minority Students Have Less Access to Resources

Research shows that school districts where low-income, high-minority students are educated consistently receive less state and local money to educate them than do the districts serving the smallest number of minority students. They received approximately $614 less per student per year in 2003 (Education Trust 2006).

Students in low-income, high-minority schools appear to have less access to computers and computer staff, science laboratories, and related resources. They also lack access to science classes and rigorous science curriculum.

Inequities of technology access exist in America's schools. School access, however, does not always mean classroom access, and a digital divide between rich and poor schools still exists. Schools with high-minority enrollment have less access to the Internet than do schools with low-minority enrollment. Access to technology is more of a given for White students than for minority students. Data from the U.S. Department of Education, National Center for Education Statistics (NCES 2000) revealed that Internet access in classrooms varies according to school characteristics. For example, in 1999, 39% of instructional rooms had Internet access in schools with a high percentage of low-income students or high concentrations of poverty compared with 62–74% in schools with low concentrations of poverty.

Minority Students Have Less Access to Classroom Opportunities

Teachers of low-income and minority students place less emphasis on essential curriculum goals such as developing inquiry and problem-solving skills. In low-ability tracks, almost all goals are less emphasized, expectations are lower, and instruction is less engaging. There are inequities in school funding. Students from non-White ethnic groups, with the exception of Asian Americans, appear more likely to attend a disadvantaged school, in terms of affluence and resources. The disadvantaged schools are more likely to have low teacher morale, deteriorating school facilities, fewer materials, lower-quality or nonexistent laboratory opportunities, lower student motivation, and fewer certified teachers—especially for science. Nationwide, only about 65% of eighth-grade teachers report adequate facilities for laboratory science (NSB 1996). Performance on the 1996 NAEP in science was higher for students from well-equipped classrooms.

Minority Students Have Less Access to Information

Low-income, rural, and minority parents have less access to information regarding educational opportunities for their children.

In summary, some of the major factors that contribute to the achievement gap in mathematics and science include inequity in access to qualified teachers, facilities, resources, challenging science and mathematics curricula for minority students, and too few students taking advantage of advanced coursework. School characteristics (such as courses offered and teacher education and experience), student characteristics (such as family income), and mathematics and science course-taking all correlated with academic achievement (U.S. ED/NCES 2000c). In addition, national, state, and school district policies regarding teacher qualifications and curricula vary, resulting in differences in access to high-quality teachers and higher-level mathematics and science courses.

The State of Mathematics and Science in the United States

International Comparisons of Student Science and Mathematics Performance

There is a growing concern that the United States is not preparing a sufficient number of students in mathematics and science. The U.S. falls behind other countries in mathematics and science. Although the most recent NAEP results show improvement in U.S. students' knowledge of mathematics and science, the large majority of students fail to reach adequate levels of proficiency. For example, among the 40 countries participating in the 2003 Program for International Student Assessment (PISA), the U.S. ranked 24th in science literacy and 28th in mathematics literacy. Compared with students in other countries, U.S. students are not achieving at high levels, and U.S. students fare worse in international comparisons at higher-grade levels than at lower-grade levels.

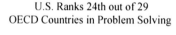

U.S. Ranks 24th out of 29
OECD Countries in Problem Solving

Source: Organisation for Economic Co-operation and Development (OECD), PISA 2003 Results, data available at http://www.oecd.org/

U.S. Ranks 24th out of 29
OECD Countries in Mathematics

Source: Organisation for Economic Co-operation and Development (OECD), PISA 2003 Results, data available at http://www.oecd.org/

Two mathematics and science assessments place U.S. student achievement in mathematics and science in an international context: the Trends in International Mathematics and Sciences Study (TIMSS; called the Third International Mathematics and Science Study in 1995) and the Program for International Student Assessment (PISA). PISA assesses the performance of 15-year-olds in mathematics and science literacy every three years. Most countries participating in PISA are members of the Organisation for Economic Co-operation and Development (OECD), although the number of participating non-OECD nations and regions is increasing. Most OECD countries are economically advanced nations (NSB 2012).

PISA is a literacy assessment, not a curriculum-based assessment. It measures how well students apply their knowledge and understanding to real-world situations. The term *literacy* indicates its focus on the application of knowledge learned in and outside of school.

The results from the two assessment programs paint a critical picture. In 1995, on the TIMSS, U.S. students performed slightly better than the international average in mathematics and science in grade 4, but by grade 8 their relative international standing had declined, and it continued to erode through grade 12. Of the 25 other countries participating in the fourth-grade component of the assessment, 12 had lower average mathematics scores and 19 countries had lower scores in science than the U.S. (NSB 2004). The eighth-grade students in the U.S. scored below the international average in mathematics but above the international average in science (NCES 1997b; NSB 2004).

The fourth- and eighth-grade results from the 1995 TIMSS study suggest that U.S. students perform less well on international comparisons as they advance through school. Four years later, a repeat study focused on the (TIMSS-R) performance of eighth-grade students in 38 countries. In 2000, the PISA assessed 15-year-olds from 35 countries in reading, mathematics, and science. TIMSS and TIMSS-R measured mastery of curriculum-based scientific and mathematical knowledge and skills. PISA assessed students' scientific and mathematical literacy, with the aim of understanding how well students can apply scientific and mathematical concepts. U.S. 12th-grade students performed below the 21-country international average on the TIMSS test of general knowledge in mathematics and science (NCES 1998; NSF 2012).

Despite recent improvement, U.S. PISA scores in mathematics remain consistently below the OECD average and also below those of many non-OECD countries. In the most recent PISA test in 2009, the U.S. average score of 487 fell below the OECD average of 496 and was lower than 17 of the 33 other OECD nations, including Republic of Korea (546), Finland (541), Switzerland (534), Japan (529), Canada (527), and the Netherlands (526). The U.S. score was also lower than the scores in several non-OECD regions/countries/economics, such as Shanghai, China (60); Singapore (562); and Hong Kong (555). In 2009, U.S. students demonstrated higher mathematical literacy than did students in only 5 out of 34 OECD countries (Greece, Israel, Turkey, Chile, and Mexico; NSF 2012).

U.S. students performed relatively better in the PISA science assessment. The average science literacy score of U.S. 15-year-olds improved by 3 points from 2006 to 2009. Whereas U.S. students scored lower than the OECD average in 2006 (489 versus 498), this gap was not evident in 2009 (502 versus 501). The U.S. gains in science since 2006 were driven mainly by improvements at the bottom of the performance distribution; performance at the top remained unchanged (OECD 2010b).

Despite improvement, the 2009 U.S. score (502) was below that of 12 OECD nations (512–554). U.S. students scored lower than students in the five top-performing OECD nations (Finland, Japan, Republic of Korea, New Zealand, and Canada) by 27–52 points. U.S. students also lagged behind their peers in (non-OECD) Shanghai, China; Hong Kong; and Singapore (by 40–73 points). The U.S. 90th percentile score in scientific literacy was 629, below the corresponding scores in 7 of 33 other OECD nations (642–667) (OECD 2010b; *The Chronicle of Higher Education,* Dec. 4, 2007).

According to a report by *The Washington Post* (December 10, 2008), U.S. students are doing better on TIMSS than they were in the mid-1990s.

TIMSS results released in December 9, 2008 show how fourth- and eighth-grade students in the U.S. measure up to peers around the world. The U.S. students made notable strides in mathematics. Since 1995, the average score among fourth-grade students has jumped 11 points, to 529. However, students in Hong Kong, Singapore, Taiwan, Japan, Russia, and England were among those with a higher average. Hong Kong topped the list with an average score of 607.

Eighth-grade students also had a higher average score than in 1996 and bested their counterparts in 37 countries. However, they lagged behind peers in Taiwan, South Korea, Singapore, Hong Kong, and Japan, among other peers.

In summary, the results from the two assessment programs paint a complex picture. U.S. students scored above the international average in the TIMSS assessment and below the international average in the PISA assessment. The two programs are designed to serve different purposes, and each provides unique information about U.S. student performance relative to other countries in mathematics and science (Scott 2004). TIMSS provides data on mathematics and science achievement of students in primary- and middle-school grades (grades 4 and 8 in the U.S.). PISA reports the performance of students in secondary schools by sampling 15-year-olds. TIMSS measures student mastery of curriculum-based knowledge and skills. PISA places emphasis on student's ability to apply scientific and mathematical concepts and thinking skills to problems they might encounter, particularly in situations outside the classroom.

In both 2006 and 2009, U.S. 15-year-olds scored below those of many other developed countries in the PISA, a literacy assessment designed to test mathematics and science. Nonetheless, U.S. scores improved from 2006 to 2009. The average mathematics literacy score of U.S. 15-year-olds declined by 9 points from 2003 to 2006, and rose by about 13 points in 2009, placing the United States below 17 of 33 other members of the OECD. The average science literacy score of U.S. 15-year-olds was not measurably different from the 2009 OECD average, though it improved by 3 points from 2006 to 2009. The U.S. score was lower than the score of 12 out of 33 other OECD nations participating in the assessment.

National data also indicate that the achievement gap among subgroup of students has not been closed. Not only do American students lag behind their international peers, but also, when student achievement is disaggregated by race, the scores of minority students, who are underrepresented in mathematics and science, are below those of their European and Asian-American peers. TIMSS measures student performance in science topics/content and cognitive skills of knowing, applying, and reasoning.

"While it is good news that fourth grade students have made significant gains in mathematics, it is troubling that our students are still behind their international peers in both mathematics and science—fields that are key to our country's economic vitality and competitiveness," said Representative George Miller (D-California), Chairman of the U.S. House Education and Labor Committee, and "It is increasingly clear that building a world-class education system that provides students with a strong foundation in mathematics and science must be part of any meaningful long-term economic recovery strategy" (Washington Post 2008).

The scores on the two international assessment tests led to renewed calls to bolster mathematics and science in the nation's schools by increasing the ranks of well-prepared teachers and providing other support.

The Policy Framework

Over the last few years, the problems in the nation's schools have rightly risen to the top of the national policy agenda. This year is no exception. The nation seems to understand that its schools are not adequately preparing its students, particularly poor and minority students, for college and careers in the twenty-first century. There is also increased awareness of the long-term social and economic implications of an inadequate education for individuals, the communities in which they live, and the nation as a whole. There is a growing consensus that there is a stronger federal role needed in addressing these issues. Given the severity of the crises, the nation cannot afford to let another generation of students pass through the system unprepared for college and careers.

The nation must ensure that K–12 schools—and their students—are no longer left behind in the education system. Policies must drive systemic reforms to help low-performing students. Numerous studies indicate that schools in the United States are failing to adequately prepare all students for a world that depends more and more on rapidly changing technology. Many students leave American schools without a basic understanding of science, mathematics, and technology. The students most affected are from underrepresented minority and low-income communities. The demand for scientists and engineers is not being met, and schools are not preparing future citizens with an adequate background of knowledge necessary to make decisions about their lives.

The nation's efforts to address the achievement gap have a long history. Expectations to address the achievement gap increased with the *Brown v. Board of Education* desegregation decision in 1954 and with the passage of the Elementary and Secondary Act (ESEA) in 1965, which focused on the inequality of school resources. ESEA authorized grants for elementary- and secondary-school programs for children of low-income families; school library resources, textbooks, and other instructional materials; supplemental education centers and services; strengthening of state education agencies; education research; and professional development for teachers.

The Civil Rights Act of 1964 speared optimism for progress in society as a whole. In 2004, the 50th anniversary of *Brown vs. Board of Education* was observed. On May 17, 2009, its 55th anniversary was observed. It reminded us of how far and how little our education system has actually come. It is astonishing that, more than 50 years after *Brown vs. Board of Education*, a large achievement gap persists. In a statement marking the 57th anniversary (2011), Representative George Miller (D-CA), the ranking Democrat on the House Education and the Workforce Committee, pointed to both progress and obstacles on the road toward equity: "Our federal education laws are rooted in the effort to uphold this promise, but sadly, education inequalities still exist on many levels in this country." Miller said, "They exist when children in the poorest schools are denied access to great teachers and they exist when school districts allow dropout factories to fail our students."

All children are entitled to a solid education in the United States. There is a good reason for this: For generations, education has been the most reliable path to a better quality of life, including access to good jobs and careers. Ensuring that every child gets a solid education will go a long way toward fulfilling America's promise of equal opportunity for all (Education Trust 2001).

The Black community has long recognized the central importance of education. That is why Black Americans have fought so hard for educational opportunities throughout this country's history. Although Black Americans won the right to equal access in public schools more than 50 years ago, the struggle for educational excellence and equity did not end with the victory in *Brown vs. Board of Education*. There is still much work to be done to ensure that Blacks and other minority children get the best education. Schools serving minorities often lack the money, qualified teachers, textbooks, and other instructional materials needed to serve their students. Even when minority students attend "better" schools, they often are not given the best teachers, not assigned to the most challenging courses, and not educated to their full potential (Education Trust 2001).

Educational Policies and Reform Initiatives

Numerous studies indicate that schools in the United States are failing to adequately prepare all students for a world that increasingly depends on rapidly changing technology. Many students leave American schools without a basic understanding of STEM education. Most of such students are from minority and low-income communities. The demand for scientists and engineers is not being met, nor are schools preparing future citizens with an adequate background of knowledge necessary to make decisions about their lives.

In an effort to improve the quality of mathematics and science in our nation's schools and to make mathematics and science accessible to all students, major national reform initiatives have been designed. These initiatives have gained wide distribution and have been or are being implemented by a wide range of U.S. schools, universities, industries, and science organizations. These comprehensive initiatives are the No Child Left Behind Act (NCLB), America COMPETES Act, and Race to the Top.

Increasing overall student achievement, especially lifting the performance of low achievers, is a central goal of education reform. This goal is reflected in the federal NCLB Act of 2001, which mandates that all students in each state reach the proficient level of achievement by 2014. This goal is also highlighted in the more recent federal Race to the Top program, which calls for states to design systematic and innovative educational reform strategies to improve student achievement and close performance gaps. The federal government also targets funds directly to low-performing schools through the School Improvement Grants program, for example, to support changes needed in the lowest-achieving schools across the nation.

Among the many factors that influence student learning, teacher quality is critical. To ensure that all classrooms are led by high-quality teachers, NCLB mandates that schools and districts hire only highly qualified teachers, defining "highly qualified" as having attained state certification and a bachelor's degree and having demonstrated subject area competence. Teaching quality has remained in the national spotlight. The Race to the Top program, a component of the American Recovery and Reinvestment Act of 2009, called for applications from states to compete for more than $4 billion for education innovation and reform, including recruitment, professional development, compensation, and retention of effective teachers. Salaries, working conditions, and opportunities for professional development contribute to keeping teachers in the profession and keeping the best teachers in the classroom (Berry et al. 2008; Brill and McCartney 2008; Hanushek and Rivkin 2007; Ingersoll and May 2010).

No Child Left Behind Act (NCLB)

The United States set a national goal of ensuring that each student receives an equitable, high-quality education, and that no child is left behind in this quest. President Bush signed the No Child Left Behind Act (NCLB) on January 8, 2002. This Act focuses on standards and aligns tests and school accountability to ensure that all students in all groups eventually perform at the grade level in all tests, and that schools show continual improvement toward this goal or face sanctions. As written, NCLB required states to immediately set standards in mathematics and reading as well as science and language arts, by 2005. The law requires that every school be held individually accountable for the progress of all students. It expects schools to close all achievement gaps in 12 years. It is a huge expectation given the size of the gap that divides many Whites and middle-class students from those who are poor or minority.

NCLB reauthorizes the Elementary and Secondary Act of 1964. It represents the president's education reform plan and contains changes to the ESEA that was enacted in 1965. This reform gives districts flexibility in how they spend their federal education dollars, in return for setting standards for student achievement and holding students and educators accountable for results. NCLB changes the federal government's role in K–12 education by focusing on school success as measured by student achievement. The Act also contains the president's four basic education reform principles: (1) stronger accountability for results, (2) increased flexibility and local control, (3) expanded options for parents, and (4) emphasis on teaching methods that have been shown to work.

NCLB is federal legislation that enacts the theories of standards-based education reform, formerly known as outcome-based education, which is based on the belief that setting high expectations and establishing measurable goals can improve individual outcomes in education. Prompted by the publication of *A Nation at Risk* in the 1980s, many states initiated a broad set of education policy reforms, including increased course credit requirements for graduation, higher standards for teacher preparation, teacher tests for certification, state curriculum guidelines and frameworks, and new statewide student assessments (CCSSO 2003). The NCLB Act of 2001 reaffirmed the key role of states by requiring them to report on school and district performance using state assessments aligned with state standards in mathematics, science, and language arts. NCLB also required states to ensure that all classrooms have highly qualified teachers in core academic subjects. The Act required states to develop assessments in basic skills to be given to all students in certain grades, if those states are to receive federal funding for schools. NCLB does not assert a national achievement standard; standards are set by each individual state.

NCLB is an ambitious law. The law sets deadlines for states to expand the scope and frequency of student testing, revamp their accountability systems, and guarantee that every teacher is qualified in their subject area. NCLB required states to make demonstrable the annual progress in raising the percentage of students' proficiency in reading and mathematics and in narrowing the test-score gap between advantaged and disadvantaged students. At the same time, the law increased funding in several areas, including K–3 reading programs and before- and after-school programs, and provided states with greater flexibility to use federal funds as they see fit.

The effectiveness and desirability of NCLB's measures generated much discussion and remain debatable.

America COMPETES Act

The America Creating Opportunities to Meaningfully Promote Excellence in Technology, Education, and Science Act (America COMPETES Act) was signed into law on August 9, 2007. The America COMPETES Act was a bipartisan legislative response to recommendations contained in the 2005 National Academies "Raising

Above the Gathering Storm" report and the Council on Competitiveness "Innovate America" report. A wide range of U.S. industries, universities, and science organizations supported it. COMPETES seeks to ensure that U.S. students, teachers, businesses, and workers will continue leading the world in science, innovation, research, and technology.

The law presents a balanced set of policies to improve the country's short- and long-term competitiveness. COMPETES invests in long-term science and research, as well as short-term technology development and innovation. Legislation is directed at increasing research investment, improving economic competitiveness, developing an innovation infrastructure, and strengthening and expanding science and mathematics programs at all points on the educational pipeline. It ensures not only that our nation will produce the world's leading scientists and engineers, but also that all students will have a strong grounding in mathematics and science and are prepared for technical jobs in every sector of the economy. The Act focuses on three primary areas of importance to maintaining and improving U.S. innovation in the twenty-first century: (1) increasing research investment; (2) strengthening educational opportunities in STEM, from elementary through graduate school; and (3) developing an innovation infrastructure. The Act provided research investments in several federal agencies to improve mathematics and science education. The agencies included the National Science Foundation (NSF), Department of Energy's Office of Science, National Institute of Standards and Technology (NIST), National Aeronautics and Space Administration (NASA), and National Oceanic and Atmospheric Administration (NOAA).

The America COMPETES Act was approved by the U.S. Congress as a measure to strengthen the U.S. position within the world's scientific and engineering communities. Although the USA has traditionally been a world leader in these two areas, other countries (e.g., China, India, and Japan) are quickly closing the gap in higher education, scientific knowledge, and technical abilities. The bill was also intended to encourage people to study and teach mathematics and science, along with supporting research into emerging technologies and increasing funds for federal science-based organizations. The America COMPETES Act was a significant step toward a national innovation agenda.

Race to the Top

In 2009, the President Obama administration instituted the Race to the Top (RTTT) program. This program provides financial incentives to states to produce measurable student gains. The primary goals of the program are improving student achievement, closing achievement gaps, and improving high school graduation rates. This initiative is similar to the NCLB Act in that it has many of the same goals, though it places greater emphasis on closing the achievement gap between high- and low-performing schools. The major difference between the two educational reform programs is that RTTT is a competitive grant program that provides incentives for schools to change, whereas NCLB mandated changes in state and local education systems (Lohman 2011).

RTTT is a competitive grant program funded by the U.S. Department of Education as part of the American Recovery and Reinvestment Act of 2009. This program is designed to encourage and reward states creating the conditions for education innovation and reform, achieving improvement in student outcomes, and implementing reform plans in four core areas: (1) adopting standards and assessments that prepare students to succeed in college and the workplace; (2) building data systems that measure student growth and success and inform teachers and principals how to improve instruction; (3) recruiting, developing, rewarding, and retaining effective teachers and principals; and (4) turning around the lowest-performing schools (NSB 2012).The American Recovery and Reinvestment Act of 2009 established a broad four-point framework to improve the K–12 education system. The framework—implemented through the creation of new competitive grant programs and the realignment of existing federal funds—focuses on developing rigorous standards and assessments, improving the effectiveness of teachers and principals, using data to improve performance, and turning around low-performing schools (Executive Office of the President 2010).

Success Stories

The concern for "raising the bar and closing the gap" in educational outcomes is widespread throughout the United States and around the world. Some schools and districts have confronted the inequities in the education system and are providing evidence that improvements are being made in student performance. These schools and districts have provided ways to improve the performance of underserved low-income and minority students and have observed how their performance has changed. They have raised their test scores and graduation rates by providing resources and making community-wide and long-term investments in poor children; creating better early-childhood programs; and using clear, ambitious goals for all students and curricula aligned to those goals.

Gains in reading, mathematics (higher than the national average), and other subjects have been made in the District of Columbia, Virginia, Maryland, Louisiana, South Carolina, Mississippi, Texas, and other states. Some of the success stories are described below.

Virginia

A new U.S. Department of Education (ED) report recognizes Virginia for narrowing achievement gaps between Black and White students in reading and mathematics. The report, *Achievement Gaps: How Black and White Students in Public Schools*

Perform in Mathematics and Reading on the National Assessment of Education Progress (NAEP), compares student achievement in 2007 with the performance in previous years. In comparing the performance of students nationwide with state-level achievement in the national fourth- and eighth-grade reading tests, the ED National Center for Educational Statistics reports the following:

- Virginia is one of only five states with achievement gaps in reading smaller than the nation's in both grades.
- Virginia is one of three states where the achievement gap in grade-4 reading narrowed between 2005 and 2007 because of increased Black achievement.
- Virginia's achievement gap in grade-4 reading is 7 points smaller than the nationwide gap, and in grade 8, the gap is six points smaller.
- Virginia is one of 13 states where fourth-grade reading achievement is higher for both Black and White students than it was in 1992, the first year of NAEP reading tests in grade 4.

The report also credits Virginia for narrowing achievement gaps in mathematics:

- Virginia is one of only four states where fourth-grade mathematics scores increased for both Black and White students between 2005 and 2007.
- Virginia is one of 15 states to narrow the achievement gap in fourth-grade mathematics as a result of Black students outpacing the gains of White students since 1992, when grade-4 NAEP mathematics testing began.
- Virginia is one of 26 states where mathematics scores for both Black and White eighth-grade students increased since grade-8 NAEP mathematics testing began in 1990.

Patricia I. Wright, Superintendent of Public Instruction, said "Closing these gaps will require the continued commitment of educators, parents and community leaders to high standards and accountability," and "The progress cited in today's report provides encouragement that we can eliminate historic disparities even as we seek to raise the achievement of all students" (Virginia Department of Education [VDOE] 2009).

The Virginia Board of Education recently honored two school divisions and 92 schools for raising the academic achievement of economically disadvantaged students. The awards are based on student achievement on state assessments during the 2009–2010 and 2010–2011 school years. Highland County and West Point schools earned the designation of "Distinguished Title 1 School Division" by exceeding all federal ESEA achievement objectives in reading and mathematics for two consecutive years (VDOE 2012).

Maryland

In September 2011, the Maryland State Education Association (MSEA) and public school advocates held a statewide forum on closing the achievement gaps for Maryland students, with educators, students, parents, and community partners. They shared successes and identified strategies for closing the gaps. An outcome of this meeting, local associations, school districts, and communities formed countywide Closing the Achievement Gap committees (Maryland State Association 2009).

Louisiana

Louisiana has made notable progress in their effort to close the achievement gap between races and socioeconomic groups. Based on NAEP data, Louisiana is one of only two states to narrow the achievement gap between Black and White students in both fourth-grade reading and eighth-grade mathematics from 2003 to 2011. Additionally, since the state implemented its accountability system in 1999, the performance gap between Black and White students on state assessments has narrowed by 11.6% in English language arts (ELA) and 11.2% in mathematics. At the same time, from 1999 to 2011, the gap between economically disadvantaged students and their peers also narrowed by 4.4% in ELA and 5.5% in mathematics (Louisiana Department of Education 2011).

These and other states have shown that poor students and minority students can perform well above norms and that the achievement gap can be narrowed if the appropriate instruction, curriculum, and resources are provided. Minority and low-income students in these states have made strides in narrowing achievement gaps and attaining the proficiency level that exceeds the averages in their states.

Each year The Education Trust, a Washington-based research and advocacy organization, identifies and honors high-performing, high-poverty, and high-minority schools (http://www.edtrust.org/dc/resources/success-stories). All of the "Dispelling the Myth" schools, as they are called, have made strides in narrowing achievement gaps, attaining proficiency levels that significantly exceed the averages in their states, or improving student performance at an especially rapid pace. These schools do not offer simple answers or easy solutions, but several common strategies emerge from their practices. They provide a rich curriculum coupled with strong, focused instruction. They have high expectations for all students. They use data to track student progress and individual student needs. They also employ purposeful professional development to improve teachers' skills (The Education Trust 2003). One of these schools includes the Longfellow School, Mount Vernon, New York. This school, with 98% African-American and 83% low-income students, outperformed three-quarters of other New York State elementary schools in mathematics and language arts for 2 years in a row. In 2001, it performed as well or better than 97% of New York schools in mathematics and 88% of New York schools in language arts. Other schools making gains in closing the achievement gap are Norview High School in Norfolk, Virginia, and DC Key Academy in the District of Columbia, where the first-year student gains were double the national average. A school district that is raising achievement for all students while narrowing gaps is Aldine, Texas (The Education Trust 2003). Some states are making gains in closing the gaps. African-American, eighth-grade students are achieving better than the national average in mathematics in Louisiana, Virginia, South Carolina, Mississippi, Texas, and the District of Columbia (The Education Trust 2003).

Concluding Statement

In this chapter, the *achievement gap* refers to the persistent disparity in achievement in mathematics and science between minority (Black, Hispanic, and Native American) and low-income students and White students as measured by standardized test scores obtained from the NAEP, the Nation's Report Card. At each precollege grade level, in both mathematics and science, minority students and students from low-income families have lower scores and are less likely to reach the proficient level than are White students and students from wealthy families. As a result, these students represent only a small proportion of scientists and engineers in the United States. Collectively, Blacks, Hispanics, and Native Americans constitute 24 % of the total US population and 7 % of the total STEM workforce.

Research shows that minorities, particularly Blacks, Hispanics, and Native Americans, are underrepresented and underserved in several areas in STEM. They are underrepresented in the scientific workforce. They are underserved in the education provided, educational resources, and school funding. The most compelling factors are inequity in access to qualified teachers, facilities, resources, and challenging mathematics and science curriculum for minority students. These deficiencies have contributed to an achievement gap. Despite policies calling for "equal opportunities to learn," minority students often do not have a chance to study as rigorous a curriculum as do more privileged students, and they are less likely to be taught by teachers with high levels of experience and expectations.

Education is the key to developing the intellectual capacity of our children—the next generation of innovators, consumers, and citizens. If the United States is to maintain its global preeminence, students must be taught the fundamentals necessary to prepare them. To increase the participation of minority students in mathematics and science and to ensure that all students receive an appropriate, high-quality mathematics and science education, measures should be taken to ensure that minority and underserved students have improved opportunities and greater encouragement to participate fully in mathematics and science education.

Educating all of its students is of critical importance to America's future. Closing the achievement gap will require a national commitment. The promise made by America and articulated by Franklin D. Roosevelt over a century ago must be reclaimed: *"We seek to build an America where no one is left out."*

References

American Educational Research Association. (2004). Closing the gap: High achievement for students of color. *Research Points*, Fall 2004.

Barnett, W. S., & Yarosz, D. J. (2007). Who goes to preschool and why it matter? Revised. *Preschool Policy Brief,* 15.

Barton, P. E. (2009). *Chasing the high school graduation rate: Getting the data we need and using it right.* Princeton: Educational Testing Service.

Building Engineering and Science Talent (BEST). (2004). *A bridge for all.* San Diego: BEST

Clark, J. V. (1996). *Redirecting science education: Reform for a culturally diverse classroom.* Thousand Oaks: Corwin Press.

Council of Chief State School Officers (CCSSO). (2009). *Effects of teacher professional development on gains in student achievement: How meta analysis provides scientific evidence useful to education leaders. Report by Rolf K. Blank, June 2009, under a grant to CCSSO, #REC-0635409.* Washington, DC: CCSSO.

Darling-Hammond, L. (2000). Teacher quality and student achievement: A review of state policy evidence. *Education Policy Analysis Archives, 8*(1).

Davenport, L. R. (1993). The effect of homogeneous grouping in mathematics. Editorial Projects in Education Research Center. 2010. Quality *Counts 2010: Education Week, 29*(17).

Dickens, W. T. (2005). *Genetic differences and school readiness.* New York: Worth.

Educational Testing Service (ETS). (2005). *Affirmative student development: Closing the achievement gap by developing human capital.* Princeton: ETS.

Education Testing Service (ETS). (2009). *Parsing the Achievement Gap II.* Princeton: ETS.

Education Trust (2001). The other gap: Poor students receive fewer dollars. *Education Trust Data Bulletin,* March 6, 2001.

Education Trust. (2003). *African American achievement in America.* Washington, DC: Education Trust.

Education Trust. (2006). *Teaching inequality.* Washington, DC: Education Trust.

Flynn, J. R. (1980). *Race, IQ, and Jensen.* London: Routledge.

Goldhaber, D. D., & Brewer, D. J. (1996). Evaluating the effect of teacher degree level on educational performance. In W. Fowler (Ed.), *Developments in Finance (pp. 197–210). NCES 97–535.* Washington, DC: U.S. Department of Education, National Center for Education Statistics.

Goldhaber, D. D., & Brewer, D. J. (2000). Does teacher certification matter? High school teacher certification status and student achievement. *Educational Evaluation and Policy Analysis, 22*(2), 129–145.

Hallinan, M. (1994). Tracking: From theory to practice. *Sociology of Education, 67*(2), 78–91.

Hanushek, E. A. (1992). The trade-off between child quantity and quality. *Journal of Political Economy, 100,* 84–117.

Hanushek, E. A., Peterson, P. E., & Woessmann, L. (2010). *U.S. math performance in global perspective: How well does each state do at producing high-achieving students?* Cambridge: Harvard University Program on Education Policy & Governance, Harvard Kennedy School.

Haycock, K. (1998). Good teaching matters: How well-qualified teachers can close the gap. *Thinking K–16, 3*(2).

Hernstein, R. J., & Murray, C. (1994). *The bell curve: Intelligence and class structure in American life.* New York: Free Press.

Hirsch, E., Koppich, J. E., & Knapp, M. S. (2001). *Revisiting what states are doing to improving the quality of teaching: An update on patterns and trends,* Seattle: University of Washington Center for the Study of Teaching and Policy.

Hodgkinson, H. L. (2003). *Leaving too many children behind: A demographer's view on the neglect of America's youngest children.* Washington, DC: Institute for Educational Leadership.

Hyunsook, K. S. (2006). *Urban teachers' beliefs on teaching, learning, and students: A pilot study in the United States of America.*

Jencks, C., & Phillips, M. (Eds.). (1998). *The black–white test score gap*. Washington, DC: Brookings Institution Press.

Johnson, L. B. (1965). Speech before the national conference on education legislation. March, 1, 1965.

Ladson-Billings, G. (2006). From the achievement gap to the education debt: understanding achievement in U.S. schools. *Educational Researcher, 35*(7), 3–12.

Lareau, A. (1987). Social class differences in family–school relationships: The impact of cultural capital. *Sociology of Education, 60,* 73–85.

Lohman, J. (2011). Comparing No Child Left Behind Act and Race to the Top. Retrieved April 9, 2011.

Lynch, S. (2000). *Equity and science education reform*. Mahwah: Erlbaum.

Magnuuson, K., & Waldfogel, J. (2005). Early childhood care and education, and ethnic and racial test score gaps at school entry. *The Future of Children, 15,* 169–196.

McKinsey & Co. (2009). *The economic impact of the achievement gap on America's schools*. New York: McKinsey & Co.

Minority Achievement Report, Trends in Subgroup Performance. (2001). Raleigh.

National Assessment Governing Board (NAGB). (2000–2001). *Science framework for the 1996 and 2000 National Assessment of Education Progress (NAEP)*. Washington, DC: NAGB.

National Assessment Governing Board (NAGB). (2002). *Mathematics framework for the 2003 National Assessment of Education Progress (NAEP)*. Washington, DC: NAGB.

National Assessment Governing Board (NAGB). (2008). *Science framework for the 2009 National Assessment of Educational Progress*. Washington, DC: NAGB.

National Center for Education Statistics (NCES). (1996). *Pursuing excellence: A study of U.S. eighth-grade mathematics and science teaching, learning, curriculum, and achievement in international context*. Washington, DC: U.S. Department of Education, Office of Educational Research and Improvement.

National Center for Education Statistics (NCES). 1999. *Educational Statistics Quarterly, 1*(4).

National Center for Education Statistics (NCES). (2000). *Highlights from the Trends in International Mathematics and Science Policy (TIMSS) 203, NCES 2005-005*. Washington, DC: U.S. Department of Education.

National Center for Education Statistics (NCES). (2001a). *The nation's report card: Mathematics 2000. NCES 2001-517*. Washington, DC: U.S. Department of Education.

National Center for Education Statistics (NCES). (2001b). *The condition of education 2001. NCES 2001-072*. Washington, DC: U.S. Department of Education.

National Center for Education Statistics (NCES). (2011). *The nation's report card: Science 2009. NCES 2011-451*. Washington, DC: U.S. Department of Education.

National Center for Education Statistics (NCES). (2000a). *Highlights from the Third International Mathematics and Science Study—Repeat (TIMSS-R). NCES 2001-027*. Washington, DC: U.S. Department of Education.

National Center for Education Statistics (NCES). (2000b). *Pursuing excellence: Comparisons of international eighth-grade mathematics and science achievement from a U.S. perspective, 1995 and 1999. NCES 2001-028*. Washington, DC: U.S. Department of Education.

National Commission on Excellence in Education. (1983). *A nation at risk: The imperative for educational reform*. Washington, DC: National Commission on Excellence in Education.

National Commission on Teaching & America's Future (NCTAF). (1996). *What Matters Most. Teaching for America's Future*. New York: NCTAF.

National Commission on Teaching & America's Future (NCTAF). (1996). *What matters most: Teaching for America's future*. New York: NCTAF.

National Commission on Teaching & America's Future (NCTAF). (1997). *Doing what matters most: Investing in quality teaching*. New York: NCTAF.

National Council of Teachers of Mathematics (NCTM). 2000. *Principles and standards for school mathematics*. Reston: NCTM.

National Research Council (NRC). (1996). *National science education standards*. Washington, DC: National Academy Press.

National Science Board (NSB). (2003). *Report of the National Board Committee on Education and Human Resources Task Force on National Workforce Policies for Science and Engineering.* Arlington: National Science Foundation.

National Science Board (NSB). (2004). *Science and engineering indicators 2004.* Arlington: National Science Foundation.

National Science Board (NSB). (2006). *Science and engineering indicators 2006.* Arlington: National Science Foundation.

National Science Board (NSB). (2008). *Science and engineering indicators 2008.* Arlington: National Science Foundation.

National Science Board (NSB). (2012). *Science and engineering indicators 2012.* Arlington: National Science Foundation. (NSB 12-01).

National Science Board Commission on Pre-college Education in Mathematics, Science and Technology. (1983). *Educating Americans for the 21st Century.* Washington, DC: National Science Foundation.

Nisbett, R. (1998). Race, Genetics, and IQ. In C. Jencks, & M. Phillips (Eds.), *The black–white test score gap* (pp. 86–102). Washington, DC: Brookings Institution Press.

No Child Left Behind (NCLB) Act of. (2001). *Public Law No. 107–110, 115 Stat. 1425 (2002).* Washington, DC: U.S. Congress.

Oakes, J. (1990). Opportunities, achievement, and choice: Women and minority students in science and mathematics. In C. B. Cazden (Ed.), *Review of research in education* (Vol. 16, pp. 153–221). Washington, DC: American Educational Research Association.

Ogbu, J. U., & Fordham, S. (1986). Black students' success: Coping with the "burden of 'acting white." *The Urban Review.*

Organization for Economic Co-operation and Development (OECD). (2003). *Education at a glance: OECD indicators.* Paris.

Organization for Economic Co-operation and Development (OECD). (2007). *PISA 2006: Science competencies for tomorrow's world*, Vol. 1. Paris.

Organization for Economic Co-operation and Development (OECD). (2010a). *Education at a Glance 2010: OECD Indicators.* Paris.

Organization for Economic Co-operation and Development (OECD). (2010b). *Lessons from PISA for the United States: Strong performers and successful reformers in education.* Paris.

Peske, H. G., & Haycock, K. (2006). *Teaching inequality: How poor and minority students are short-changed on teacher quality.* Washington, DC: Education Trust.

President's Council of Advisors on Science and Technology. (2010). *Prepare and inspire: K–12 education in science, technology, engineering, and math (STEM) for America's future.* Washington, DC: Office of Science and Technology Policy, Executive Office of the President.

Researchers bemoan lack of progress in closing education gaps between the races. *The chronicle of higher education*, March 26, 2008.

Sanders, W. L., & Rivers, J. C. (1996). 1986 Cumulative and residual effects of teachers on future student academic achievement. *Research Progress Report.* University of Tennessee Value-added Research and Assessment Center.

Sax, L. (2005). *Why gender matters: What parents and teachers need to know about the emerging science of sex differences.* New York: Doubleday.

Simmons, T. (1999). *The News & Observer.* Raleigh

Steele, C., & Aronson, J. (1998). Stereotype threat and the test performance of academically successful African Americans. In C. Jencks & M. Phillips (Eds.), *The black–white test score gap* (pp. 401–430). Washington, DC: Brookings Institution Press.

U.S. Census Bureau. (2007). *Current population survey, annual social and economic supplement.* Washington, DC: U.S. Census Bureau.

VanderHart, P. G. (2006). Why do some schools group by ability? *American Journal of Economics and Sociology, 65*, 435–462.

White House. (n.d.). *Educate to innovate.* Accessed May 2011

Workforce. (2000). Work and Workers for the Twenty-First Century. 1987. Johnston & Packer.

Chapter 5
Closing the Science, Mathematics, and Reading Gaps from a Canadian Perspective: Implications for Stem Mainstream and Pipeline Literacy

Larry D. Yore, Leslee Francis Pelton, Brian W. Neill, Tim W. Pelton, John O. Anderson and Todd M. Milford

Introduction

What constitutes an achievement gap and how to close the gap when it is identified and verified are critical issues facing Canadian educational systems. Lee (2004) identified eight gaps in education, three of which are the focus of this chapter:

The gap between the current status of student achievement and idealized expectations;

The gap between (a) curricular standards developed at the provincial or territorial level and (b) the goals articulated, instruction delivered, and classroom outcomes; and

The gap between the levels of achievement of two or more groupings of students within an educational system.

Explorations of type 1 gaps provide benchmarks for and insights into future planning, because they are uncoupled from current curricula and instructional practices. Explorations of type 2 gaps address accountability issues related to the status of the current curricula, instruction, and learning. Explorations of type 3 gaps frequently consider differential performance and social justice issues within an educational system. These three gaps are explored as achievement performances of Canadian students on international, national (interprovincial/territorial), and intraprovincial/territorial stages using secondary analyses and systematic reviews of large data sets and published reports.

Public education is responsible for equitably teaching children to achieve. This means that some mastery-teaching approaches were designed to enhance learning and reduce gaps in achievement in order to move everyone to established, prescribed levels of achievement, thereby increasing the average performance and redu-

L. D. Yore (✉) · L. F. Pelton · B. W. Neill · T. W. Pelton · J. O. Anderson
University of Victoria, Victoria, BC Canada
e-mail: lyore@vic.ca

T. M. Milford
Griffith University, Mt. Gravatt, QLD Australia

J. V. Clark (ed.), *Closing the Achievement Gap from an International Perspective*,
DOI 10.1007/978-94-007-4357-1_5, © Springer Science+Business Media B.V. 2014

cing the group variance. Other outcome-based teaching approaches were designed to enhance learning of all students, but the top students tended to outgain the bottom students, thereby increasing both the average performance and group variance. Both of these general instructional approaches are criticized for different reasons; therefore, informed teachers use hybrid approaches where teaching strategies of the second type are used as the basic approach, which is supplemented with individual tutorials and small-group support to bring the low-performing students to minimum standards or achievement levels. These hybrid approaches do not necessarily eliminate group variance, but they reduce it and bring the majority of students to acceptable achievement levels without limiting the performance of the outstanding students.

A concern of Canadian public education is that of achievement differences in national, provincial/territorial, or group performance in which these differences are relatively stable over time and located within specific groups and achievement domains—science, mathematics, and reading. These gaps become serious education and social justice issues when they are large and specifically associated with group membership over which the student has no control or influence; for example, status as a new Canadian, Indigenous person, male or female, English or French language learner, urban or rural dweller, or membership in a particular socioeconomic status (SES). Furthermore, gaps in some domains (e.g., reading or mathematics) may produce associated gaps in other domains (e.g., science) or serve as gatekeeper functions that deny access to future study or employment. These types of gaps are inherently unfair, and substantial efforts ought to be expended to reduce or eliminate them.

This chapter explores Canadian achievement gaps—performance differences in comparisons with other countries (Program of International Student Assessment [PISA]), across provinces and territories (PISA 2009; Pan-Canadian Assessment Program [PCAP]), and within a specific province (British Columbia [BC] Foundation Skills Assessment [FSA]; grade 10 and 12 provincial examinations and course marks)—and potential strategies or approaches that capitalize on achievement patterns and address these gaps. Understanding the Canadian context is essential to making sense of any international, national, or provincial/territorial comparisons and associated claims.

Context

Canada, the United States of America's (USA) northern neighbor, is not well-known other than as the source of "cold air masses," "maple syrup," "hockey," "Cirque du Soleil," and "Celine Dion." Canada is like the USA in many ways but distinctly different in many others. The population of Canada, mostly located within 150 km of the shared border, when compared to the USA's population would lead one to believe that Canada should be less diverse than the USA, but in actuality this is far from being true. Canada, like the USA, is a country of immigrants and ever-increasing diversity.

Canada—A Country of Diversity and Consistency by Design

Whereas the USA is a melting pot that integrates and blends diversities into an ethnic and cultural alloy to strengthen the national character, Canada has chosen a multicultural mosaic that retains individual characteristics and accentuates diversity. From the time of confederation (1867), Canada maintained two official languages (English and French) for its (mostly European) immigrants joining the original Indigenous people. Canada's population subsequently expanded with Asian, Caribbean, and South Asian immigrants, as well as (most recently) African and South American arrivals. Shopping centers and schools are enlivened with a spectrum of languages and skin colors, where Cantonese, Mandarin, Punjabi, and Spanish frequently drown out the official languages.

The cultural, ethnic, and linguistic diversities continue into the organization and control of the Canadian educational systems. Canada does not have a federal office of education because education policy, funding, and organization are the responsibility of the ten provincial (Alberta, BC, Manitoba, New Brunswick, Nova Scotia, Newfoundland and Labrador, Ontario, Prince Edward Island, Québec, and Saskatchewan) and three territorial (Northwest Territories, Nunavut, and Yukon) governments; therefore, it is difficult to make accurate generalizations about education. This complexity has increased with the recognition of Indigenous self-governance. Nonetheless, generally each of the 13 jurisdictions has a Ministry of Education (MoE) headed by an appointed member of the majority party of the government as Minister, an appointed nontenured bureaucrat as Deputy Minister, a senior civil servant as Assistant Deputy Minister, and numerous civil servants as staff responsible for specific functions and programs. The continuity in education policy and procedures comes from the ongoing tenure of the Assistant Deputy Minister and long-serving civil servants. Within these ministries are a variety of systems and programs mandated by the Articles of Confederation or provincial or territorial laws, such as parallel English- and French-language Catholic, private, and public systems, recently expanded to include First Nations, Métis, and Inuit systems in some jurisdictions. These systems employ different organizations involving elementary (K–5, K–6, or K–7), middle/junior secondary (grades 6–8, 7–9, or 8–10), and secondary/high (grades 9–12, 10–12, or 11–12) schools. Language arts, mathematics, and science curricula are developed by the separate jurisdictions (some territories adopt and modify a neighboring province's curricula) that have general focus for K–10 and specialized focus in grades 11–12 (biology, chemistry, calculus, earth sciences, environmental education, mathematics, and physics).

The strong centralized provincial or territorial MoE has a top–down organization vested with funding, curriculum, and licensing powers. MoE provides a consistency within jurisdictions. Unlike some states in the USA, there is little variation in instructional focus, prescribed learning outcomes, class sizes, funding per student, and instructional resources within a specific province or territory. Funding is set on a student basis or scale, with additions from various agencies for special-needs students, English/French language learners, and Indigenous students.

Teacher qualifications and salaries are likewise consistent across most school districts in the jurisdictions. Teacher preparation and licensing are also provincial and territorial responsibilities; however, there are recent employment mobility agreements that facilitate transition of teachers across provincial/territorial boundaries as part of internal trade in Canada (http://www.ait-aci.ca/index_en/labour. htm). Most elementary teachers are educated as generalists with strong language arts background, graduating with a Bachelor of Education (B.Ed.) degree; middle teachers are educated as both generalists and specialists, graduating with a B.Ed. or Bachelor of Arts or Sciences (B.A. or B.Sc.) degree with a post-baccalaureate teaching diploma (B.A. or B.Sc. with a diploma in education), with most being generalists; and most secondary teachers are educated as specialists in one or two disciplines, having a B.A. or B.Sc. with a diploma in education. Teachers are members of a labor organization or federation for personnel, working conditions, and financial interests and may belong to teaching specialist associations for pedagogical reasons. Most of the professional memberships are with provincial or territorial associations, but some teachers belong to the International Reading Association, National Council of Teachers of Mathematics (NCTM), and National Science Teachers Association (NSTA), all headquartered in the USA.

There are informal and formal factors that influence consistency regarding language, mathematics, and science education in Canada. Since most textbook publishers are located in Toronto (English versions) or Montréal (French versions), there appears to be a homogenization of instructional resources developed for provincial and territorial curricula. An inspection of author, editor, and production teams indicates some degree of common membership and an inspection of textbooks would indicate similar commonalities in content. The only formal attempts to influence and coordinate curriculum frameworks are (1) the Western and Northern Canadian Protocol (WNCP), involving BC, Alberta, Saskatchewan, Manitoba, and the Yukon and Northwest Territories, and (2) the Council of Ministers of Education, Canada (CMEC), involving all territories and provinces except Québec. The WNCP for mathematics and the CMEC's Pan-Canadian Framework for science have been and continue to be reasonably successful in influencing interjurisdictional cooperation on mathematics and science curricula while recognizing provincial and territorial rights over the last 12–15 years (McAskill et al. 2004; Milford et al. 2010). The Council of Atlantic Ministers of Education and Training (CAMET) 2008 has developed common K–12 curricula for mathematics, science, and language arts, and they have adopted the WNCP mathematics curricula for use in the four Maritime provinces.

Table 5.1 Canadian participation numbers in PISA 2000–2009

Year of survey	Literacy emphasis	Schools participating	Students participating
2000	Reading	1,117	16,489
2003	Mathematics	1,087	27,953
2006	Science	896	22,646
2009	Reading	978	23,207

Reading, Mathematics, and Science Literacy Achievement—Canada on an International Basis

Canada has been a consistently high-performing participant in PISA over the first complete cycle and through the start of the second cycle: 2000–2009 (Organisation for Economic Co-operation and Development [OECD] 2002, 2009, http://www. pisa.oecd.org). PISA, unlike the Trends in International Mathematics and Science Study, has uncoupled its assessments from curriculum and instruction common across all participants and has coupled the assessments with an idealized interpretation of adult literacies for an information-rich society and knowledge economy. PISA seeks to measure how well prepared 15-year-olds are to meet these challenges (Knighton et al. 2010). It measures three domains, one of which is emphasized during each administration: reading literacy based on information text, mathematics literacy based on problem solving, and science literacy based on application to socio-scientific contexts. The Canadian PISA samples for schools and students across all 4 years were large and relatively consistent (Table 5.1).

The published league tables (i.e., the mean achievements of participating countries in the principal testing domains of each period) provide an initial reference point for summarizing performance patterns. Canada has been one of the high-performing participants, but its rankings have varied as more non-OECD participants joined the international surveys. Table 5.2 summarizes the Canadian students' performance in PISA 2000–2009 relative to other participants.

Table 5.3 summarizes the mean performances and standard errors for each of the literacies measured. OECD countries' scores were standardized to 500 for each literacy measure. These results paint a consistent and relatively strong image of the performance of Canadian students across all literacy domains for each iteration of PISA and the reasonably consistent performance of the provinces and territories around the national means, which may be related to top–down development and management of educational policy, curriculum, and instructional expectations within the provinces and territories.

However, these performance data require secondary analyses to establish clearer understanding of any gaps (Anderson et al. 2007). Anderson et al. (2010) and Milford et al. (2011) found that the correlation coefficients for 2000–2009 student performances for OECD nations in reading literacy, mathematics literacy, and science literacy ranged from 0.75 to 0.88, indicating 56–77% shared variance (Table 5.4). Correlation analyses of the PISA 2006 performance for sampled schools within participating nations and city-states revealed that the variance between schools ranged

Table 5.2 Canadian rankings by literacy in PISA 2000–2009

Year of survey	Number of participating nations	Literacy measure		
		Reading	Mathematics	Science
2000	43	2nd	6th	5th
2003	41	3rd	3rd	11th
2006	57	4th	7th	3rd
2009	65	6th	9th	8th

Table 5.3 Estimated grand mean for Canada in PISA 2000–2009

Year	Estimated grand mean (standard error)		
	Reading literacy	Mathematics literacy	Science literacy
2000	531.44 (1.86)	530.26 (1.32)	521.15 (2.12)
2003	527.21 (1.99)	531.27 (1.33)	517.59 (2.28)
2006	524.60 (2.12)	524.58 (2.58)	531.90 (2.35)
2009	522.08 (1.91)	524.72 (2.04)	526.73 (1.95)

Table 5.4 Correlations between literacy domains in PISA 2000–2009

Year of survey	Reading and mathematics	Science and reading	Science and mathematics
2000	0.78	0.80	0.88
2003	0.75	0.77	0.80
2006	0.78	0.84	0.85
2009	0.78	0.83	0.87

from 7 to 88% (Iceland had the lowest variance and The Netherlands had the highest), while Canada had 17% variance between the schools (Milford et al. 2011).

Data from the first complete cycle (2000–2003–2006) of PISA surveys about student achievement of learning outcomes coupled with student, school, and home characteristics offer information directly related to the performance patterns of the educational system and to achievement gaps for groups defined by these characteristics (Anderson et al. 2010). A contemporary approach widely used in education is multilevel or hierarchical linear modeling (HLM) to reveal such relationships. HLM (Raudenbush and Bryk 2002) is a regression-based analysis that explicitly incorporates into the analysis the hierarchical structure common to many educational data sets—students nested within schools within countries. The data required for these analyses consist of both the achievement (performance) and personal measures of students (level 1) and the measures of school traits for each school (level 2) attended by the students (Anderson et al. 2007). PISA data sets are well suited to multilevel modeling as the national samples were collected through a stratified random sampling technique with schools as primary sampling units. Once schools were collected at random, students aged 15 years were drawn randomly from each sampled school (OECD 2002). This design established a data hierarchy of students nested in schools; thus, a multilevel analysis becomes necessary.

Milford et al. (2011) discussed a number of case studies that explored the differences in students' educational outcomes across several nations participating in PISA. Secondary analyses using these data sets for Canada and other comparators—like the USA, Asia (Hong Kong–China, Japan, and Korea), and other areas of high immigration (Australia, Germany, New Zealand, and Spain)—revealed interesting and informative relationships and unique features. The specifics of these case studies are detailed elsewhere; however, some observations from them are worth noting here. Comparisons between these differing nations in PISA 2003 and 2006 results uncovered similarities in final models on student-level variables such as self-concept, self-efficacy, SES, and positive academic achievement. Students with higher levels of motivation and self-concept tended to achieve higher scores than students with lower levels of these attributes from all nations sampled. Additionally, immigrant status was predictive of negative academic achievement, with some exceptions (e.g., Australia, New Zealand, and the USA). Some differences were uncovered among demographic variables (i.e., gender, SES, family structure, and immigrant status). For example, larger proportions of girls predicted higher achievement in mathematics in Canada only, and SES (although positive) was a larger predictor in Canada than in Asia. Overall, there is a good degree of homogeneity in final models across the nations selected for these case studies, which potentially lend themselves to more universal recommendations.

Reading, Mathematics, and Science Achievements in Canada

The exploration of the potential performance gaps in Canada will consider two perspectives: national and interprovincial/territorial. First, the performance on an international survey (PISA) will be considered on a national basis, and second, the performance on a national survey (PCAP) will be considered on an interprovincial/territorial basis.

Canada's Performance on a National Basis

PISA 2009, which emphasized reading literacy, begins the second cycle of a series of 3-year assessments of knowledge and skills related to the three literacies. The HLM analyses and statistical models identify and describe relationships among important variables between and among student-level measures (level 1) and school-level traits and measures (level 2). PISA assessed students' reading, mathematics, and science literacies with tests and collected student-level and school-level information about school, home, and personal factors through questionnaires. These data can be used to offer a number of scales that describe the background of students

Table 5.5 School-level variance for Canada in PISA 2000–2009

Year	Proportion of school-level variance		
	Reading literacy	Mathematics literacy	Scientific literacy
2000	0.18	0.19	0.26
2003	0.16	0.18	0.17
2006	0.20	0.22	0.25
2009	0.19	0.21	0.19

and the school environment. Much of this information, combined with the outcome measures of reading, mathematical, and science literacies, can be used to explore the differences—gaps—in educational outcomes among students within and across schools.

Often the objective of research on school effects at the national level is to uncover how the background variables (both student and school climate) influence student outcomes and performance patterns across participating schools. This can be achieved with a series of specific multilevel models that address questions such as why schools differ from each other in student outcomes. The initial outcome of HLM analyses is the intra-class correlation coefficients generated by running an unconditioned model (the null model; Raudenbush and Bryk 2002, p. 24). This model contains only an outcome variable with no independent variable except an intercept. The null model offers two bits of information: the estimated grand mean of the outcome measure adjusted for clustering of students in schools and differing sample sizes and the estimate of variance at the school level (O'Connell and Mc-Coach 2008). The first offers a relative standing compared to other nations in PISA (the outcome measures are scaled with mean 500 and standard deviation 100), and the second offers evidence that variance exists and can be modeled at the school level (i.e., above and beyond student-level differences). The results of the null models from Canadian students in PISA 2000–2009 are provided in Table 5.5.

These results indicate reasonable consistent school-level variance in Canada for the four iterations of PISA across all literacies. Additionally, the fact that school-level variance is consistently close to, if not slightly below, 20% (the data from PISA 2003 shows a range of school-level variance from a low of 6% in Iceland to a high of 63% in The Netherlands) supports two things: First, although much of the achievement variance is at the student level (i.e., ~80%), a fair amount does exist at the school level; the statistically significant between-school variance in the national sample indicates that the average measures of literacy achievement do vary across schools (i.e., science 2009 for Canada; $\tau_{00} = \text{var}(\mu_{0j}) = 1605.92$, $\chi^2(977) = 7248.27$, and $p < .001$). Second, when this school-level variance is compared to other nations, Canada is among the group with the lowest levels (typically, this group includes Finland, Norway, Sweden, Denmark, and Iceland). This suggests that, despite the significant between-school differences in PISA, it is unlikely that sizable gaps exist between the participating schools at the provincial or territorial level. Therefore, parents might be well advised to worry less about their children's school and more about their children's friends!

The relationship between the measures of science, mathematics, and reading lite-racies is strong and positive (~0.80$^+$; Table 5.3). These high correlation coefficients indicate strong associations and much shared variance between the literacies that need to be considered seriously, especially when these correlation coefficients are compared to other high-stakes test results (e.g., Iowa Test of Educational Develop-ment, Stanford 9, and other statewide tests), in which the coefficients are 0.35–0.45, with a range of 10–20% shared variance. The correlation between mathematics literacy and science literacy is not surprising, as mathematics is frequently consi-dered to be the language of science and the gatekeeper to success in science. The associations between reading and mathematics literacies and between reading and science literacies were pleasantly surprising, in that they support fundamental lite-racy in mathematics and science proposed by the interactive dynamic of reading and disciplinary understanding (Alberts 2010; Yore et al. 2007). We believe these dif-ferences and high associations may be in part due to PISA's reading literacy focus on informational text rather than narrative text. The abilities to comprehend and use common informational genre (form–function) were likely better indicators of ma-king meaning *of* and *with* mathematical and scientific texts to solve problems and address socio-scientific issues. Furthermore, with the high associations and small likelihood of gaps between schools, it is reasonable to speculate that achievement gaps in reading, mathematics, and science literacies will be similar and small—but taken together, they point to the potential impact of explicit disciplinary literacy instruction embedded in the study of mathematics and science (Anthony et al. 2010; Pearson et al. 2010; Tippett 2011).

Some of the more complex results from the multilevel models explored differen-ces in students' educational outcomes across the years when PISA was administered in Canada. Simply stated, the most fundamental multilevel procedures for estima-ting school effects indicate the relationships between student-level and school-le-vel characteristics and outcomes. The value of such analyses have been explored in several studies using literacy measures from PISA as the educational outcomes and background data from students and principals—again from PISA—as student characteristics and school characteristics. Despite the complexity of determining the difference that males and females have in terms of educational outcomes, inter-national data suggest that females are advantaged in reading (OECD 2009). In fact, this difference has been documented to be on the rise across PISA from 2000 to 2006. In contrast to females, males show advantages in both mathematics and sci-ence.

These observations were also observed for Canada in PISA 2009 (Knighton et al. 2010). Females continued to outperform males in reading nationally and across the provinces. Canadian females outperformed males by 34 points (similar to the ave-rage gender gap of 33 points in OECD countries). However, on average, there was a much smaller difference with male students outperforming females by 12 points in mathematics and by 5 points in science. The reasonable performance, relationships among and between student characteristics and achievements, and the constrained variance among Canadian schools provide a sound foundation for moving **science** literacy, **mathematics** literacy, and **technology, engineering** literacy (STEM; we

Table 5.6 Standardized achievement gaps for the ten Canadian Provinces

Province	Reading	Mathematics	Science
Alberta	−0.09	0.01	0.24
BC	−0.14	−0.16	−0.12
Manitoba	−0.28	−0.21	−0.24
New Brunswick	−0.36	−0.39	−0.35
Newfoundland	−0.36	−0.22	−0.15
Nova Scotia	−0.29	−0.43	−0.20
Ontario	0.02	0.06	−0.01
Prince Edward Island	−0.40	−0.51	−0.36
Québec	0.26	0.17	0.11
Saskatchewan	−0.29	−0.39	−0.20

The gaps are based on the difference between the provincial mean score and the Pan-Canadian composite mean score divided by 100

chose T/E to illustrate our belief that these disciplines are on a single continuum) toward a fuller and balanced consideration of the citizenship (STEM mainstream literacy) and career needs (STEM pipeline literacy) of Canada (Let's Talk Science & Amgen Canada Inc. (SSL) 2012; Yore 2011, 2012; Yore et al. 2007). These performance indicators should encourage and allow ST/EM teachers to incorporate more demanding outcomes and challenging teaching approaches into the instructional agendas.

Performance on Interprovincial Basis

The PCAP, administered by the CMEC, provided data suitable to investigate the reading, mathematics, and science achievements of 13-year-old students in Canada, which complemented the PISA in defining achievement domains, age of students, and year of administration. PCAP surveys students on a tri-annual basis; and the 2007 data (latest available) were used here to report gaps (i.e., the standardized mean differences) for provinces, gender, and Indigenous status. The mean differences between the target groups or a specific province and the overall group average are expressed in terms of the PCAP-scaled standard deviation of 100.

The ten provincial performances vary somewhat across the discipline being assessed and administration of assessment. (N.B. The only territory to participate was the Yukon; therefore, the following analysis focuses only on the provinces to simplify the discussion.) The general trend is that Québec, Ontario, and Alberta were top performers and Manitoba, Newfoundland and Labrador, Nova Scotia, and Prince Edward Island were bottom performers, with slight variations across reading, mathematics, and science (SSL 2012). The 2007 PCAP achievement gaps associated with provincial-level results (Table 5.6) indicated the relative performance among provinces, with Québec showing the highest mean performance on reading (0.26 standard deviations above the Pan-Canadian mean) and mathematics (0.17) and Alberta having the highest relative performance in science (0.24). BC, the focus of

Table 5.7 Standardized gender and Indigenous status achievement gaps

Gap	Reading literacy	Mathematics literacy	Science literacy
Gender	0.24[a]	0.00	0.00
Indigenous status	0.30*	0.47[b]	0.28[b]
Indigenous status × male	0.24[b]	0.43[b]	0.27[b]
Indigenous status × female	0.35[b]	0.52[b]	0.30[b]

The gap is the difference between group means divided by the PCAP standard deviation
[a] Denotes that the gap favors females
[b] Denotes that the gap favors non-Indigenous students

a case study reported later, performed consistently below the Pan-Canadian average in all three disciplines (−0.12 to −0.16).

The basic PCAP results for student gender and Indigenous status are reported in Table 5.7. There is a gender gap in reading that favors females by about a quarter of a standard deviation, whereas there are no gender gaps in performance in mathematics or science. The results indicated a consistent gap in achievement between Indigenous and non-Indigenous students in all three domains with non-Indigenous students outperforming Indigenous students by approximately a quarter standard deviation on reading and science and almost a half standard deviation on mathematics. When the data were further aggregated by Indigenous status and gender, the gaps for Indigenous males tended to be less than the gaps for Indigenous females.

Aggregating provincial-level results by student gender and Indigenous status yields more variation in the gaps (Table 5.8). The gender gaps in reading have different magnitudes but consistently favor females. The gender gaps in mathematics and science that were nonexistent for the Pan-Canadian data (Table 5.7) vary across the provinces. For example, four gaps favor females and five gaps favor males in mathematics; five gaps favor females and three gaps favor males in science. The male students in Alberta outperformed females in mathematics (0.10), whereas Québec females outperformed males by the same margin (0.10). Only New Brunswick in mathematics and Alberta and Nova Scotia in science did not demonstrate gender gaps at the provincial level.

In general parlance, gaps imply separation or space between at least two entities. It should be noted that we have represented the achievement gap as the standardized difference between group means, suggesting separation in terms of achievement. However, if we represent the distributions of these data as histograms, the gaps are not as distinctive. The gender gap of 0.24 standard deviations in reading achievement (Fig. 5.1) is more of an offset of overlapping patterns, with the range of achievement of male students (bottom distribution) covering the full range of female (top distribution) achievement (i.e., no separation). Even the larger Indigenous status achievement gap in mathematics (0.47) can be viewed as an offset of overlapping distributions of Indigenous (top distribution) and non-Indigenous (bottom distribution) students' performances (Fig. 5.2).

In interpreting achievement gaps, the magnitude and direction of the gap constitutes the main focus; however, attention needs to be paid to how the gaps are actually calculated. For example, the Indigenous status gap in reading for Québec

Table 5.8 Standardized achievement gaps for gender and Indigenous status for the ten Canadian Provinces

Province	Reading gender/ indigenous	Mathematics gender/ indigenous	Science gender/ indigenous
Alberta	0.20[a]/0.28[c]	0.10[b]/0.68[c]	0.00/0.48[c]
BC	0.15[a]/0.22[c]	0.08[b]/0.50[c]	0.05[a]/0.40[c]
Manitoba	0.17[a]/0.27[c]	0.02[a]/0.48[c]	0.10[a]/0.30[c]
New Brunswick	0.29[a]/0.32[c]	0.00/0.25[c]	0.03[b]/0.26[c]
Newfoundland	0.34[a]/.30[c]	0.12[b]/0.12[c]	0.03[b]/0.07[c]
Nova Scotia	0.21[a]/0.23[c]	0.06[b]/0.23[c]	0.00/0.14[c]
Ontario	0.21[a]/0.18[c]	0.05[b]/0.41[c]	0.01[b]/0.11[c]
Prince Edward Island	0.23[a]/0.21[c]	0.04[a]/0.31[c]	0.02[a]/0.26[c]
Québec	0.32[a]/0.45[c]	0.10[a]/0.17[c]	0.01[a]/0.13[c]
Saskatchewan	0.16[a]/0.38[c]	0.06[a]/0.55[c]	0.17[a]/0.55[c]

The differences between the student gender or Indigenous status mean scores divided by PCAP standard deviation
[a] Denotes an advantage to female students
[b] Denotes an advantage to male students
[c] Denotes an advantage to non-Indigenous students

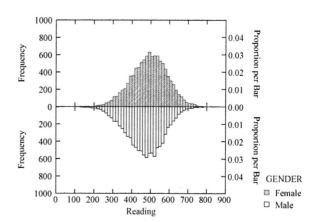

Fig. 5.1 The 2007 PCAP gender gap in reading

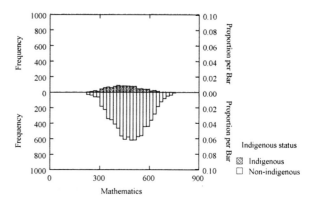

Fig. 5.2 The 2007 PCAP indigenous status gap in mathematics

(0.45) is the largest of the ten provinces, whereas for Ontario the reading gap is the lowest (0.18). However, the actual achievements of Indigenous students in Québec and Ontario are not only equivalent but also the highest in Canada. The reason the gap is high for Québec compared with Ontario is that the non-Indigenous students of Québec obtained higher scores, thereby yielding the larger gap. So although the achievement of Québec-Indigenous students is the highest in the nation (along with Ontario students), the gap suggests issues of equity. These perspectives indicate similarities rather than differences in potentialities, where both gender groups and Indigenous status groups have the ability to perform equally.

Reading, Mathematics, and Science Achievements on an Intraprovincial Basis—The Case of BC

The BC Assessment Branch of the MoE developed the FSA to address accountability issues involving the teaching--learning effectiveness of core domains (i.e., reading comprehension, writing, and numeracy) for identifiable groups. The FSA is conducted annually for students in grade 4 (end of primary or lower-level elementary school) and grade 7 (end of elementary school) in specific school districts and schools to document the effectiveness of early- and middle-year schooling within schools and districts. (N.B. The Assessment Branch does not explicitly publish league tables for school districts and schools.) The FSA results can be used to identify gaps for specific groups of students and monitor the progress of these groups, which have been partially addressed for females and males and for dominant language learners. However, one group has not enjoyed these enhancements—Indigenous students have not been served well by schools in Canada dating back to the residential schools of the late 1800s to the mid-1900s. These experiences have left many Indigenous peoples questioning the value of public education.

Performance of BC Schools

The FSA results provide indications of school and student performances that must be interpreted in the context of other factors, but these results can help school districts, schools, parents, and students to plan and monitor achievement trends. The MoE (n. d.-a) states, "Attempting to rank schools or districts based on FSA results invites misleading comparisons that ignore the particular circumstances that affect achievement in each school" (para 7). However, the Fraser Institute used synthesis procedures that involved several factors, including grade 4 and 7 FSA results for rating elementary schools. The Fraser Institute (Cowley et al. 2012a) reported that private and urban/suburban schools were ranked as top performers, whereas rural and mainly Indigenous schools were bottom performers among the 860 BC elementary

Table 5.9 Percentage of BC Indigenous and non-Indigenous students with specific identified special needs (2010–2011)

Special need	Indigenous (%)	Non-Indigenous (%)	Gap (%)
Behavior disability	6.40	1.90	4.50
Sensory deficit	0.41	0.25	0.16
Learning disability	0.48	0.29	0.19
Gifted	0.32	1.40	1.08

schools considered. The Fraser Institute (Cowley et al. 2012b) used similar procedures to rank 280 BC secondary schools based on several factors, including performance in grade 10 and 12 provincial examination results. The results indicated consistent findings in which private schools ranked the highest and rural and mainly Indigenous schools ranked the lowest. The Fraser Institute reports implied that there were significant differences across the schools, but, in fact, the school-level differences address about 20 % of the variance in achievement (see earlier-reported PISA and PCAP results).

Performance of BC Female, Male, and Indigenous Students

The proportion of BC public school students self-identifying as Indigenous (aboriginal, First Nation, Inuit, and Métis) is about 11 % of the total student population (BC MoE n. d.-b). The total BC public school population (K–12) in the 2010/2011 school year was 515,206, with 63,899 identified as Indigenous (9,908 students were classified as on-reserve and 53,991 as off-reserve). More than half of these schools reported an Indigenous enrollment of less than 10 %, whereas about 6 % (87 of 1,408) indicated that the majority of their students were Indigenous. Approximately 2 % of the public school students are in alternative programs designed to meet the special needs of students who may be unable to adjust to the requirements of regular schools (timetable, schedules, and traditional classroom environment); these programs are normally offered in separate facilities. Enrollment data show that 5.8 % of the Indigenous students (3,680/63,899) were attending alternative programs, whereas only 1.3 % of non-Indigenous students (6,637/515,216) were in alternative programs. The profiles of Indigenous and non-Indigenous students with special needs (disabilities and gifted) are substantially different (Table 5.9). The participation in alternative and special programs suggests that the normal school program does not address the preferences and needs of a substantial number of Indigenous cultures, students, and families.

There is a provision in the FSA for parents to "opt out" of having their children participate, as well as one for schools to not administer the test to students with severe learning disabilities who could not meaningfully participate. The student participation rates (in public schools) are somewhat lower for Indigenous students (76–79 %)

Table 5.10 Grade 4 and 7 gender and Indigenous gaps (differences in the proportions of female and male or Indigenous and non-Indigenous students) for students meeting or exceeding expectations in the 2010–2011 FSA

Grade	Reading gender/Indigenous	Writing gender/Indigenous	Numeracy gender/Indigenous
4	0.06[a]/0.20[b]	0.09[a]/0.22[b]	0.02[a]/0.24[b]
7	0.08[a]/0.19[b]	0.10[a]/0.18[b]	0.01[a]/0.27[b]

[a] Denotes an advantage to female students
[b] Denotes an advantage to non-Indigenous students

than for non-Indigenous students (83–86%) but relatively consistent across topics and grade levels (differences of 6–8%). These participation rates would suggest that their results are reasonably representative of the grade 4 and 7 populations.

Student responses were reported as one of the four performance categories: unknown (i.e., assessment not completed), not yet meeting expectations, successfully meeting expectations, and exceeding expectations (Edudata Canada 2011). The performance gaps (the differences in the proportions of students in comparison groups) in the top two performance categories (i.e., successfully meeting or exceeding expectations) in each of the assessments are reported for female and male students and Indigenous and non-Indigenous students in Table 5.10. These results illustrate that the differences in the proportions of male vs. female students achieving success are small (0.01–0.10) and that the differences between non-Indigenous and Indigenous students are larger (0.18–0.27). The inspection of the differences in proportions of students across domains indicates that gender gaps are the smallest for numeracy and the largest for writing, whereas the Indigenous status gaps are the largest for numeracy and are about the same for reading and writing. The gaps remain relatively consistent across grade levels.

The other sources of information for secondary school achievement are the required formal provincial examinations (BC MoE 2011a) and optional examinations (BC MoE 2011b) for grade 10 and 12 courses in the BC graduation program. The MoE provides examination scores and blended marks comprising examination and course marks. We have used both sets of data to explore Indigenous and non-Indigenous students' achievements.

The comparison of 2010–2011 enrollment and required examination data demonstrated that Indigenous students are substantially underrepresented, with low participation rates in university preparation courses (English 10, Mathematics 10—Foundations or Pre-calculus, Science 10, Social Studies 11, and English 12) and are overrepresented in alternative and specialized First Nations courses (Mathematics 10–Apprentice and Workplace, First Nations Studies, Communications 12). Generally, a smaller percentage of Indigenous than non-Indigenous students participated in the required examinations for the standard and alternative courses. The comparison of these participation percentages revealed gaps of about 18–20% for a single course or combined alternative courses focused on the same graduation requirement. The success (C−/pass or better) rates across the various examinations revealed a consistent pattern where the percentage of Indigenous students achieving a

Table 5.11 Percentage of Indigenous and non-Indigenous students earning good (C+ or better) ratings in required course examinations (2010–2011)

Course	Indigenous (%)	Non-Indigenous (%)	Gap (%)
English 10	42.5	64.2	21.7
Mathematics 10 Foundations and pre-calculus	33.8	59.6	25.8
Apprentice and workplace[a]	12.1	16.9	4.8
Science 10	29.5	57.6	28.1
Social studies 11	36.2	57.8	21.6
BC First nations studies 12[a]	35.7	46.0	10.4
English 12	43.2	61.1	17.9
Communications 12[a]	48.2	51.4	3.2

[a] Denotes alternative course to normal required course

pass or better rating is lower than the percentage of non-Indigenous students achieving the same rating. Whereas the gaps were moderate in English and social studies courses (~5–11 %), the gaps widen for university preparation mathematics and science courses (~14–17 %). However, gaps in the alternative courses were small (<5 %), indicating higher achievement by Indigenous students in these culturally sensitive and responsive courses than in the standard courses.

A better predictor of preparation that allows further study and predicts success requires a somewhat higher level of achievement in these required courses. When the cutoff criterion was raised to indicate the percentage of students earning a good rating (C+ or better) in these examinations, the gaps increased slightly for the alternative courses but nearly doubled for the standard courses (Table 5.11). These data were likely better indicators of the performance gaps between Indigenous and non-Indigenous students and predictors of acceptance into postsecondary studies.

The comparison of 2010–2011 enrollment and examination data for elective academic grade 12 courses demonstrated that Indigenous students are underrepresented, with low participation rates in noncompulsory university preparation courses: English 12 (English Literature), Modern Foreign Languages (French and Spanish), Mathematics 12 (Applications or Principles), Science 12 (Biology, Chemistry, Geology, Geography, and Physics), and Social Studies 12 (History and Geography). Small percentages of students (<1–16.5 % for Indigenous and 2–39.1 % for non-Indigenous) participate in these elective courses. There are variations in participation rates between Indigenous and non-Indigenous students: ~2–5 times higher percentages of non-Indigenous than Indigenous students enrolled in some elective courses. For both Indigenous and non-Indigenous students, the least popular elective courses are modern foreign languages, mathematics applications, geology, geography, chemistry, and physics. The most popular elective courses are biology, principles of mathematics, and history. Generally, small percentages of Indigenous and non-Indigenous students participated in the provincial examinations for these elective courses as the provincial requirement for blended marks was dropped in 2009. The low participation in provincial examinations necessitated the use of course grades

Table 5.12 Percentage of Indigenous and non-Indigenous students earning good (C+ or better) ratings in grade 12 elective course examinations (2010–2011)

Course	Indigenous (%)	Non-Indigenous (%)	Gap (%)
Biology 12	51.4	68.6	17.1
Chemistry 12	61.9	76.9	15.0
Geology 12	51.9	68.0	16.2
Geography 12	65.1	77.9	12.8
Physics 12	68.4	78.6	10.2
Principles of mathematics 12	55.3	72.4	17.1
Applications of mathematics 12	42.6	44.9	2.3
English literature 12	72.4	85.3	12.9
History 12	74.2	74.2	22.2
French 12	76.9	90.2	13.2
French immersion 12	60.2	77.7	17.4
Spanish 12	75.7	87.2	11.6

consisting of marks in teacher-made assignments and tests as achievement indicators, which are reported to the MoE. The pass rates (C− or better) across the various elective courses revealed a consistent moderate pattern where the percentage of Indigenous students achieving a pass or better rating is slightly lower than the percentage of non-Indigenous students achieving the same rating (2.7–8.1 %).

However, if a more rigorous cutoff criterion of the students earning good marks (C+ or better) in these elective courses was used, the gaps favoring non-Indigenous students increased to approximately 10–22 % (Table 5.12). These data were likely better indicators of the performance gaps between Indigenous and non-Indigenous students and, therefore, predictors of acceptance into postsecondary studies and future success in the STEM disciplines. The variation in participation rates makes it difficult to directly compare results; however, as it is likely that a much larger spectrum of non-Indigenous students take these university preparation elective courses, it is likely that the achievement gaps reported here would underestimate the gaps of the larger populations of Indigenous as well as the non-Indigenous students.

Mathematics—the gatekeeper of science, technology, and engineering—is often linked with science, as numeracy skills can impact success in science courses, particularly in the physical sciences (see Table 5.4 for mathematics literacy and science literacy correlations reported in PISA 2000–2009). The WNCP research project reported that many Indigenous students struggle in mathematics (McAskill et al. 2004). Epp (2007) examined this continuing gap using the regression analysis of grade 7 FSA scores, school size, gender, Indigenous status, and ESL status as predictor variables to develop equations to predict student performance on the grade 10 mathematics provincial examination. Indigenous status was the only predictor variable that had a negative coefficient for all three mathematics courses (principles, applications, and essentials). The success in Chemistry 11 and 12 and Physics 11 and 12 courses in particular is highly correlated with success in mathematics prerequisites (Mathematics 10 and 11). Furthermore, these participation and success rates in mathematics appear to be linked to graduation rates.

BC MoE 5-year completion (complete grade 9–12 requirements) data for the years 1995–2002 and 2006–2011 reported graduation rates for Indigenous students of about 36–42% (BC MoE 2011c; Snively and Williams 2006, 2008). An examination of the 6-year completion data for the years 2006–2011 showed a continuing increase in graduation rates for Indigenous students from 48% (2006/2007) to 54% (2010/2011) while participation rates in all of the subjects remained relatively stable. Although the upward trend is positive, it is still a concern that this rate significantly lags behind the 83% completion rate of non-Indigenous students, a graduation rate gap of ~29%. Hargreaves (2011) found that overall the 6-year completion rate for students in an apprenticeship program in 2008–2010 (77%) was comparable to the completion rate for students in all programs (76.5%). The average 6-year completion rate for Indigenous students in apprenticeship programs was 59.2%, which was substantially higher than the overall Indigenous graduation rate of 48.2%. He suggested that provisions for educational programs, which the Indigenous students viewed as more relevant, played a role in the increased completion rates.

Indigenous students are greatly underrepresented in enrollment in most postsecondary programs and institutions. Mendelson (2006) found that 27% of the total population in Canada had attained university graduation, whereas 11% of the Indigenous population had attained a university degree. In particular, Indigenous people are highly underrepresented in science (especially the physical sciences) and engineering occupations (Canadian Council on Learning [CCL] 2007). Although this could indicate a lack of interest in these careers, enrollment and performance in the necessary prerequisites at the secondary school level may be a significant barrier to pursuing postsecondary mathematics- or science-related programs and careers. However, the completion rates were reversed for community colleges and technical institutions where graduation rates are 49% for non-Indigenous and 64% for Indigenous students. Mendelson (Mendelson 2006) stated that, "Failure to complete high school explains 87.8% of the variation in PSE [postsecondary education] completion rates among provinces and territories. This is an extremely strong correlation and is further evidence that [success in postsecondary programs starts with success in the K–12 programs]." Collectively, these participation and success rates for Indigenous students appear to support the need to build STEM programs in relevant, authentic, supportive, and culturally responsive environments.

Building on Successes and Addressing the Gaps

Explorations of science and mathematics performance within the international, national, and provincial/territorial arenas revealed a good news/bad news story: interesting trends and perplexing achievement gaps. Canada has done well on international surveys (PISA 2000–2009), with very high associations at the student level among reading, mathematics, and science literacies. These findings suggest that any discussion about future STEM curriculum and instruction innovations designed to take advantage of the successes while closing the gaps needs to consider the process of planned change, a constellation of interrelated disciplines, and cultural as well as pedagogical factors. Case studies revealed informative relationships among science or mathematics literacies, student-level traits, and school-level characteristics. However, some of these student- and school-level relationships varied in magnitude and direction for the comparators. Interestingly, students with higher science self-efficacy and science self-concept tended to have higher science literacy achievement; however, at the country/city-state level, science self-concept and science literacy were negatively associated, suggesting that jurisdictions with higher science self-concept tended to achieve lower on scientific literacy. It is easy to become overconfident.

Secondary analyses of the PISA 2009 and PCAP 2007 data sets revealed moderate to large differences among provinces, moderate differences among participating schools, small to moderate differences in gender, and moderate to large differences between Indigenous and non-Indigenous students for reading, mathematics, and science achievement at the national level. The interprovincial achievement gaps favored Québec, Ontario, and Alberta over the Maritime provinces and Manitoba, whereas BC was a middle-level performer. Similarly, the analysis of BC's FSA data sets revealed persistent school and slight gender differences for reading comprehension, writing, and numeracy. Unfortunately, there were substantial gaps between Indigenous and non-Indigenous students in FSA results and other examination, success, and participation measures. These persistent differences favored non-Indigenous over Indigenous students. Similar differences have been documented for immigrant student groups, but they have not been as persistent; gender difference has decreased in science and mathematics and persisted in reading. The sources of the Indigenous gaps appear to be the ongoing and long-lasting negative effects of residential schools and assimilationist pedagogies and the lack of cultural awareness, social capital, and respect among Indigenous groups, governments, and other Canadians. Fortunately, the comparison of the distributions in PCAP scores indicated similarities in performance potentials between non-Indigenous and Indigenous students and between females and males.

Collectively, the PISA, PCAP, FSA, and BC case study results illustrate that enhanced STEM literacy and achievement might be as much about culture as they are about pedagogy! The female gender gaps in mathematics and science have been reduced drastically over the last 30 years, but their participation in postsecondary mathematics, physical sciences, and some engineering programs continues to lag behind that of males. Reading gaps for males have persisted, but these gaps appear to be greater for traditional measures of reading than for measures involving information texts and modern sources. Likewise, some success has been demonstrated for Indigenous students in extracurricular internships, postsecondary apprenticeship, and university programs, but much is left to be done. We believe that these efforts must address cultural factors as well as curricular and instructional factors within gender- and Indigenous-appropriate STEM learning environments.

The development of culturally sensitive and responsive learning environments and interdisciplinary teaching approaches has been, in part, the central foci of the Centres for Research in Youth, Science Teaching, and Learning (CRYSTAL) projects funded by the Natural Sciences and Engineering Research Council (http://education2.uvic.ca/pacificcrystal; this website provides links to the other centers and their results). Five regional, multiagency, collaborative centers were located in Eastern Canada (Maritime provinces, University of New Brunswick), Québec (University of Sherbrooke), Manitoba (Prairie provinces, University of Manitoba), Alberta (University of Alberta), and BC (Pacific province, University of Victoria). These centers addressed issues related to STEM and disciplinary literacy, student learning, curriculum, teacher preparation, and professional development and have published the lessons learned in various journals and books. Pacific CRYSTAL established a mission statement as the Centre for Scientific and Technological Literacy that promotes ST/EM literacy for responsible citizenship and ecological sustainability focused on underrepresented and underserved populations using community-based partnerships of universities, faculties within the university, school districts, First Nations, and nongovernmental organizations (NGOs; Yore et al. 2011).

STEM Leadership and Policy Influence

Strategies to take advantage of success and to address STEM gaps flow partially from the Pacific CRYSTAL (Yore et al. 2011) and SSL (2012) reports. Some of these ideas (leadership and policy influence, revitalization of national and regional STEM curricular frameworks, interprovincial and school gaps, gender, and Indigenous status) will be connected to other remediation approaches and may be applicable to other settings beyond Canada. Capitalizing on these reasonable performances and addressing the identified gaps need to consider people's natural reluctance to change—because on the surface, it appears that Canadian STEM education "is not broken, so do not mess with it!" But the time to consider, plan, and manage change is before (not after) it becomes necessary. Planned change is like engineering design and development; it requires defining problems, visualizing solutions, and considering leadership—sponsors, advocates, and change agents.

Canada has an inconsistent history of identifying STEM needs, achievable goals, and focused actions to utilize knowledge, design curricula, and implement innovations in science and mathematics classrooms (Yore et al. 2011). Several efforts and provincial agencies have identified needs and goals before, but they lacked long-term sustained actions to influence policy with evidence-based insights and to implement changes in order to achieve those goals. As an example, the BC MoE recently eliminated its research and curriculum implementation functions. Frequently, leadership has been fragmented by national sponsors and advocates who are not effectively connected to provincial and local leaderships. Recent efforts have identified the STEM mainstream (citizenship) and pipeline (career) needs with potential remediation strategies at the national level (SSL 2012), but it is yet to be seen if provincial-level and local-level sponsors, advocates, and change agents will extend these strategies into schools and classrooms. Based on our experiences, the major concerns are to find influential high-profile sponsors (politicians, professional associations, corporations, stakeholder groups, individuals, and universities/colleges) to promote the innovations and local advocates (professionals, parents, decision makers, superintendents, and school board members) who will stimulate actions. Canada lacks or has underutilized national, provincial, and local professional and educational associations to sponsor and advocate for STEM reforms. But, most importantly, it is necessary to find school- and school district-level STEM change agents (principals, directors of instruction, and lead teachers) who will work with the targets of the change to adapt and implement robust evidence-based innovations that address local resources and constraints. Unfortunately, leadership has been transient and change agents have been lacking—no supervisors of STEM instruction, public advocates for STEM, or leadership-grooming programs. Teachers, the potential change agents, lack a national STEM organization to support their efforts, and those who are members of NSTA and NCTM have their goals, needs, and solutions overwhelmed by the weighted priorities of the USA membership, thereby reducing the impact of these organizations on Canadian problems.

The updated and modernized national and regional STEM curriculum frameworks need to address disciplinary literacy and career awareness for all students; future studies for some students; the communicative, epistemic, and rhetorical functions of language (learning with and from informational text, constructive-interpretative language arts pairs—speaking–listening, writing–reading, and representing–viewing); core ideas; crosscutting concepts; disciplinary practices; and domain-specific self-concept, efficacy, and identity. The education communities need to clarify and establish shared contemporary visions of STEM literacies to guide the healthy tension between current successes and future expectations and to engage curriculum and instructional strategies that enhance future learning performance and close identified gaps.

Science and mathematics literacies for all have enjoyed worldwide cachet for several decades without associated, shared definitions and applications; technology and engineering literacies may be facing similar problems. The experience, research, and scholarship with first-generation curricular reforms and classroom practices need to be used to develop visions of STEM literacies that illustrate the interactive and symbiotic relations among fundamental disciplinary literacy while understanding the big ideas, unifying themes, and crosscutting concepts that will allow people to more fully participate in the public debates about socio-scientific/technical issues resulting in informed decisions and sustainable actions, which will allow a seamless trajectory to address STEM-related careers (Yore 2011, 2012; Yore et al. 2007). Canada, like many countries and city-states, must address the quality of their citizens' STEM literacies (mainstream issue) and its need for STEM-related professionals (pipeline issue). The mainstream and pipeline issues involve social justice; a democratic country must prepare its citizens for full participation, and it is inappropriate for developed countries to recruit highly qualified STEM personnel from less developed countries. The recent expert panel report *Spotlight on Science Learning: A Benchmark of Canadian Talent* (SSL 2012) stated:

We need a robust science culture in this country that goes beyond the classroom, one that's evident in a broader interest in, awareness of and involvement with science. For Canada, strong interest and abilities in STEM is critical. We require it to fill and create rewarding jobs across all sectors. We also need those probing and problem-solving STEM traits to grow a thriving twenty-first-century economy, foster innovative processes and discoveries and keep Canada competitive. As other countries invest more heavily in their STEM learning, we can't afford to be left behind. (p. 5)

The report, recognizing the pattern of success within PISA, went on to state:

However, there is a huge drop-off in the uptake of science courses once they're no longer compulsory, usually after grade 10. By the end of high school, the vast majority of students are taking no science at all. At the post-secondary level, enrolment in STEM fields is up—but that's not as encouraging as it might seem, as enrolment is up in all fields, and the proportion of students studying STEM has not moved. (p. 5)

The expert panel identified 11 dimensions to address the STEM pipeline issue within a knowledge-based economy and society, six of which apply to the mainstream issue central to this chapter: attitudes toward these disciplines and awareness of related careers; encouraging enrollment in optional high school science courses, the monitoring of ongoing student performance on international and national science and mathematics tests, apprenticeships in STEM-related disciplines, and enhanced science culture. The panel made eight recommendations to monitor the outcomes related to elementary, middle, and secondary school STEM programs (SSL 2012):

1. Establish a national forum for ongoing multi-stakeholder discussion related to STEM talent development.
2. Support and scale effective STEM-teaching and -learning programs, in and outside school to revitalize young people's love of science with compelling programming and help youth see how science education is relevant, i.e., it will serve them well no matter what career they envision (and in life, too).
3. Establish or improve tracking and reporting systems required for effective data collection, around participation in high-school STEM programs, and postsecondary applications, registrations and graduation in STEM programs.
4. Build better connections between job forecasts and STEM learning demands— and make this information available to schools in a relevant way—so youth and parents are more aware of future employment opportunities.
5. Build awareness about the breadth of career opportunities that are available with STEM learning.
6. Conduct a system-wide review of STEM curricula across Canada to develop programs that increase interest and participation in STEM studies (optional high-school courses and postsecondary programs).
7. Assess the factors that affect the capacity of universities and colleges to support and maintain STEM studies.
8. Determine a suite of benchmarks, with public input, that can be used to measure the state of the science culture in Canada.

These recommendations reinforce earlier requests to renew and revitalize the national and regional frameworks for STEM curricula (McAskill et al. 2004; Milford et al. 2010). Canada would be well advised to follow the examples of the NCTM (2000) updated mathematics framework and the recent NRC (2012) framework for science. The science, mathematics, and technology/engineering content, core ideas, and practices outlined in these documents could be incorporated into the next generation of Canadian science and mathematics curricula.

Instructional programs should include explicit STEM language and literacy instruction that capitalizes on the communicative, constructive, and persuasive functions of language in doing and learning in these disciplines. Interdisciplinary pedagogies that capitalize on the common grounds of literacy, science, technology, engineering, and mathematics should be central to instruction. Such approaches have been used by Willers (2005) as expanded mathematics 10/11 courses (engineering, science, and mathematics); by Carruthers et al. (2011) as modules focused on fundamental, but complex, ideas in computing sciences; by Tippett (2011) as projects developed to enhance disciplinary knowledge, practices, and literacy; and by Francis Pelton and Pelton (2011) as high-motivation approaches and devices to enhance technological design and learning.

Provincial-level and School-level Gaps

The development and implementation of national and regional STEM frameworks will do much to address the gaps among provinces and territories and among schools within a jurisdiction. We believe that these gaps are likely unique to differences in curriculum, instructional resources, and instructional time assigned to STEM courses. Collaborations such as the Pan-Canadian framework in science and the WNCP and CAMET in mathematics have developed common curricula, instructional resources, and professional development to reduce these differences. One of the guiding principles of the renegotiated WNCP statement (2011) states:

> In providing high quality K–12 education for all students, the WNCP leadership and management explores new opportunities for innovation that focuses on continuous improvement in fostering student learning, the attainment of strategic priorities and goals, and in creating solutions to opportunities and challenges, that yield new and/or improved curriculum, project management models and processes, and research.

Similarly, the CAMET 2009–2012 Strategic Direction document (2008) established its main objective of having Atlantic Canadian students graduate from high school performing at the same level or above that of students from across Canada. Two goals specific to numeracy are "To develop classroom-based assessment tools to assist teachers with monitoring numeracy skills of students; and to provide students with access to an effective next generation math curricula" (CAMET, p. 2).

Gender Gaps

Reading programs and instruction have long had a preferential bias toward females, leaving many male students disinterested and underachieving. This may well be a side effect of instructional resources and measurements that favor narrative texts over information text and testing situations that lack authenticity. Newer reading programs have attempted to include materials of interest to both females and males. Continued effort is needed to explore reading as purpose inquiry and to search for solutions embedded within a broader context of problem-focused learning (Alberts 2010). STEM literacies position reading as a fundamental literacy component and suggest that explicit strategic reading comprehension instruction with meta-cognition consideration should be embedded within ongoing inquiry, design, and problem-solving activities (Tippett 2011).

Most STEM programs have been developed with a narrow but overwhelming consideration of dominant groups: males and university/college-preparation students. Little has been done to address underserved and underrepresented groups until recently as once underachieving female students are now narrowing or eliminating the gap with male students in science and mathematics. Still, there is much to do in some areas of science and mathematics to address the culture and pedagogy; identity experiences and career awareness need continued efforts to engage and challenge females. Science internships in university research laboratories (Hsu 2008) and mapping possible science selves (Marshall et al. 2011) for middle school and high school females have demonstrated growth and success in career awareness and self-concept/self-efficacy. Hsu explored and documented the experiences and perceptions of high-school biology students, mostly females, in a university water-quality laboratory. She found that these females developed improved images of scientists as people and the work that scientists do. Marshall et al. explored junior secondary students' science identities using the possible selves mapping procedure and found that repeated and comparative mapping of "self" helped students develop more informed career awareness and understandings of educational requirements.

Continued progress in addressing gaps in females' science and mathematics achievement needs to coordinate and use both formal and informal learning opportunities. Science Venture (http://www.scienceventure.ca/), a nonprofit program started by the University of Victoria Engineering Students' Society, has a 20+ year history of success offering informal summer programs and teacher workshops, some of which focus on females. Venture Girls Club (http://www.scienceventure.ca/weekend-clubs/) is a unique opportunity for grade 3–6 girls to learn about science and engineering in a safe and fun environment. Other organizations for women in science and engineering seek to engage K–12 female students in science and mathematics to promote interest in STEM careers. The Society for Canadian Women in Science and Technology (http://www.scwist.ca/) sponsors conferences for girls in grades 9–12, hands-on workshops for girls in grades 1–12, eMentoring for girls in grades 11–12, and scholarships for women in science and technology. The University of Victoria Women in Engineering and Computer Science (WECS) is a faculty-sponsored program designed to encourage women and girls to consider engineering or computer science as a career and to support them in their decision (http://www.csc.uvic.ca/Outreach/wecs.htm). WECS provides information and role models and promotes changes to the established culture and teaching methods within schools and universities. WECS holds events that inspire, educate, and build a sense of community, such as mother–daughter workshops, guest speakers, and Lego robotics festivals and workshops. Faculty and staff also provide informal learning outreach efforts to elementary school students across BC with challenging computer science concepts and robotic-programming missions (Carruthers et al. 2011).

Indigenous Status Gaps

Successful remediation of the gaps between Indigenous and non-Indigenous students may be as much cultural as pedagogical, with attitudes, beliefs, and identity playing important roles. Traditional schools have not been welcoming and safe environments for many older Indigenous people. This has had a long-lasting negative effect on the educational engagement and success of recent generations of Indigenous peoples. Fortunately, many forward-thinking Indigenous leaders have decided to give schools and universities/colleges another guarded chance to address the dislocation of Indigenous peoples. We need to use this new trust to make a difference for Indigenous students.

The persistent science and mathematics gaps for Indigenous students involve establishing two-way border crossings between Indigenous and non-Indigenous cultures and mutual benefits (Aikenhead 2001, 2002). Respectful engagement, sensitive use, and honest appraisal of traditional STEM knowledge and wisdom by both cultures (Chinn et al. 2008) and the use of technology and technological design as initial engagement with western techno-science for Indigenous students have demonstrated practical advantages in several Indigenous cultures with highly developed place-based technologies (Yore and Guo 2008). Technology and mathematics, unlike science, do not involve conflicting ontological assumptions that separate spiritualism from physical causality while capitalizing on highly developed and rigorous epistemologies (Aikenhead and Elliott 2010; Francis Pelton 1995; Lewthwaite and McMillan 2007; Lewthwaite and Renaud 2009; Neel 2011; Snively and Williams 2008; Sutherland and Dennick 2002; Sutherland and Henning 2009; Yore 2008).

Overcoming the science and mathematics gaps and moving forward with responsive STEM education require better understanding of the different views of learning and scaling-effective small-scale projects. Indigenous people view learning as holistic, lifelong, experiential, spiritually oriented, and community-based and as rooted in Indigenous language and as integrating Indigenous and western knowledge (Cappon 2008). This view does not fully align with Western views of learning and of science and mathematics and does not capture the potential of STEM education. "The…conventional reporting on learning success of [indigenous] people provides only a partial picture and therefore does not support effective policy [and program] development" (Cappon, p. 61). However, using deficit indicators like special needs, gifted, alternative programs, examination scores, course marks, and participation and completion rates as predictor variables indicated clearly that many policy makers are ill-informed and that there is a lack of alignment between Indigenous students' learning resources and the traditional "one-size-fits-all" school system expectations. Effective models of Indigenous learning and constructivist-oriented approaches need to consider these students' prior experiences and knowledge and the formal and informal opportunities across early, elementary, middle, secondary, postsecondary, adult, and intergenerational learning within the home, school, community, land, and workplace. As well, external assessments need to "underline the critical connection between community regeneration and well-being and individual learning…[and] the relationships that contribute to [indigenous] learning" (Cappon, p. 64). CCL's (http://www.ccl-cca.ca/CCL/Home.html) dialogs were implemented in a First Nations community, using the First Nations Holistic Lifelong Learning Model as a focus for the roles and interactions of language, culture, and history. The following shared priorities were revealed:

1. Becoming a healthy community—spiritually, socially, intellectually, and physically.
2. Increasing parental involvement in learning through teaching of traditional values and virtues in the home, at school, and in the community.
3. Increasing the understanding and awareness of kinship and genealogy.
4. Improving the fluency of [indigenous] language among learners of all ages.
5. Improving the engagement of elders within all learning environments across the community.
6. Creating a learning space (resource centre or library) to facilitate the transfer of cultural and indigenous knowledge.
7. Learning to work together by building relationships across all agencies, organizations, and families within the community.
8. Increasing opportunities for the community to understand, develop respect for, experience and learn from the land.
9. Developing entrepreneurship and employment opportunities within [local] and with neighboring communities. (Cappon 2008, p. 66)

These priorities could serve as tentative design principles—there will be variations across Indigenous groups—for developing future STEM programs and clearly illustrate the Indigenous side of the two-way border crossing needed to facilitate movement between cultures.

Addressing career awareness and identities with self-exploration experiences, STEM internships, and established models of STEM professionals to which Indigenous females and males can aspire has been successful in small-scale projects. Some school districts and First Nations have started to demonstrate advancement and success that provide promising foundations for STEM education. The T'Sou-ke First Nation has developed a 75-kilowatt solar photovoltaic installation on its traditional territory (Sooke, BC) with solar power units on the band office, other buildings, and community houses. This large grid-connected photovoltaic solar energy system reflects Indigenous people's continued reliance on renewable, nonfossil energy sources and provides internship opportunities for active participation and career awareness of Indigenous youth, which could result in a decrease in the gaps currently observed in education (BC Ministry of Small Business, Technology, and Economic Development 2009). The Sooke School District, which serves the T'Souke First Nation, reported an improvement in Indigenous engagement in schools and graduation rates: 38–73 % over the last 5 years (largest aboriginal graduate class ever in SD 62 2012). Again, these gains occur only with effort and changes on behalf of the communities, schools, teachers, and students.

Fisher (2010) explored secondary school Indigenous students' mathematics identities using the *possible selves mapping procedure.* She found that this self-exploration experience enhanced Indigenous students' views of themselves, specifically in relation to mathematics. Many students, but especially Indigenous students, have limited awareness of career choices and insights into how traditional ecological knowledge and wisdom (TEKW) and western knowledge about nature and naturally occurring events integrate into authentic problem solving. Wright et al. (2011) provided experiences to Indigenous students in TEKW and a university water-quality research laboratory within an authentic ecological restoration project sponsored by SeaChange Marine Conservation Society, an NGO. They found that these internships allow border crossings between the knowledge bases and informed one another; however, to be successful, there needed to be two-way border crossings and mutual recognition and respect.

The Heiltsuk First Nation (Bella Bella, BC) has an innovative approach to increase interest in science and technology of their youth. They have collaborated with Pacific Wild, an NGO, to provide an 8-week place-based learning experience for student interns focused on ecology and natural resource management. The Supporting Emerging Aboriginal Stewards (http://www.pacificwild.org/site/great-bear-live/seas-community-initiative.html) project has Indigenous interns working with a field crew and researchers from Simon Fraser University to perform stream and juvenile salmonid assessments, conduct crab surveys, and carry out other research projects in the Koeye watershed. The fieldwork and scientific studies use remote cameras, hydrophones, and other sampling methodologies to collect data and Smart-Board technology to present data. Interns are presented with an array of authentic information and communication technology experiences and information-gathering techniques to study their territorial ecosystems.

The University of Victoria's Aboriginal Connections with Computing, Engineering, and Software Systems (ACCESS, http://www.csc.uvic.ca/Outreach/access.htm) has investigated ways in which computer science and engineering education can be established, connected, maintained, and nurtured in Indigenous communities. ACCESS activities (e.g., technology camps, games with Indigenous content, workshops on algorithms and revolutionary engineering, and Indigenous women acquiring knowledge through engineering) have been held on campus and in local and remote communities. Early results indicate high levels of engagement and success. These place-based education efforts have provided voice to the participants and offer tangible experiences that have the potential to narrow the gaps in the present school experience.

Concluding Remarks

The promises of addressing the interprovincial, gender, and Indigenous status gaps and improved performance in reading, science, and mathematics are central to Canada's social justice and economic future dependent on STEM advancements. There are promising efforts by some NGOs, provinces, and schools, but the overall efforts at the national and provincial/territorial levels are questionable. The *Spotlight on Learning Science* effort by Amgen Canada and the Let's Talk Science Foundation indicates promise among corporations and NGOs, but this promise has been counterbalanced by the cancelation of federal government funding for the CCL and the Natural Sciences and Engineering Research Council's funding of the CRYSTAL initiatives, the reduced funding in provincial ministries of education for STEM education research and curriculum development, and the lack of educational reform efforts by the CMEC.

The critical issue appears to be the lack of local STEM education leadership, bottom–up "grass roots" advocacy, long-term reform effort, and use of STEM education research and scholarship to inform education policy and decisions. Local, provincial/territorial, and national efforts are needed to develop and encourage lead STEM teachers to take on the roles of advocates and change agents. STEM stakeholders, sponsors, and advocates need to realize that reform is not an event but rather a long-term vector process where small advances in the desired direction need to be consolidated before the next advance can be planned and enacted. Likewise, academic organizations and universities/colleges need to encourage and facilitate their scholars and researchers to disseminate research findings and scholarship to various levels of government and education systems. STEM stakeholders need to become aware of the political process and to influence priorities within political parties and governments. In closing, we believe that successful efforts to address mainstream STEM literacy for all citizens will do much to address the STEM pipeline issue by increasing the pool of interested and literate Indigenous people and young females and males.

Acknowledgement The authors would like to acknowledge and thank Shari Yore, SAY Professional Services, Saanichton, BC, for her thorough editorial work and professional and helpful comments on format, style, references, and internal consistency.

References

Aikenhead, G. S. (2001). Integrating western and aboriginal sciences: Cross-cultural science teaching. *Research in Science Education, 31*(3), 337–355. doi:10.1023/a:1013151709605

Aikenhead, G. S. (2002). Cross-cultural science teaching: Rekindling traditions for aboriginal students. *Canadian Journal of Science, Mathematics and Technology Education, 2*(3), 287–304. doi:10.1080/14926150209556522.

Aikenhead, G. S., & Elliott, D. (2010). An emerging decolonizing science education in Canada. *Canadian Journal of Science, Mathematics and Technology Education, 10*(4), 321–338. doi:10.1080/14926156.2010.524967.

Alberts, B. (2010). Prioritizing science education (Editorial). *Science, 328*(5977), 405. doi:10.1126/science.1190788.

Anderson, J. O., Chiu, M.-H., & Yore, L. D. (2010). First cycle of PISA (2000–2006)—International perspectives on successes and challenges: Research and policy directions. Special issue. *International Journal of Science and Mathematics Education, 8*(3), 373–388. doi:10.1007/s10763-010-9210-y.

Anderson, J. O., Lin, H.-S., Treagust, D. F., Ross, S. P., & Yore, L. D. (2007). Using large-scale assessment datasets for research in science and mathematics education: Programme for International Student Assessment (PISA) special issue. *International Journal of Science and Mathematics Education, 5*(4), 591–614. doi:10.1007/s10763-007-9090-y.

Anthony, R. J., Tippett, C. D., & Yore, L. D. (2010). Pacific CRYSTAL project: Explicit literacy instruction embedded in middle school science classrooms Special issue. *Research in Science Education, 40*(1), 45–64. doi:10.1007/s11165-009-9156-7.

British Columbia Ministry of Education. (2011a). *Provincial optional examinations–2010/11; Province—Public and independent schools combined.* Victoria: Author. http://www.bced.gov.bc.ca/reports/pdfs/exams/opt/prov.pdf.

British Columbia Ministry of Education. (2011b). *Provincial required examinations–2010/11; Province–Public and independent schools combined.* Victoria: Author. http://www.bced.gov.bc.ca/reports/pdfs/exams/req/prov.pdf.

British Columbia Ministry of Education. (2011c). *Summary of key information–2010/11.* Victoria: Author. http://www.bced.gov.bc.ca/reporting/docs/SoK_2011.pdf.

British Columbia Ministry of Education. (n. d.-a). *Foundation Skills assessment (FSA)—Results.* Victoria, BC: Author. http://www.bced.gov.bc.ca/assessment/fsa/results/.

British Columbia Ministry of Education. (n. d.-b). *Provincial reports.* Victoria: Author. http://www.bced.gov.bc.ca/reporting/.

British Columbia Ministry of Small Business, Technology and Economic Development. (2009). *First nation caps clean energy success. (News release 2009STED0004-000110).* Victoria: Author. http://www2.news.gov.bc.ca/news_releases_2009-2013/2009STED0004-000110.htm.

Canadian Council on Learning. (2007). *Lessons in learning: The cultural divide in science education for Aboriginal learners.* Ottawa. http://www.ccl-cca.ca/CCL/Reports/LessonsInLearning/LinL20070116_Ab_sci_edu.html.

Cappon, P. (2008). Measuring success in first nations, inuit, and métis learning. *Policy Options,* (May), 60–66. http://www.irpp.org/po/archive/may08/cappon.pdf.

Carruthers, S., Milford, T. M., Coady, Y., Gibbs, C., Gunion, K., & Stege, U. (2011). Teaching problem solving and computer science in the schools: Concepts and assessments. In L. D. Yore, E. Van der Flier-Keller, D. W. Blades, T. W. Pelton & D. B. Zandvliet (Eds.), *Pacific CRYSTAL centre for science, mathematics, and technology literacy: Lessons learned* (pp. 99–112). Rotterdam: Sense.

Chinn, P. W. U., Hand, B., & Yore, L. D. (2008). Culture, language, knowledge about nature and naturally occurring events, and science literacy for all: She says, he says, they say. *L1—Educational Studies in Language and Literature, 8*(1), 149–171. http://l1.publication-archive.com/show?repository=1&article=220.

Council of Atlantic Ministers of Education and Training. (2008). *2009–2012 strategic direction.* Halifax, NS. http://camet-camef.ca/images/eng/pdf/2009-2012%20Strategic%20Direction%20SUMMARY%20FINAL%20English.pdf.

Cowley, P., Easton, S. T., & Thomas, M. (2012a). *Report card on British Columbia's elementary schools 2012 edition.* Vancouver, BC: Fraser Institute. http://www.fraserinstitute.org/research-news/display.aspx?id = 2147484266.

Cowley, P., Easton, S. T., & Thomas, M. (2012b). *Report card on secondary schools in British Columbia and Yukon 2012.* Vancouver: Fraser Institute. http://www.fraserinstitute.org/research-news/display.aspx?id=18259.

Edudata Canada. (2011). *FSA 2011 item level response reports* (Vol. 2012). Vancouver: University of British Columbia. http://www.edudata.ca/apps/fsa_item.

Epp, B. A. (2007). Examining the predictive validity of the FSA on the provincial mathematics 10 examinations (Master of Arts thesis). University of Victoria, Victoria, British Columbia, Canada. http://hdl.handle.net/1828/2464.

Fisher, K. (2010). Aboriginal students' high school mathematics experiences: Stories of opportunities and obstacles (Master of Arts thesis). University of Victoria, Victoria, British Columbia, Canada. http://hdl.handle.net/1828/3103.

Francis Pelton, L. (1995). Multicultural mathematics: Non-European examples of common mathematical relations. In G. J. Snively & A. MacKinnon (Eds.), *Thinking globally about mathematics and science education* (pp. 107–121). Vancouver: Centre for the Study of Curriculum and Instruction, University of British Columbia.

Francis Pelton, L., & Pelton, T. W. (2011). Outreach workshops, applications, and resources: Helping teachers to climb over the science, mathematics, and technology threshold by engaging their classes. In L. D. Yore, E. Van der Flier-Keller, D. W. Blades, T. W. Pelton & D. B. Zandvliet (Eds.), *Pacific CRYSTAL centre for science, mathematics, and technology literacy: Lessons learned* (pp. 113–129). Rotterdam: Sense.

Hargreaves, R. (2011). Youth appenticeship programming in British Columbia (Master of Education thesis). University of Victoria, Victoria, British Columbia, Canada. http://hdl.handle.net/1828/3634.

Hsu, P.-L. (2008). Understanding high school students' science internship: At the intersection of secondary school science and university science (Ph.D. dissertation). University of Victoria, Victoria, British Columbia, Canada. http://hdl.handle.net/1828/1096.

Knighton, T., Prochu, P., & Gluszynski, T. (2010). *Measuring up: Canadian results of the OECD PISA study—The performance of Canada's youth in reading, mathematics and science; 2009 first results for Canadians aged 15*. Ottawa: Statistics Canada. http://www.statcan.gc.ca/pub/81-590-x/81-590-x2010001-eng.pdf.

Largest aboriginal grad class ever in SD 62. (2012, May 16). Sooke news mirror. http://www.sookenewsmirror.com/news/151585385.html.

Lee, J. (2004). Multiple facets of inequity in racial and ethnic achievement gaps. *Peabody Journal of Education, 79*(2), 51–73. doi:10.1207/s15327930pje7902_5.

Let's Talk Science & Amgen Canada Inc. (2012). Spotlight on science learning: A benchmark of Canadian talent. http://www.letstalkscience.ca/our-research/spotlight.html.

Lewthwaite, B., & McMillan, B. (2007). Combining the views of both worlds: Perceived constraints and contributors to achieving aspirations for science education in Qikiqtani. *Canadian Journal of Science, Mathematics and Technology Education, 7*(4), 355–376. doi:10.1080/14926150709556740.

Lewthwaite, B., & Renaud, R. (2009). Pilimmaksarniq: Working together for the common good in science curriculum development and delivery in Nunavut. Special issue. *Canadian Journal of Science, Mathematics and Technology Education, 9*(3), 154–172. doi:10.1080/14926150903118334.

Marshall, E. A., Guenette, F. L., Ward, T., Morley, T., Lawrence, B., & Fisher, K. (2011). Adolescents' science career aspirations explored through identity and possible selves. In L. D. Yore, E. Van der Flier-Keller, D. W. Blades, T. W. Pelton & D. B. Zandvliet (Eds.), *Pacific CRYSTAL centre for science, mathematics, and technology literacy: Lessons learned* (pp. 47–65). Rotterdam: Sense.

McAskill, B., Holmes, G., Francis Pelton, L., & Watt, W. (2004). *WNCP Mathematics research project: Final report*. Victoria: Hold Fast Consultants.

Mendelson, M. (2006). *Aboriginal peoples and postsecondary education in Canada*. Ottawa: Caledon Institute of Social Policy.

Milford, T. M., Anderson, J. O., & Luo, J. (2011). Modelling of large-scale PISA assessment data: Science and mathematics literacy. In L. D. Yore, E. Van der Flier-Keller, D. W. Blades, T. W. Pelton & D. B. Zandvliet (Eds.), *Pacific CRYSTAL centre for science, mathematics, and technology literacy: Lessons learned* (pp. 187–201). Rotterdam: Sense.

Milford, T. M., Jagger, S., Yore, L. D., & Anderson, J. O. (2010). National influences on science education reform in Canada. *Canadian Journal of Science, Mathematics and Technology Education, 10*(4), 370–381. doi:10.1080/14926156.2010.528827.

National Council of Teachers of Mathematics. (2000). *Principles and standards for school mathematics*. Reston: Author.

National Research Council (2012). *A framework for K–12 science education: Practices, crosscutting concepts, and core ideas*. In H. Quinn, H. A. Schweingruber & T. Keller (Eds.), *Board on Science Education, Division of Behavioral and Social Sciences and Education*. Washington, DC: National Academies Press.

Neel, K. (2011). Factors that motivate aboriginal students to improve their achievement in school mathematics. In D. J. Brahier & W. R. Speer (Eds.), *Motivation and disposition: Pathways to learning mathematics* (73rd yearbook) (pp. 113–126). Reston: National Council of Teachers of Mathematics.

Norris, S. P., & Phillips, L. M. (2003). How literacy in its fundamental sense is central to scientific literacy. *Science Education, 87*(2), 224–240. doi:10.1002/sce.10066.

A. A. O'Connell & D. B. McCoach (Eds.). (2008). *Multilevel modeling of educational data*. Charlotte: Information Age.

Organisation for Economic Co-operation and Development. (2002). PISA 2000 technical report. In R. J. Adams, & M. L. Wu (Eds.). Paris, France: Author. http://www.oecd.org/education/preschoolandschool/programmeforinternationalstudentassessmentpisa/1841899.pdf.

Organisation for Economic Co-operation and Development. (2009). *Equally prepared for life? How 15-year-old boys and girls perform in school*. Paris: Author. http://www.oecd.org/document/51/0,3746,en_32252351_32236191_42837811_1_1_1_1,00.html#TOC.

Pearson, P. D., Moje, E. B., & Greenleaf, C. (2010). Literacy and science: Each in the service of the other. Special issue. *Science, 328*(5977), 459–463. doi:10.1126/science.1182595.

Raudenbush, S. W., & Bryk, A. S. (2002). *Hierarchical linear models: Applications and data analysis methods* (2nd ed.). Thousand Oaks: Sage.

Snively, G. J., & Williams, L. B. (2006). The aboriginal knowledge and science education research project. *Canadian Journal of Native Education, 29*(2), 229–244.

Snively, G. J., & Williams, L. B. (2008). "Coming to know": Weaving aboriginal and western science knowledge, language, and literacy into the science classroom. *L1–Educational Studies in Language and Literature, 8*(1), 109–133. http://l1.publication-archive.com/public?fn=enter&repository=1&article=218.

Sutherland, D., & Dennick, R. (2002). Exploring culture, language and the perception of the nature of science. *International Journal of Science Education, 24*(1), 1–25. doi:10.1080/09500690110067011.

Sutherland, D., & Henning, D. (2009). Ininiwi-Kiskānītamowin: A framework for long-term science education. *Canadian Journal of Science, Mathematics and Technology Education, 9*(3), 173–190. doi:10.1080/14926150903118359.

Tippett, C. D. (2011). Exploring middle school students' representational competence in science: Development and verification of a framework for learning with visual representations (Ph.D. dissertation). University of Victoria, Victoria, British Columbia, Canada. http://hdl.handle.net/1828/3250.

Western and Northern Canadian Protocol (WNCP) for Collaboration in (Kindergarten to Grade 12) Education. (2011). Protocol. http://www.wncp.ca/english/wncphome/protocol.aspx

Willers, M. (2005). Enriched mathematics 10/11: Focus on the NCTM process standards (Unpublished master's thesis). University of Victoria, Victoria, British Columbia, Canada.

Wright, N., Claxton, E. Jr., Williams, L., & Paul, T. (2011). Giving voice to science from two perspectives: A case study. In L. D. Yore, E. Van der Flier-Keller, D. W. Blades, T. W. Pelton & D. B. Zandvliet (Eds.), *Pacific CRYSTAL centre for science, mathematics, and technology literacy: Lessons learned* (pp. 67–82). Rotterdam: Sense.

Yore, L. D. (2008). Science literacy for all students: Language, culture, and knowledge about nature and naturally occurring events. *L1—Educational Studies in Language and Literature, 8*(1), 5–21. http://l1.publication-archive.com/show?repository=1&article=213.

Yore, L. D. (2011). Foundations of scientific, mathematical, and technological literacies—common themes and theoretical frameworks. In L. D. Yore, E. Van der Flier-Keller, D. W. Blades, T. W. Pelton & D. B. Zandvliet (Eds.), *Pacific CRYSTAL centre for science, mathematics, and technology literacy: Lessons learned* (pp. 23–44). Rotterdam: Sense.

Yore, L. D. (2012). Science literacy for all—More than a slogan, logo, or rally flag! In K. C. D. Tan & M. Kim (Eds.), *Issues and challenges in science education research: Moving forward* (pp. 5–23). Dordrecht: Springer.

Yore, L. D., Bisanz, G. L., & Hand, B. (2003). Examining the literacy component of science literacy: 25 years of language arts and science research. *International Journal of Science Education, 25*(6), 689–725. doi:10.1080/09500690305018.

Yore, L. D., & Guo, C.-J. (2008, March). Cultural beliefs and language practices in learning and teaching science–symposium: The intersection of the influence of schooling, culture, and nature on the motivation of Hawaiian and Taiwanese indigenous children Taiwanese perspective. Paper presented at the annual meeting of the National Association for Research in Science Teaching, Baltimore, MD.

Yore, L. D., Pimm, D., & Tuan, H.-L. (2007). The literacy component of mathematical and scientific literacy. Special issue. *International Journal of Science and Mathematics Education, 5*(4), 559–589. doi:10.1007/s10763-007-9089-4.

Yore, L. D., & Van der Flier-Keller, E. (2011). Epilogue of pacific CRYSTAL—Lessons learned—About science, mathematics, and technology literacy, teaching, and learning. In L. D. Yore, E. Van der Flier-Keller, D. W. Blades, T. W. Pelton, & D. B. Zandvliet (Eds.), *Pacific CRYSTAL centre for science, mathematics, and technology literacy: Lessons learned* (pp. 237–252). Rotterdam: Sense.

Yore, L. D., Van der Flier-Keller, E., Blades, D. W., Pelton, T. W., & Zandvliet, D. B. (Eds.). (2011). *Pacific CRYSTAL centre for science, mathematics, and technology literacy: Lessons learned*. Rotterdam: Sense.

Chapter 6
Achievement Gap in Mexico:
Present and Outlook

Armando Sanchez-Martinez

> *It is better to light one candle than to curse the Darkness.*
> Adage (cited by Carl Sagan in *The Demon-Haunted World).*

Mexico is an expansive country with enormous inequality and this is reflected in education no matter the government efforts, which are summarized in a former section. At the end of the chapter, two successful experiences at schools in poor communities, constructed from the bottom, are discussed and two more, the first one of the Mexican Academy of Science and the other an innovative proposal of technology in the classroom, are presented. All of them point out what else there is to do. The conclusions are a series of reflections to foster the discussion, mainly about the need to innovate and promote the autonomous curricular development, considering learning achievements as the fundamental educative purpose and viewing the school as a learning community in interaction with the neighborhood.

Mexican Education Consultant, with teaching experience in the superior level and Master's degrees on education, and on teachers' training for all education levels. He worked in the Federal Public Education Ministry (SEP) in curriculum design, held training programs, coordinated national textbooks and educative materials of Science for Basic Education from 1993 to 2004, and participated in international meetings. Since 2005, he has been working in Editorial Santillana where he is the textbook High School Manager since 2008.

A. Sanchez-Martinez (✉)
Mexican Education Consultant, Editorial Manager, Santillana, Mexico, North America
e-mail: amartinezs@santillana.com.mx

J. V. Clark (ed.), *Closing the Achievement Gap from an International Perspective,*
DOI 10.1007/978-94-007-4357-1_6, © Springer Science+Business Media B.V. 2014

What Kind of Country is Mexico?[1]

Mexico is a country situated in North America, just south of the USA, and it is organized into 32 federative states.

According to the 2010 Population Census (INEGI 2011), Mexico is the 11th most populated country, with 112,336,538 inhabitants[2], 51.22% women, 76.9% in urban and 23.1% in rural population, 38.8% (43,541,908) aged 0–15 years, and 6.2% over the age of 65 (6,938,913).

Approximately 20.1 million people live in the metropolitan zone of Mexico Valley (16 districts in Mexico City—called Federal District—, 59 municipalities in Mexico State, and one in the state of Hidalgo), whereas 10.5 million live in localities of less than 500 inhabitants. As much as 13.8% of the total population is concentrated in 11 of the 2,456 municipalities and districts. In 1930, Mexico's population was about 16,552,722; some 80 years later, it has increased sevenfold. Approximately, 52 million people or 46.2% of the total population live in poverty; of these, 11.7 million or 10.4% live in extreme poverty (Coneval 2010).

In the global context, Mexico has the 14th largest GDP (FMI 2010), and ranks 57th on the world human development index (HDI), with a value of 0.77 (UNDP 2011) and a life expectancy of 77 years. In conclusion, Mexico is an expansive country with enormous inequality.

Education in Mexico

Approximately, 4.8% of the GDP is allocated to educational expenditures. Mexico's education system is organized into: *basic education* [preschool or kindergarten (1, 2, and 3), primary (1–6), and secondary (7–9)], *middle education* (10–12), *superior or tertiary education*, and *training for the workforce*.

From 1990 to 2010, the population grew from 25,091,966 students to 34,384,971—an annual growth rate of 1.6%. The percentages of growth in decreasing order per level were as follows: middle 99.3%, training for work 74.7%, preschool 69.7%, secondary 46.5%, superior 38.1%, and primary 3.4% (SEP 2011a). Information about the Mexican education system by level is shown in Table 6.1.

Of the country's 32 federative states, 12 enroll more than a million students or 66.4% of the total. Of these 12, only 4 have more than 2 million students (see Table 6.2).

---

[1] For a wider, more critical and propositional vision of Mexico, I recommend "Mexico frente a la crisis: hacia un nuevo curso de desarrollo" (Cordera et al. 2009), a document written by 16 academics, investigators, intellectuals, and first-level politicians with the participation of ten guest expositors, with the same profile as the former.

[2] *International Human Development Indicators* (UNDP 2011) reports for Mexico 114,793 thousand inhabitants in 2011, with 78.1% in urban towns or cities.

Table 6.1 Students, teachers, and schools of the Mexican education system, school year 2010–2011. Source: SEP (2011). Mexican United States Education System, main numbers, school year 2010–2011. Mexico: Planning and Programming General Direction, Public Education Ministry (Dirección General de Planeación y Programación, Secretaría de Educación Pública)

Education level	Roll	% per level	Teachers	Schools
Total education system	34,384,971	100	1,801,793	252,061
Basic education	25,666,451	74.6	1,175,535	226,374
Preschool	4,641,060	13.5	222,422	91,134
Primary	14,887,845	43.3	571,389	99,319
Secondary	6,137,546	17.8	381,724	35,921
Middle education	4,187,528	12.2	278,269	15,110
Technical professional	376,055	1.1	27,557	1,399
High school	3,811,473	11.1	250,712	13,711
Superior education	2,981,313	8.7	308,061	4,689
Technical superior	113,272	0.3	11,121	256
Degree	2,659,816	7.8	366,032	4,127
Postgraduate course	208,225	0.6	38,026	1,906
Training for the workforce	1,549,679	4.5	39,928	5,888

Table 6.2 Students of federative states with more than two million enrollments School year 2010–2011. Source: SEP (2011). Mexican United States Education System, main numbers, school year 2010–2011. Mexico: Planning and Programming General Direction, Public Education Ministry (Dirección General de Planeación y Programación, Secretaría de Educación Pública)

State	Students	% of the total
Mexico state	4,284,974	12.5
Mexico city–D.F.	2,798,110	8.1
Veracruz	2,220,728	6.5
Jalisco	2,196,662	6.4

In contrast, two states have less than 200,000 students: Colima (186,276) and South Baja California (187,640). These data point to the complexity of the Mexican education system because there are striking variations at the federation level (SEP 2011a).

Information on the coverage and terminal efficiency of students is presented in Table 6.3 for preschool, primary, secondary, and middle high education levels.

As shown in Table 6.3, in 15 years, there has been a substantial increase in both the coverage and the terminal efficiency at all education levels, except primary, which was close to 100 % in coverage in 1995–1996. Nevertheless, the challenge to widen the coverage in preschool and middle education is more daunting. So, too, is the case of increasing secondary and middle superior terminal efficiency. The national average school grade is 8.7, with differences between Mexico City (D.F.) and Chiapas of 10.6 and 6.3, respectively.

The main modality in preschool and primary is called general, but some schools are indigenous and others communitarian, so they have in general three modalities. Indigenous is for localities with a majority population of native indigenous language speakers, and communitarian is for localities with the smallest populations, so a middle education student is training as a "communitarian teacher" for teaching all

Table 6.3 Coverage and terminal efficiency of students of different education levels School years 1995–1996 and 2010–2011. Source: SEP (2011). Mexican United States Education System, main numbers, school year 2010–2011. Mexico: Planning and Programming General Direction, Public Education Ministry (Dirección General de Planeación y Programación, Secretaría de Educación Pública)

Level	Coverage		Terminal efficiency	
	1995–1996	2010–2011	1995–1996	2010–2011
Preschool[a]	45.6	80.9	n.a.	n.a.
Primary[b]	95.2	100.6	80.0	95.0
Secondary[c]	74.9	95.9	75.8	86.5
Middle education[d]	40.5	66.7	55.5	63.3

[a] Range of 3–5 years
[b] Range of 6–12 years
[c] Range of 13–15 years
[d] Range of 16–18 years. n.a. Information not available

students of the level. In preschool, the general modality takes care of the 86.8 % of the roll, indigenous 8.4 %, and communitarian 1.4 %; the rest 3.4 % attend the children development centers (Centros de Desarrollo Infantil – Cendi). In primary, the percentages of the three modalities are 93.6 %, 5.7 %, and 0.7 % (SEP 2011a), and the marginalization conditions increase from the first to the last. If the general regular primary schools were differentiated (with at least a teacher per group and per grade) from those called "multi-grade" (from one to five teachers for the six grades), we would find differences of marginalization.

In secondary, the main modalities are general (51.2 %), technical (28.2 %), and telesecondary[3] (20.6 %) (SEP 2011a). Except for the last one, all groups are taken care of by different teachers, one for each subject. The difference between general and technical is that the latter prepares the students in some technical activity. Although there are communitarian secondary schools and others for workers, its participation is minimal because they respectively take care of just the 0.2 % and 0.5 % of the total enrollment (INEE 2009, p. 34). However, there are states where the percentage of communitarian schools is larger, such as Guerrero with 10.5 % and Campeche with 9.8 % of its public schools. These states and those where the percentage of telesecondary schools is larger than the national average provide insight into the educational achievements of the states. Zacatecas, San Luis Potosí, Guanajuato, Veracruz, Puebla, Chiapas, Hidalgo, and Oaxaca stand out as states in which at least seven of each ten public secondary schools are telesecondary, whereas that for Mexico City (D.F.) is only 5 % (INEE 2009, p. 35). To further understand the context of telesecondary schools, it is important to know that 64 % of its population is

---

[3] Modality designed to give service to a small rural town; it has operated since 1968 with a printed guide for the student, TV programs via satellite, and teacher's books. In general, they operate with one teacher per group and a monitor, although, in some very small schools, there is only one teacher for the three grades.

Table 6.4 Percentage distribution of secondary schools according to the marginalization level of the locality where they are situated, school year 2007/2008. Source: INEE (2009). Learning in Mexico in the third year of secondary. *Report about the results of Excale 09, 2008 application. Spanish, Mathematics, Biology, and Civical and Ethical Training*

Degree of marginalization	Public secondary				Private secondary
	General[a]	Technical	Telesecondary	Communitarian	
Very high	0.4	2.1	8.5	38	0.1
High	8.3	14.5	52.3	45	0.9
Medium	8.7	11.2	18.1	5.7	2.6
Low	19.9	21	14.2	3.2	10.2
Very low	62.1	50.7	5.8	0.9	85.1
Lost[b]	0.6	0.5	1.1	7.2	1.2

[a] Includes the secondary schools for workers
[b] Corresponds to the schools where it was not possible to identify the marginalization level of the locality in which they are located

registered in the Popular Insurance System[4] and goes to the health-care center, drug store, or clinics[5]; and 7% has no medical care. For the general secondary schools, the percentages are 32 and 10 (INEE 2009, p. 40).

Information related to marginalization, with indicators such as potable water, electricity, telephones, and other kind of services, is generalized in Table 6.4, which presents the percentage distribution of secondary schools according to the marginalization degree of the locality. Such is a reference to understand the later data analysis related to the gap in educational achievement for this level.

The Ministry of Education (Secretaría de Educación Pública in Spanish [SEP]) defines plans and study programs for basic education; therefore, all the students who attend preschool, primary school, and secondary school have the same plan and study programs (SEP 2011b).

Since 1993, secondary education has been obligatory. Nevertheless, universal coverage has yet to be reached. In primary education, however, it was reached at the beginning of this century (SITEAL 2011). In 2010, the National Institute for the Education Evaluation (INEE)[6] estimated that 5% of the primary graduates failed to continue their studies and that approximately 80% of those who do continue their studies finish it in 3 years (INEE 2010a).

Middle education is divided into technical professional and high school; the first one is a medium terminal option, with the possibility of continuing the degree studies, and the second one is basically preparatory. The middle education is complex; it is estimated that there are more than 300 study plans divided in several systems

---

[4] Program created by the federal government this century for citizens who are not registered in the health government institutions, so they can have the basic services going to the health centers and clinics.

[5] These options are the only possible ones for the most marginalized sectors of society.

[6] All the INEE publications can be consulted in its page http://www.inee.edu.mx/index.php/english-version

Table 6.5 Roll, schools, and study plans for middle education, school year 2010–2011

Roll 4,187,528 & 15,110 schools (More than 300 study plans).	Technical professional: roll 376,055 & 1,399 schools	*National School of Professional Education* [Conalep[a]]: roll 287,927 & 501 schools (42 technical degrees)
		Others: roll 88,128 & 898 schools
	High school (HS): roll 3,811,473 & 13,711 schools	*General*: roll 1,631,003 & 7,390 schools (24 HS public universities & incorporated ones, HS autonomous private universities, private HS DGB-SEP, federative systems & art schools)
		Technological: roll 1,288,749 & 2,798 schools (191 technical degrees)
		High school DGB-SEP [Colegio de bachilleres[b]]: roll 717,733 & 1,463 schools
		TV–high school: roll 173,988 & 2,060 schools

[a] Colegio Nacional de Educación Profesional Técnica
[b] There is one of these high schools in each federative state and they have a lot of schools

and subsystems. Table 6.5 summarizes the composition of the middle education systems as is works up to date as a personal interpretation, in special HS DGB-SEP.

Nevertheless, the Middle Education Integral Reform (RIEMS) has tried to homogenize the study plans of the basic subjects of the most important subsystems such as the National School of Professional Education (Colegio Nacional de Educación Profesional [Conalep]), DGB-SEP (state high schools, and cooperative and private schools incorporated to SEP), and for all the technological high schools.

In 2010, 3.65 million or 53 % young people studied the middle education while 3.23 million or 47% did not. This pattern was even more accentuated in rural towns (36.4 %) than in urban towns and cities (60.3 %). In the extreme, there are the young speakers of indigenous languages, for whom the percentage is only 29.4 % (INEE 2011a)[7]. As a matter of fact, in the homes where an indigenous language is spoken, 52% of young people between the ages of 15 and 29 have not finished basic education compared with 27.6 % for other homes (INEE 2011b). On February 9, 2012, the Mexican Constitution reaffirmed that middle education is obligatory (DOF 2012). In short, making clear that the State obligation is that of guaranteeing "a place to study it for those who having the typical age[8], would have concluded the basic education and it will be realized in a gradual and growing way starting from the school year 2012–2013, until achieving the total coverage in the country in its different modalities at the latest in the school year 2021–2022" (DOF 2012, second transitory article).

---

[7] Information taken by INEE (2009) of INEGI (2009).

[8] In Mexico, 80 % of EMS students are between the ages of 15 and 17.

The Education Evaluation and the Achievement Gap
in Mexico

In Mexico, the issue of institutional evaluation on the education achievement applied to students is recent. Since the end of the twentieth century, Mexico has participated in the Trends in International Mathematics and Science Study (TIMSS) and the Program of National Standards of SEP.

There are three kinds of evaluations applied to students, two national and one international:

- *Exams of Quality and Education Achievement (Exámenes de la Calidad y el Logro Educativos* [Excale]): These are administered by the National Institute for the Education Evaluation (Instituto Nacional para la Evaluación de la Educación [INEE]) since 2005, to key school years of the basic education (BE): 3rd year of preschool, 3rd and 6th years of primary, and 3rd year of secondary (INEE 2011c). The periodicity of application for each grade is every 4 years. In preschool, the test evaluates the formative fields of language and communication and mathematical thought, whereas that in primary and secondary evaluates Spanish, mathematics, natural sciences, and social sciences. The tests are based on the curriculum. However, they are administered only to a representative sample of students from each federative stage based on matrices similar to those of the *Programme for International Student Assessment* (PISA). In order to evaluate the most important curricular content, the exams are divided into units.
- *National Evaluation of Academic Achievement (Evaluación nacional del logro académico* [Enlace]): This is a standardized test based on the curriculum and with multiple-option questions, administered by the General Direction of Evaluation of SEP's Policies (Dirección General de Evaluación de Políticas de la SEP). Since 2006, this test has been applied to all students from third to sixth grades of primary education and to the students of the third grade of secondary. Beginning in 2008, it will also be applied in the 1st and 2nd years of secondary and in the last year of middle education. In the first three applications, reading comprehension and mathematics were evaluated, but, since 2008, a third subject for basic education was introduced, which is repeated every 5 years (natural sciences in 2008, civical and ethical training in 2009, history in 2010, and geography in 2011, for which sciences have been repeated in 2012).
- *Programme for International Student Assessment (PISA)* is an international evaluation based on the concept of literacy, which evaluates reading, mathematical, and scientific skills. It was first applied in 2000, and it is repeated every 3 years. At the beginning, 43 countries participated, with 68 participating in 2012, including OECD members and partner countries/economies[9].

---

[9] Because this test is internationally known, for bigger affairs related to it, please consult: http://www.pisa.oecd.org/pages/0,2987,en_32252351_32235731_1_1_1_1_1,00.html

In the following, some of the results of the three evaluations concerning Excale will be analyzed with added information published by INEE regarding the sixth grade of primary and third grade of secondary, to compare the information related to the education gap in places with certain marginalization[10]. From Enlace, more information will be presented because it is for everybody and annual, so there are indicators of education achievement since 2006 for basic education and since 2008 for middle education, but with different marginalization indicators regarding Excale. In particular, there is information about middle education presented so that it can be related to PISA results besides those of the 3rd year of secondary of Enlace and Excale. Because PISA has become the international reference of the evaluation of education achievements, we will analyze more information on students of the middle education provided that 72.6% of the students who present the test are in that education level, most of them (71.9%) in the 1st year.

Case-study information about Chiapas and Mexico City (D.F.) are highlighted because the former has the lowest HDI (0.7395)[11] (PNUD 2011), especially with a higher percentage of the young population out of the standard educational level (50.3% in 2008), with minor results of educational achievement and associated with the poorest states in Mexico; however, the latter has the biggest HDI of the country (0.9176; UNDP 2011), the lowest percentage out of the standard educational level (14.8%), the best achievement level, and better opportunities for its inhabitants.

According to the World Economic Forum (WEF 2011, p. 259), Mexico ranks 121th among 142 countries in primary education, 107th in superior education, and 126th in mathematics and sciences. Overall, Mexico ranked 58th in this report (pp. 11 and 15).

Excale

These exams have four levels of achievement: under the basic, basic, medium, and advanced. To illustrate the gap of educational achievement in these exams, only the first level of achievement is used because it measures the students who do not reach the basic level. In Table 6.6, the results of Excale 2007 in mathematics and Spanish for sixth grade of primary are presented.

These data are revealing. For example, in the indigenous stratum, of the biggest marginalization, 42% and 37% of the students do not reach the basic level in Spanish and mathematics, respectively. This is in sharp contrast with the private stratum, where the corresponding percentages are 2% for each.

---

[10] It is important to clarify that the INEE exams have a higher coverage regarding the main contents of the national curriculum of sciences, Spanish, and mathematics, the reason why there are many elements to analyze. The previous data are the result of its matrix design that has sense because it is applied to student samples. For those who are interested, we recommend consulting http://www.inee.edu.mx/explorator (English) or http://www.inee.edu.mx/explorador (Spanish).

[11] This information is from 2008, although the report from Mexico is from 2011.

Table 6.6 Achievement level percentage for students of sixth grade of primary, according to the school stratum. Spanish and Mathematics Excale 2007. Source: INEE (2008). Comparative study of the learning in sixth grade of primary in Mexico 2005–2007

School stratum	Underneath the basic		Basic		Medium		Advanced	
	Spanish	Mathe-matics	Spanish	Mathe-matics	Spanish	Mathe-matics	Spanish	Mathe-matics
Indigenous education	42	37	50	53	7	9	1	1
Rural public	21	20	56	56	20	20	3	4
Urban public	11	12	50	51	31	28	8	9
Private	2	2	23	31	45	43	30	23

In Table 6.7, data are presented for Excale 2008 secondary mathematics (INEE 2009) for rural populations, urban of big marginalization, and urban of low marginalization in the general and technical modalities, and for students in private schools.

As shown, there are differences between rural schools and urban schools of high marginalization but they are considerable between the former and urban schools of low marginalization. However, the differences that stand out even more are between the technical ones of the first stratum (rural) and the last one (ULM; 20 points) and more of the technical rural with the privates (42 points). The results for the UHM technical and general are similar to that among ULM schools of both modalities. Nevertheless, the differences between the UHM and the ULM of both modalities are 10 and 11 points, respectively.

On the other side, the percentage of telesecondary students under the basic level for mathematics in Excale 2008 was 62%. Data were unavailable to differentiate the achievement of students in rural, urban of high marginalization, and urban of low marginalization telesecondary schools, but, according to Table 6.4, 52.3% of its schools are in towns of high marginalization, and 8.5% in very high ones; therefore, that 62% is also an indicator of how the gap in the education achievement is larger regarding the less-favored students.

In Table 6.8, the information focuses on Chiapas and Mexico City (D.F.)

As mentioned earlier, the differences are significant and will be confirmed in subsequent discussions of the Enlace and PISA results.

Enlace Test[12]

The Ministry of Public Education (SEP) warns that the results among the different years of application are not comparable because of technical reasons. Consequently, only the information of the 2011 application will be used, with the clarification that there are four grades of academic achievement defined as unsatisfactory,

---

[12] The Enlace information comes from www.enlace.sep.gob.mx/ms/estadisticas_de_resultados/

Table 6.7 Percentage of students of the third year of secondary below the basic level for mathematics in Excale 2008, of the stratums in rural populations, urban of high marginalization (UHM), and urban of low marginalization (ULM), for general schools, technical, and private. Source: INEE (2009). Learning in Mexico in the third year of Secondary. *Report about the results of Excale 09, 2008 application. Spanish, Mathematics, Biology, and Civical and Ethical Training*

Education modality	Mathematics		
	Rural	UHM	ULM
Technical	67	57	47
General	57	56	45
Private	–	–	25

Table 6.8 Percentage of students from the third year of secondary below the basic level for Chiapas and Mexico City (D.F.) Spanish, Mathematics, and Biology, Excale 2008. Source: INEE (2009). Learning in Mexico in the third year of Secondary. *Report about the results of Excale 09, 2008 application. Spanish, Mathematics, Biology, and Civical and Ethical Training*

State	Spanish	Mathematics	Biology
Chiapas	48	64[a]	40
Mexico City (D.F)	26	39	16

[a] In this case, Guerrero (68%), Tabasco (67%), and Michoacan (65%) are over that

elementary, good, and excellent, provided the second one represents the basic level established in the curriculum.

Enlace in the Basic Education

Data in Table 6.9 present the percentage of students for basic education in the national level with unsatisfactory results in Spanish and mathematics in 2011.

The high percentages of students with unsatisfactory results in both subjects stand out, but especially in secondary mathematics. The fact that a student has an unsatisfactory result means that the expected achievements established in the curriculum have not been achieved. These data are consistent with the low levels of performance on PISA.

As shown in Table 6.10, the percentages of students with the unsatisfactory level in primary in the state of Chiapas exceed that of Mexico City (D.F.). In secondary, mathematics is in the reversed pattern. This information is inconsistent with the results obtained by both states in Excale.

Some education experts have been critical of the Enlace test (Ramírez 2010). Now that the test is the standardized measure of the curriculum, education authorities and schools direct their actions toward student enhancement tests. Mexico's results in the PISA test suggest that the schools are not achieving the best education results and performance of competencies that are defined in all curriculums. The data regarding primary education in both states in the last three of them public and using the government budget are summarized in Table 6.11.

Table 6.9 Percentages of students with unsatisfactory results at national level in Enlace 2011. Source: http://www.enlace.sep.gob.mx/ba/

Level	Mathematics	Spanish[a]
Primary (3–6)	43	42
Secondary (7–9)	57	42

[a] In specific, reading comprehension is evaluated

Table 6.10 Percentages of students with unsatisfactory and elementary results for Chiapas and Mexico City (D.F.) Enlace 2011. Source: http://www.enlace.sep.gob.mx/ba

Level	Mathematics		Spanish[a]	
	Unsatisfactory	Elementary	Unsatisfactory	Elementary
Chiapas				
Primary (3–6)	23	36	24	40
Secondary (7–9)	48	24	44	36
Mexico City (D.F.)				
Primary (3–6)	14	48	11	40
Secondary (7–9)	52	32	39	40

[a] Reading comprehension is specifically evaluated

Table 6.11 Percentages of primary students with unsatisfactory results for Chiapas and Mexico City (D.F.) Enlace 2011. Source: http://www.enlace.sep.gob.mx/ms/ba

	Private	General	Indigenous	Communitarian
Chiapas				
Mathematics	12	13	41	54
Spanish*	9	13	44	55
(Mexico City) D.F.				
Mathematics	5	16	-	-
Spanish[a]	3	13	-	-

[a] Reading comprehension is specifically evaluated

The results are a finer sample of the education gap because the towns, cities, or villages or each one of these modalities go from a smaller to a larger degree of marginalization. In the case of Chiapas, the results are more striking because it is a state with many indigenous and communitarian schools. The difference in the mathematics grading between the private schools and the indigenous ones is 29 points, whereas, with the communitarian ones, it is 46. For Mexico City (D.F.) the favorable results are evident, and they show a great gap in the education achievements between a state with more opportunities for its inhabitants and a less-favored one.

The information about secondary education is summarized in Table 6.12. These results also show what was stated previously for both federative states. Nevertheless, the information about students with an unsatisfactory performance at telesecondary schools in Chiapas of 36% for mathematics and 35% for Spanish is surprising compared with the high percentages of the other modalities. In general, the performance of these students goes from less to more going from left to right in the modalities, that is, less in private schools, then general and technical ones,

Table 6.12 Percentages of secondary students with unsatisfactory results for Chiapas and Mexico City (D.F.) Enlace 2011. Source: http://www.enlace.sep.gob.mx/ba

	Private	General	Technical	Telesecondary
Chiapas				
Mathematics	46	56	62	36
Spanish	34	49	56	35
Mexico City (D.F.)				
Mathematics	29	59	53	64
Spanish	18	44	41	56

and at the end, with a bigger number of students, the telesecondary schools. This seems like an error if compared with the results of 2007 and 2008: 67% and 59% for mathematics and 70% and 62% for Spanish.

Enlace in the Middle Education

The unsatisfactory results for the middle education students in 2011 are presented in Table 6.13. It is clear that a larger degree of marginalization lowers performance. These are differences between the very high and the very low degrees. The differences are 25 for reading comprehension and 28 for mathematics.

Pisa

The PISA scores for Mexico are presented in Table 6.14. Based on the way in which PISA establishes the score ranks for each level, Mexico has always been in level 2. Nevertheless, when the students' percentages are differentiated according to the performance levels, it is found that, for reading in 2000, 44% of the students were under this level, with 6% of them in level 0; and in 2009 39% of students were also under the level, with a 14% in or under level 1b, the percentage equivalent to the previous level 0 (INEE 2010b; OECD 2010)[13]. If the results for Mexico City (D.F.) in 2009 are analyzed, the previous percentages are 20% and 1%, whereas that for Chiapas is 67% and 36%, respectively (INEE 2010b). It goes without saying that the students of level 0 are not capable of locating a fragment of concrete information in the text, recognizing the main topic, or recognizing simple relationships among close fragments.

For reading in 2000, 49% of the students were in levels 2 and 3, and 7% above level 4, whereas in 2009 the first percentage was 54% and the second 6%. If the re-

---

[13] Since 2009, the reading performance levels are 8 (0, 1b, 1a, and from 2 to 6), 6 being the highest grade whereas 2 is defined as "the minimum to perform in the nowadays society" (INEE 2010b, p. 37). From 2000 to 2006, the previous levels were from 0 to 5. In mathematics and sciences, they have always ranged from 0 to 6, without differentiating level 1 yet.

Table 6.13 Percentage of middle education with unsatisfactory performance Enlace 2011. Source: http://www.enlace.sep.gob.mx/ms/estadisticas_de_resultados/

Year	Degree of marginalization				
	Very high	High	Medium	Low	Very low
Reading comprehension					
2011	38	24	18	16	13
Mathematics					
2011	59	49	42	37	31

Table 6.14 Mexico in PISA. Source: OECD (2010). *Pisa 2009 Results: What Students Know and Can Do. Students Performance in Reading, Mathematics and Science* (Volume I). Paris: Organization for Economic Co-operation and Development

	2000	2003	2006	2009
Reading	422	400	410	425
Mathematics	387	385	406	419
Sciences	422	405	410	416

sults among the students who present PISA studying secondary or middle education are compared, important differences can be noticed. For the first ones, the percentages in levels above 3 in reading in 2000 and 2009 were 8% and 13%, respectively, whereas those of middle education were 41% and 37%. The results among public and private schools reflect the education gap because for the first ones; the same data are 19% and 25%, whereas that for the second ones were 58% and 44%. A possible cause of the decrease in some of the previous results in 2009 (37 and 44%) could be due to the reading competence having been evaluated more deeply because the integration of what is read was included (INEE 2010b).

Table 6.15 presents the percentages by levels of performance of PISA 2009 in Mexico, Mexico City (D.F.), and Chiapas for sciences and mathematics, compared with the averages of OECD and from Latin America (LA).

The previous data, along with what was revised by Enlace and Excale, are a sample of the low results of the three tests. These results get even lower when the marginalization conditions increase. For instance, in 2009, the percentage of middle education students of a high marginalization with unsatisfactory results in Enlace was 43% in reading comprehension and 72% in mathematics, whereas, for those of low marginalization, it was 14 and 42%, with national averages for this level of 17 and 46%, respectively.

What Has Been Done in Mexico?

In the first decade of this century, the demographic pressure that kept Mexico in a complicated race during almost all the past century to achieve the universal coverage of the primary education started to decrease. After the education reform in

Table 6.15 Percentages of Mexico for sciences and mathematics in PISA 2009 by levels of performance. Source: INEE (2010). *Mexico in PISA 2009*. Mexico: National Institute for the Education Evaluation (Instituto Nacional para la Evaluación de la Educación)

	Levels ≤ 1		2 ≤ levels ≤ 3		Levels ≤ 4	
	Sciences	Mathematics	Sciences	Mathematics	Sciences	Mathematics
Average OECD	18	22	53	46	29	32
Mexico City (D.F.)	27	32	64	57	8	11
Mexico	47	51	49	44	3	5
Average LA	52	63	43	32	5	5
Chiapas	71	72	28	26	0.4	1

1972, it was not until 1993 that a series of changes in the primary and secondary plans and programs began (e.g., the free text books are updated, secondary is declared obligatory, the education services in the federative states are decentralized, and a teaching degree is created as an incentive program based on evaluations to teachers). Additionally, the normal education was reformed with infrastructure supports for schools, and an aggressive updating program for teachers was encouraged with the creation of more than 600 teaching centers distributed along the country at the end of the last century. These various reforms were introduced in all the curricula based on competencies, and they started in 2004 for preschool, 2006 for secondary, and 2008 for middle school and primary, and they finished up in 2011 with the articulation of the basic education where PISA is explicitly considered as a referent (SEP 2011b).

There are many variables that intervene in the results of the academic achievement with a wide inequality gap:

- The great cultural and socioeconomic diversity of the Mexican population, characterized by the large differences between those more and less marginalized, combined with a very high percentage of population in poverty and a high percentage in extreme poverty, is the reason why the education efforts are not enough and are least with an education system completely centralized in the twelve grades of basic education and with bureaucratic practices of excessive control.

- Since the creation of the Sindicato Nacional de Trabajadores de la Educación (SNTE; National Union of Workers of the Education) in 1943, the government made a pact ceding the control of the teaching positions and those of all the directives (sector chiefs, supervisors, school principals, and teaching chiefs) to the union (Arnaut 1998; Barba and Arnaut 2010). As a result, reaching those positions is part of the political union race of many teachers, based on scales and on looking good with the leaders; therefore, this is not an academic degree. Besides, this situation can favor the climb to other political positions like member of parliament (state or national), senator, political party leader, and, in some cases, even governor. In this sense, for many teachers, the union race is more attractive than the academic one. The union force is such that the key positions in many education secretariats of the federative states are even negotiated with the local governments. Although there is a democratic movement inside the SNTE with a long tradition of fight (Street 2010), it has also evolved into a political movement, in spite of efforts to encourage a real pedagogical movement (Street 2001).

- The national education system is prescriptive, with detailed study programs, unique and national textbooks in primary and subject of government authorization in secondary. There has not been enough space for the education innovation and for the curricular development to be in the teachers' hands, because of the fact that the bureaucratic and administrative controls and the union politicians have limited the initiative of teachers and schools as a whole.

- Except for the teachers registered in a teaching degree, most of them and most of the directives are not subject to periodic strict evaluations. Besides, when the SEP has had hard information about deficiencies in the basic contents from the teachers, nothing has been done to improve the updating programs. The present pressure to reject the evaluation of all teachers still causes strikes and sit-ins, particularly in states with the lowest educational results. SEP announced the first massive evaluations to 541 thousand teachers of basic education for23 and 24 June and 6 and 7 July, 2012; and since the beginning of June, the CNTE protest marches have increased to the point where in some federative states (mainly Oaxaca, Guerrero, Chiapas, and Michoacan) the application of the Enlace test for this school year is almost being prevented from going forward because the results of this assessment would be part of the teachers' evaluation. This boycott was successful in these states and, as a matter of fact, in June, there were 150,000 lawsuits against the SEP and the SNTE to stop the application of the test (Blancas 2012). Finally, the Ministry of Education recently reported 54 % of attendance to the evaluation.

What Else is There to do?

First of all, there is a necessity for a national educational policy that only defines the achievements and the general standards so that the teachers can develop their curriculum. However, this must be in accordance with the school and community context, which means working in a collaborative way with the schools of all the education levels. This policy could be gradually favored, for instance, first with open contests for those interested teachers who could present projects, receive economic supports, and guarantee that the executions will be done in complete freedom, once they have been selected with transparent mechanisms and strictly academic criteria[14]. All this implies gradually making the national curriculum more flexible so that the study programs stop being "omnimonopolized" or omnicovered (Cordera et al. 2009, p. 35).

It is essential, however, that the federal and state governments rethink their relationship with the National Union so that all academic matters, including of course

---

[14] Although this kind of contest has existed since the 1990s, their number needs to be increased.

the periodic teachers and directives' evaluation, become an exclusive attribution of the state, just as the appointment of directive posts and all those of the education authorities are.

However, there are successful experiences, especially in schools in poor communities like the following two experiences in secondary, that turn out to be very significant:

- *Telesecondary schools are linked to the community*, in which 14 schools of the Puebla mountain region participate with a model of productive workshops related to the characteristics of each community. All these schools belong to a poor rural environment, and they have high migration. This is why the workshops try to provide young men with a practical and ecological training to take advantage of the land, community resources, and local productions (e.g., vegetables, edible mushrooms or medicinal plants, or elaboration of processed food). In addition, students are taught to acquire technological skills as a possible source of future work. Workshops on blacksmith handicrafts have been organized as well. In all the workshops, the students learn to make budgets. This experience is an example of the connection between theory and practice, because it relates the workshops to the official curriculum, and a part of its success has been that many of the new teachers are "graduates" of this education model, which has its foundation in the pedagogy based in projects. Its founder was Salom (2009), zone coordinator of those telesecondary schools; he created the model in 1994 and coordinated it with its members until his death in August 2011. His work was a great example of collaboration work among teachers, directives, and the community.
- *Educational Coexistence* (Convivencia Educativa) was founded by Gabriel Camara in 1996 with a model of learning communities based on tutorial networks integrated by students but with the freedom of choosing their subject of interest to prepare themselves as peer tutors. The teachers offer students a menu of contents, which the teachers know well[15], and give them personal consultations to form them, give them confidence, and let them rehearse as tutors until they are ready to do it independently and are able to give presentations to parents or other schools, teachers, and principals. This pedagogical model has favored the competence of "Learning to Learn" and that of adequately expressing themselves in different audiences to teach something since the early ages. It first started in few one or two teachers telesecondary schools in Chihuahua, Zacatecas, and San Luis Potosi. Nowadays, it is an SEP's Integral Strategy Program for the Improvement of the Educational Achievement (*Programa de Estrategia Integral para la Mejora del Logro Educativo)*; it exists in all federative states and deals primarily with the 9,000 schools of basic education that obtained the lowest results in the Enlace test, and it also coordinates the training of Spanish and mathematics teachers of the first grade in all secondary schools of the country in the first weeks of the school year 2011, to develop a preparatory course about the tutor relationship. Besides, the tutor relationship gives a new dimension to the secondary

---

[15] Remember that in telesecondary schools there is only one teacher per grade or even per school.

education reform, along with the redefinition of the technical advisor's functions, and it is already mentioned as part of the educational policy in accordance with the Agreement 592 (SEP 2011b). The participating schools have started to show substantial improvements in the results of the Enlace test (Cámara 2010; Malone 2011).

Both innovative experiences constructed from the bottom have proven to be an alternative to improve the education achievements of poor communities, and that improvement would be even greater if they were supported resolutely and without concealment. It is not a coincidence that, in a centralized, prescriptive, and authoritarian educational system like that of Mexico, the innovations are born in marginalized sectors and at the margin of the system (Barba and Zorrilla 2010). The federal and state governments need to bring about innovation in these sectors in a more compromised way and with more resources.

Moreover, it is important to consider for future actions the science and mathematics programs for basic education developed by the Mexican Academy of Science, such as *Summer in the Scientific Investigation, Teaching of Mathematics* and *Science in your School* (AMC 2010a). For instance, the last one was started in 2002, and it links the scientific community with primary and secondary teachers to improve the teaching of science and mathematics through a course. To date, 6,168 professors have been prepared (AMC 2010b).

In this chapter, I did not focus on technology in the classroom. The analysis was made in the basic achievements that are not fulfilled and the inequities between those more and less favored. The need to encourage projects linked to the ICT is unquestionable[16]. With the previous clarification, the proposal of *Classroom of the Future* of the Center of Applied Sciences and Technological Development of the Mexico National Autonomous University (Universidad Nacional Autónoma de México [UNAM]) is mentioned only as an example (Gamboa 2009). The basis of this proposal is the interactive surfaces "in which several users can collaborate without having to use a mouse or a keyboard; it's enough to put, move, or remove physical elements from the surface to do all the actions that are traditionally done with a simulator" to work business and collaborative strategies, "to support and promote the collaborative work among students" (Gamboa 2009). In particular, innovative projects of natural sciences, technology, and mathematics (STEM) should be imposed to transform those in an effective education, just as the teaching of English in the case of Mexico, both using the ICT. For instance, the secondary curriculum contemplates the realization of bimonthly projects that should take into consideration, both by students and teachers, the systematization of the projects made by Harland (2011) in his handbook.

---

[16] ICT (Information and Communication Technologies) plus LCT (Learning and Knowledge Technologies), and EPT (Empowerment and Participation Technologies); called TIC, TAC, and TEP in Spanish (http://toyoutome.es/blog/tic-tac-tep-las-siglas-del-aprendizaje-aumentado/12734).

Conclusions

The preceding discussion does not deny the need to implement wider state policies that:

- Combat poverty, especially extreme poverty, and not just give compensatory measures.
- Increase the country productivity, especially with communitarian projects that compromise and empower people, mostly involving the young, but with a strict attachment to legality and observance of human rights.
- Raise the education expense regarding the GDP, mainly with labeled budgets to take care of the school infrastructure and improve it, develop innovation and education investigation projects emphasizing the importance of the poorest communities. They are called labeled to differentiate them from those of the current expense, which would imply the betterment of the tax mechanisms to guarantee a good use of these resources.
- Set out to a bigger impulse related to the scientific and technological investigation, both basic and applied, to start with a bigger budget including the incubation of new companies with favorable cost and feasibility studies.
- Involve the scientific community in more SMET programs for basic and middle education with interdisciplinary groups where specialists in the didactics of sciences, mathematics and technologies, besides engineers, of the middle education participate.

Because these matters go far beyond what many educative actors can do, the examples expressed in the previous section must be considered as programs that can be extended. Therefore, it is imperative to foster and implement innovation and autonomous curricular development, considering learning achievements as the fundamental educative purpose and viewing school as a learning community that interacts with the school neighborhood for improving together.

Acknowledgment Special thanks to Minerva Guevara for her readings, which improved the chapter content, and Annette Santos del Real, for her critical revision of the rough draft, especially regarding the presentation of Mexican Education System data and evaluation results.

Translation and later corrections: Nelly Pérez Islas.

References

AMC. (2010a). *Programas*. Mexico: Academia Mexicana de Ciencias. http://www.amc.edu.mx/p5/index.php?option=om_content&view=article&id=72&Itemid=61. Accessed 7 June 2012.

AMC. (2010b). *La ciencia en tu escuela*. Mexico: Academia Mexicana de Ciencias. http://www.lacienciaentuescuela.amc.edu.mx/sobreprograma. Accessed 7 June 2012.

Arnaut, A. (1998). *Historia de una profesión. Los maestros de educación primaria en Mexico, 1887–1994*. Mexico: Secretaría de Educación Pública y Centro de Investigación y Docencia Económicas.

B. Barba, & M. Zorrilla (Eds.). (2010). *Innovación social en educación. Una base para la elaboración de políticas públicas*. Mexico: Siglo XXI y Universidad Autónoma de Aguascalientes.

Barba, B., & Arnaut, A. (2010). Las relaciones SEP-SNTE y el proyecto educativo de Mexico: escenario para comprender a los maestros en la política educativa. In B. Barba, & M. Zorrilla (Eds.), *Innovación social en educación. Una base para la elaboración de políticas públicas* (pp. 106–131).

Blancas, M. D. (2012). La CNTE se resiste a la evaluación con paros… y 150 mil demandas. In *Crónica Nacional*, June 4, p. 3. http://www.cronica.com.mx/nota.php?id_nota=666284. Accessed 5 June 2012.

Cámara, C. G. (2010). Un cambio sustentable. La *comunidad de aprendizaje* en grupos de maestros y alumnos de educación básica. *Perfiles educativos, 32*(13). Cfile:///Volumes/Trabajos/Trabajos/Art%C3%ADculos%20e%20información%20diversa/Educación/Experiencias%20innovadoras/Perfiles%20educativos%20-%20Un%20cambio%20sustentable:%20La%20comunidad%20de%20aprendizaje%20en%20grupos%20de%20maestros%20y%20alumnos%20de%20educación%20básica.webarchive. Accessed 28 May 2012.

Coneval. (2011). *Medición de la pobreza*. Mexico: Consejo Nacional de la Evaluación de la Política de Desarrollo Social. http://www.coneval.gob.mx/cmsconeval/rw/pages/medicion/index.es.do. Accessed 9 May 2012.

Cordera, C. R., Heredia, Z. C., & Navarrete, L. J. E. (Grupo de redacción) (2009). Mexico frente a la crisis: hacia un nuevo curso de desarrollo. *Economía UNAM, 6*(18), 7–60.

DOF. (2008). *Acuerdo número 442 por el que se establece el Sistema Nacional de Bachillerato en un marco de diversidad*. Mexico: Diario Oficial de la Federación. Secretaría de Gobernación. file:///Volumes/Trabajos/Trabajos/Acuerdos%20oficiales/EMS/RIEMS/Acuerdo%20442%20SNB.webarchive. Accessed 21 May 2012.

DOF. (2012). *Decreto por el que se declara reformado el párrafo primero; el inciso c) de la fracción II, la fracción V del artículo 3º y la fracción I del artículo 31 de la Constitución Política de los Estados Unidos Mexicanos*. Mexico: Diario Oficial de la Federación. Secretaría de Gobernación.

FMI. (2010). *Ranking de países según su economia*. Cited in as the following: http://www.mogilo.com/forum/topics/ranking-mundial-de-economia-de. Accessed 9 May 2012.

Gamboa, R. F. (2009). El Aula del Futuro: Diseñando nuevos ambientes de aprendizaje. *Boletín SUAyED, 6*. file:///Volumes/Trabajos/Trabajos/Art%C3 %ADculos%20e%20información%20diversa/Educación/Experiencias%20innovadoras/El%20aula%20del%20futuro.webarchive. Accessed 29 May 2012.

Harland, D. J. (2011). *STEM Student Research Handbook*. U.S.A.: National Science Teachers Association.

INEE. (2008). *Estudio comparativo del aprendizaje en sexto de primaria en Mexico 2005–2007: Español y Matemáticas*. Mexico: Instituto Nacional para la Evaluación de la Educación.

INEE. (2009). *El aprendizaje en tercero de secundaria en Mexico. Informe sobre los resultados del Excale 09, aplicación 2008. Español, Matemáticas, Biología y Formación cívica y ética*. Mexico: Instituto Nacional para la Evaluación de la Educación.

INEE. (2010a). *El derecho a la educación en Mexico*. Mexico: Instituto Nacional para la Evaluación de la Educación.

INEE. (2010b). *Mexico en PISA 2009*. Mexico: Instituto Nacional para la Evaluación de la Educación.

INEE. (2011a). *La Educación Media Superior en Mexico. Informe 2010–2011*. Mexico: Instituto Nacional para la Evaluación de la Educación.

INEE. (2011b). *Panorama educativo de Mexico. Indicadores del Sistema Educativo Nacional. 2009. Educación Media Superior*. Mexico: Instituto Nacional para la Evaluación de la Educación.

INEE. (2011c). *Explorador Excale*. Mexico: Instituto Nacional para la Evaluación de la Educación. http://www.inee.edu.mx/explorador. Accessed 23 May 2012.

INEGI. (2009). *Encuesta Nacional de Ingresos y Gastos de los Hogares, 2008*. Mexico: Instituto Nacional de Estadística y Geografía.

INEGI. (2011). *Censo de Población y Vivienda 2010*. Mexico: Instituto Nacional de Estadística y Geografía. http://www.inegi.org.mx/sistemas/olap/proyectos/bd/consulta.asp?p=17118 & c=27769 & s=est#. Accessed 9 May 2012.

Malone, H. J. (Ed.). (2011, November). Q & A with Gabriel Cámara. *Lead the Change Series*, *11*. http://www.aera.net/SIG155/EducationalChangeSIG155/tabid/12179/Default.aspx. Accessed 5 June 2012.

OECD. (2010). *Pisa 2009 Results: What Students Know and Can Do. Students' Performance in Reading, Mathematics and Science (Volume I)*. Paris: Organization for Economic Co-operation and Development.

Gochicoa, E. P., Raimondi y, G. M., & Colectivo docente (2008). Nuestras historias. El lugar del *trabajo en las Telesecundarias Vinculadas con la Comunidad*. Mexico: Universidad Iberoamericana.

PNUD. (2011). *Informe sobre desarrollo humano. Mexico*. Mexico: Programa de las Naciones Unidas para el Desarrollo.

Ramírez, R.. (2010, April 3). La prueba Enlace: contra el sentido de la educación. *Educación UACM. Suplemento de la Universidad Autónoma de la Ciudad de Mexico*. No. 9. *La Jornada*. http://www.jornada.unam.mx/2010/04/03/enlace.html. Accessed 9 May 2012.

Salom, G. (2009). Cómo se ha ido tejiendo mi historia. In *Decisio. Saberes para la acción en educación de adultos*, *23*, 77–85. http://atzimba.crefal.edu.mx/decisio/index.php?option=com_content & view=featured & Itemid=111. Accessed 1 June 2012.

SEP. (2011a). *Sistema Educativo de los Estados Unidos Mexicanos, principales cifras, ciclo escolar 2010–2011*. Mexico: Dirección General de Planeación y Programación, Secretaría de Educación Pública. www.dgpp.sep.gob.mx/Estadi/principales_cifras_2010_2011.pdf. Accessed 5 June 2012.

SEP. (2011b). *Acuerdo 592 por el que se establece la Articulación de la Educación Básica*. Mexico: Secretaría de Educación Pública.

SITEAL. (2011). *Dato destacado 21. El desafío de universalizar el nivel primario*. Sistema de información de tendencias educativas en América Latina (OEI, UNESCO e IIPE). http://www.cedus.cl/files/El%20desaf%C3 %ADo%20de%20universalizar%20el%20nivel%20primario_SITEAL%202011.pdf. Accessed 9 May 2012.

Schmelkes, S. (2010). El papel de la comunidad en el cambio desde la escuela. In B. Barba, & M. Zorrilla (Eds.), *Innovación social en educación. Una base para la elaboración de políticas públicas* (pp. 207–223).

Street, S. (2001). When politics becomes pedagogy: Oppositional discourse as policy in Mexican Teachers' struggles for union democracy. In M. Sutton, & B. A. Levinson (Eds.), *Policy as practice: Howard a comparative sociocultural analysis of educational policy* (pp. 145–166). Westport: Ablex Press.

Street, S. (2010). Contribución del movimiento magisterial democrático al cambio educativo en Mexico. In B. Barba, & M. Zorrilla (Eds.), *Innovación social en educación. Una base para la elaboración de políticas públicas* (pp. 132–156).

UNDP. (2011). *International human development indicators*. New York: United Nations Development Programme. http://hdr.undp.org/en/statistcs/. Accessed 21 May 2012.

WEF. (2011). *The global competitiveness report 2011–2012* Switzerland: World Economic Forum. http://www3.weforum.org/docs/WEF_GCR_Report_2011–12.pdf. Accessed 23 May 2012.

Part III
South America

Chapter 7
Racial Achievement Gaps in Another America: Discussing Schooling Outcomes and Affirmative Action in Brazil

Ricardo A. Madeira and Marcos A. Rangel

Introduction

A negative association between African ancestry and measures of socioeconomic success in regions colonized by Europeans can be considered an empirical regularity across the social sciences. In the USA, Brazil, and South Africa, for example, the intense trade of African slaves by English and Portuguese colonizers and the Dutch displacement of indigenous populations made the color of one's skin an indicator of European ancestry and made it play a key role in social stratification. Most studies document the presence of this historically rooted stratification and uncover racial differences in a variety of contexts, even in the presence of sharp differences in patterns of economic development, enforcement of civil rights, and institutional arrangements regarding racial segregation.[1]

The case of Brazil is particularly outstanding due to somewhat contradictory observations. On the one hand, widespread interracial marriages and desegregation in housing markets have helped spread the view of a Brazilian "racial democracy." Approximately one in every four heterosexual couples is the result of the union between White and Black individuals, whereas the geographic dispersion of population in major urban areas indicates that from one-fifth to one-third of the neighbors

---

[1] See Alexander et al. (2001) for discussions regarding South Africa, the USA, and Brazil. See also Herring et al. (2004) and Telles (2004) on the North American and Brazilian experiences.

M. A. Rangel (✉) · R. A. Madeira
University of Sao Paulo (USP), Sao Paulo, Brazil
e-mail: rangelm@usp.br

R. A. Madeira
e-mail: rmadeira@usp.br

M. A. Rangel
NORC/University of Chicago, Chicago, IL, USA

Abdul Latif Jameel Poverty Action Lab (MIT), Cambridge, MA, USA

J. V. Clark (ed.), *Closing the Achievement Gap from an International Perspective*,
DOI 10.1007/978-94-007-4357-1_7, © Springer Science+Business Media B.V. 2014

of a White Brazilian are Black.[2] As pointed by Reichmann (1999), these indicators led foreign observers to become fascinated with a "haven of racial reconciliation and affinity." On the other hand, and in sharp contrast with the image of tolerance portrayed by such mix, there are stark and persistent inequalities in living standards across skin color groups. In fact, a recent Human Development Report (United Nations 2005) states that skin color difference in economic achievement is one of the main social challenges facing Brazil. The report suggests that antidiscrimination (color-sighted) policies should be a central component of any poverty reduction program implemented in the country.

In this chapter, we explore the recent evidence of racial disparities in socioeconomic outcomes in Brazil. We then trace these differences to income-generating capabilities materialized in an uneven accumulation of human capital (formal education in particular) by Black and White adult Brazilians. We also explore unique and novel data on school transitions and proficiency for the case of the Brazilian southeastern state of Sao Paulo in order to establish general stylized facts in education trends among younger cohorts. The discussion that follows is centered on the assessment of color-blind and color-sighted policies that suggest a closing (but not the elimination) of racial gaps in both the quantity and the quality of education.

Data

We base the analysis in this chapter on two national sources of aggregate data on households and individuals, one source of regional longitudinal information on students built from administrative records, and two national sources of information on high school graduates and college students. National data aggregates are computed from public microdata records of the Brazilian Population Census of 2000 and the Brazilian yearly Household Surveys from 1989 to 2009, both collected and organized by the Brazilian Census Bureau (Instituto Brasileiro de Geografia e Estatistica, IBGE). Regional data are sourced from Sao Paulo's school system and covers the years 2007 to 2011. Sao Paulo is the most populous, richest, and most heterogeneous of all 26 Brazilian states. In Sao Paulo, the School Authority (Secretaria Estadual de Educacao, SEE-SP) directly manages over 5,000 schools, employs about 220,000 people (180,000 of whom are teachers), and serves 4.4 million students (2.3 million in high schools and 1.85 million in primary schools). The Secretary is also responsible for regulation of private schools in the system and maintains straight cooperation agreements with all municipality-run schools across the state. Finally, data with national coverage on high school graduates and college students come from (1) the Exame Nacional do Ensino Medio (ENEM), the Brazilian equivalent of the Scholastic Aptitude Test (SAT), and (2) the Exame Nacional de Desempenho de Estudantes (ENADE), an exam taken by a sample of college

---

[2] These figures are approximately 25 times larger than the ones observed in the USA, respectively. See Fryer (2010) on marriage markets and Massey and Denton (1988) on spatial segregation.

students in their first and last years of college. We merge these two sources of data by exploring administrative data made available by the Brazilian Ministry of Education. All data sets employed are presented in more detail below.

National Data on Households and Individuals

The first data set used in the present study is the 2000 Brazilian Census of Population (Censo Demografico, Instituto Brasileiro de Geografia e Estatistica, IBGE). The public use data, available for purchase from the IBGE website, consist of 10% samples of the population for localities with more than 15,000 inhabitants and 20% samples of the other localities. The interviews were conducted on private households. Information on dwellings' construction and general living standard measures related to access to basic public services and to ownership of assets/durable goods was collected. With respect to individual characteristics, a knowledgeable adult (most frequently the spouse of the household head) was asked to report basic demographics, migration, school enrollment, educational attainment, fertility history (for women 10 years and older), and sources of income.

The 2000 Census maintained the structure used in other editions and asked respondents to report individual members' "skin color or race," reflecting the Brazilian social norm that skin color and race are interchangeable concepts. For the skin-color question, respondents were given five options: white, black, indigenous, yellow (Asian), and brown. The indigenous population and Asians are a small fraction of the overall population (0.6%) and are geographically concentrated in the North and Sao Paulo regions, respectively. In the analysis that follows, we have dropped any household in which at least one member was reported to be in either of these two groups. Henceforth, browns and Blacks are combined in one group.

The census data are complemented using data from the Brazilian Household Survey (Pesquisa Nacional de Amostra de Domicilios, PNAD) also conducted by the Brazilian Census Bureau (Instituto Brasileiro de Geografia e Estatistica, IBGE).[3] The sampling scheme is based on a three-level multistage procedure, a successive selection of municipalities, census sectors, and households. The PNAD collects information on household demographic characteristics, income, labor supply, and human capital investments. The PNAD yearly random sample consists of approximately 65,000 observations on households.

Regional Data on Basic Education

Sao Paulo's Secretaria Estadual de Educação has agreed to share with the authors, under cooperation and confidentiality agreements, detailed information on the uni-

---

[3] Due to budgetary problems, the PNAD was not conducted in 1994.

verse of students in its education system. We merged data sets from three distinct sections of their data bank: matriculation information, standardized tests of proficiency, and transcript records.[4] In what follows, we refer to them as flow measures, standardized scores, and teacher assessments, respectively.

The Brazilian precollege educational system is organized into four levels: preschool (first grade), elementary school (second to fifth grade, ideally attended by 7–10-year-olds), middle school (sixth to ninth grade, ideally attended by 11–14-year-olds), and high school (tenth to twelfth grade, ideally attended by 15–17-year-olds). The elementary school comprises four school years. The basic disciplines offered at such an educational level are language (Portuguese), mathematics, social studies, and sciences. All the basic subjects are taught by the same teacher, but curricular activities also include physical education and the arts, which are taught by specialized teachers. For middle and high school students, teachers' subject specialization is required.

Matriculations in the entire state of Sao Paulo covering elementary, middle, and high schools are centralized by the Secretary of Education. The centralized system exists as a way to prevent parents from matriculating their children in more than one school (private or public) in order to guarantee a slot. In the past, this practice had itself led to a number of children who could not be absorbed by the system (because some had taken two or three slots). The centralization of information coupled with the generation of individual tracking numbers offers interesting ways of measuring student mobility within the school system, especially in the case of dropout and migration between or within public and private systems.

Standardized scores are collected in the context of Sao Paulo's performance evaluation system (Sistema de Avaliação de Rendimento do Estado de São Paulo, SARESP). The system consists of a statewide exam taken by students enrolled in grades 2, 3, and 5 (elementary school); in grades 7 and 9 (middle school); and in grade 12 (high school) of the public schools directly managed by the state-level authority. The test has been applied in slightly different formats since 1996. This chapter uses data from its 2007 edition onward. We have information about 1.8 million test takers in approximately 5,400 schools every year since then. All students/ parents, and a sample of teachers answer a survey that asks questions on socioeconomic status, study habits, teaching and pedagogical practices, and perceptions about the school environment, among others.

The main purpose of such an exam is to measure the students' proficiency on the subjects assigned to each specific grade according to a predetermined curriculum. The exams have two sets of questions covering mathematics and Portuguese language. The mathematics set contains up to 24 multiple-choice questions. The Portuguese language component also includes a short essay for more advanced grades. Grading is electronic for the multiple-choice questions: students use a test sheet, which is scanned and graded automatically, without human interference. This grad-

---

[4]The Secretary itself has never attempted to combine these data. There are different teams of bureaucrats in charge of each of these sections. This is the first time these data are used in an integrated format.

ing procedure assures that a completely blind score (relative to a child's identity) is obtained.[5] We combine matriculation and test score data in order to follow proficiency gains over time for individual students.

The SARESP exams are taken in late November (spring), close to the end of the academic year, during class time and in the same place where the students take regular classes. Students take the exam on two consecutive days, one for each subject. Because 5th, 7th, 9th, and 12th graders (the focus of this chapter) can attend classes in the morning, in the afternoon, or at night, a different exam (yet similar in difficulty) has to be prepared for each group. All students who usually attend classes during the same school shift take the same test. The State Secretary of Education hires an independent institution to prepare the exam, according to predetermined guidelines. To oversee the students during the test, teachers from other schools are mobilized, such that students are supervised by a teacher different from their regular ones. External observers are also assigned to each school to guarantee the strict fulfillment of all rules.

Microdata on these tests' results are provided in the format of percentage of correct answers and proficiency scores in each subject after application of item response theory (IRT) methods. These scores are also converted into a (grade-subject-specific) four-step classification system that reflects educators' consensus regarding levels of proficiency (below basic, basic, sufficient, and more than sufficient) after the statistical definition of anchor items. Proficiency in the essay portion of the language exam is reported in a separate four-level scale. Individual-level results from SARESP are not made publicly available to children, parents, or schools. Until 2008, school-level results were not used in any explicit accountability system either, and they have been serving the sole purpose of "diagnosing" the entire educational system. From that year onward, the education authority has implemented a bonus payment scheme that rewards schools' personnel based on test performance by their students.

Transcript data have been based on a uniform criterion-referenced rule for teacher evaluations adopted by the Secretary's directly managed schools since September 2007. According to these guidelines, all teachers have to attribute numeric integer grades ranging from 0 to 10, and the passing grade is set at 5 points. As part of the official records, teachers also compute attendance rates on a 0–100 (percentage) scale. Teachers and other school administrators were not given instructions on how to attribute grades as a function of a student's observed proficiency level beyond the ones implicitly imposed by a uniform school curriculum. The state administration provides pedagogical material aligned to such a curriculum, and teachers are supposed to evaluate students according to proficiency in such material. Nonetheless, no explicit guidance regarding the design of evaluations (except for questions included at the back of the teacher's booklet) is given. Therefore, the uniformization of grading scales occurs with respect to format but not necessarily in terms of meaning.

---

[5] For students in grades 2 and 3, scoring is not blind. Either their own teacher or a committee (formed within the school) grades the exams.

National Data on High School and College Performance

Starting in 1998, the Ministry of Education implemented a low-stakes exam focused on measuring scholastic abilities for individuals who were graduating or had previously graduated from high school. The original objective was to offer some sort of certification of high school knowledge for those entering the labor market. The so-called Exame Nacional do Ensino Médio (ENEM) is now a comprehensive yearly test designed to assess several subjects. Participation is still voluntary, and students from both public and private schools are eligible to take it. Its popularity and importance increased after 2004, when it became the main criterion to select the recipients of the newly created federal scholarships program, the ProUni (College for All), which awarded full or partial scholarships to low-income students who studied in tuition-free high schools (public or private). At the same time, some colleges, including the prestigious federal universities, began to use the scores obtained in the ENEM as one of the criteria to select students in competitive admission processes.[6]

The test format changed over time, comprising 63 objective, multiple-choice questions, and a writing sample in the 2006 edition used in this chapter. The exam was taken in just 1 day in October, and it had the same questions nationwide. In this edition, the test covered four subjects: mathematics, Portuguese language, natural sciences (chemistry, biology, physics), and social studies (history and geography). The exam is not explicitly divided by subject. Usually, the questions require the understanding of more than one subject. Scores were simply the percentage of correct answers obtained by the candidate.[7]

We merge ENEM data with the data of the 2007 Exame Nacional de Desempenho de Estudantes (ENADE), taken by college students at the end of their first and last years of college attendance. The ENADE exam applied to a sample of students from college matriculation records. The exam is applied by the Ministry of Educations and is used for college accreditation. ENADE 2007 was taken by approximately 250,000 students in their first year of college. We exclude from our analysis students who scored zero on ENADE, resulting from boycott by organized student organizations. ENADE evaluates two areas, major-specific material (covering material delivered in the first year of college) and a general formation material (which basically reflects, once more, material that should be mastered by the end of high school).

---

[6]Thanks to a provision in the Federal Constitution, public institutions of higher education cannot charge tuition, independent of the socioeconomic standing of the student. As a consequence, there is excess demand to enroll, and candidates have to excel in highly competitive entrance exams in order to be admitted. Private colleges charge tuition and generally have lower quality compared with public institutions. Further, state universities are not allowed to charge tuitions either, nor are they allowed to discriminate against out-of-state applicants by giving more weight to state residents.

[7]In 2009, the exam experienced another major change, when it became the only admission criteria to enroll in several federal universities. The number of questions jumped from 63 to 180, the exam is taken in two consecutive days, and item response theory has been used since then to calculate the scores.

Background

Brazil was colonized by Portugal starting in the year 1500. Colonization followed extractive institutions, and Portuguese familial settlements were rare. After an initial period of enslavement of the indigenous populations, expansion of economic activities toward sugarcane plantations required more laborers and led the colonizers into one of the most profitable activities of the colonial times: the trafficking of enslaved Africans. For over 200 years, until the middle of the nineteenth century, approximately 3.6 million Africans were sent to Brazil as slaves. The excessive dependence on such a labor force made Brazil the last country in the whole Western hemisphere to abolish slavery in 1888.

During this early period, migration flows from Europe were composed mostly of male colonizers. This created a clear sex-ratio imbalance in the colony. As a result, the mixing of Whites and Blacks was set in motion, explaining a more diffuse concept of race in Brazil than in the USA. In practice, ancestry was substituted by a phenotype-based perception of racial groups. In this sense, beyond a Black–White dichotomy, Brazil ended up heading toward a racial debate with many shades of gray. Current census counts indicate a population of self-declared African origin only smaller than that in Nigeria, corresponding to approximately half of the 180 million Brazilian inhabitants. This is most likely an underestimate nonetheless. Genetic research has recently uncovered that a large proportion of Brazilian self-declared Whites have mitochondrial DNA (maternal lineage) that can be traced to an African origin.[8]

Large rates of miscegenation have led most observers to conclude that, in the absence of racial conflict, Brazil had simply avoided the consequences of enslavement on socioeconomic outcomes and mobility.[9] That is not the picture emerging from a careful study based on sociodemographic data, however. There is now overwhelming evidence that such racial tolerance indicators coexist with pertinent differences between Whites and Nonwhites in terms of wages and other measures of living standards (see Arias et al. 2004; Campante et al. 2004; Telles 2004). A recent publication by the World Bank (see Perry et al. 2006) extended that analysis and presented evidence that even returns to schooling (in terms of wages) among dark-skinned individuals are lower than among Whites. These findings suggest that industrialization, economic progress, and modernization of the social structure have not eliminated color as a potential determinant of social inequalities (see Hasenbalg et al. 1999) more than 100 years after the abolition of slavery.

In order to illustrate these stylized facts, we reproduce such findings using microdata from the 2000 Brazilian Census of Population. Figure 7.1 presents rates of home ownership and access to public utilities. Blacks are consistently found in worse conditions when compared with Whites on all dimensions of living standards investigated. They are less likely to own their homes, even considering the loose

---

[8] See Parra et al. (2004).
[9] See Pierson (1945).

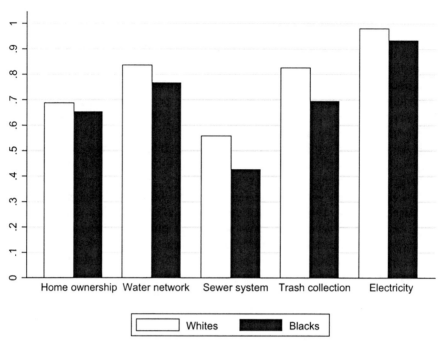

Fig. 7.1 Living standards by race, Brazil 2000. Data source: Population Census 2000, IBGE

definition of ownership used by the Census enumerators (not based on formal/legal ownership). Their homes are also less likely to be served by water, electricity, and sewer systems. They also live in areas less likely to have trash regularly collected by the public sector.

We then consider differences in the main source of income for Brazilian families: the sale of one's labor. Figure 7.2 explores the same source of data as mentioned above by looking at the distribution of hourly wages (in a log scale) commanded by workers of different racial background at 35 years of age. For both men and women, the evidence indicates that the wage distribution is shifted to the right for Whites. In general, hourly wages are approximately 40 % higher among the latter.

Such differences in income-generating capabilities are remarkably constant in the 15-year period between 1995 and 2009. Data from the Brazilian Household Surveys in the period indicate that both hourly wages and unemployment rates (for male adults aged 30–35) are less favorable for Blacks.[10] Racial differences are slightly reduced in terms of wages, but there is no sign of relative improvement in the unemployment indicator among Blacks, as seen in Fig. 7.3.

---

[10] We focus on male workers in order to avoid changes in the composition of female labor force due to time-changing participation decisions.

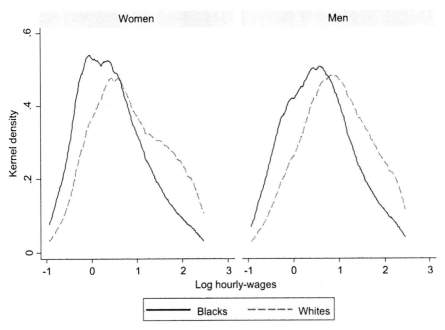

Fig. 7.2 Hourly wages by race (in logarithms), Brazil 2000. Data source: Population Census 2000, IBGE

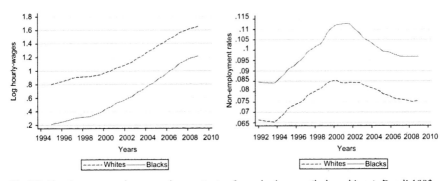

Fig 7.3 Hourly wages and non-employment rates for males by race (in logarithms), Brazil 1992–2009. Data source: PNAD, IBGE

There are at least two main factors that could explain racial differentials in those economic outcomes. It is possible that dark-skinned individuals receive lower wages, are less likely to be employed, or have limited access to certain jobs due to discrimination or prejudice among labor market actors. Alternatively, observed differences may be the result of darker-skinned individuals' relatively lower invest-ment in the accumulation of skills, which translates into a scarcity of economic op-portunities. We focus here on the latter and show how differently (in terms of human

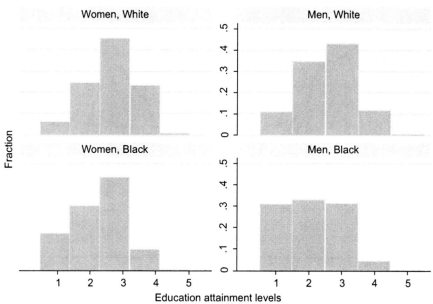

Fig. 7.4 Education attainment by race (completed degrees), Brazil 2000. Data source: Population Census 2000, IBGE

capital) Black and White Brazilians arrive in the labor market. Figures 7.4 and 7.5 reproduce the distribution of education attainment by race in the year 2000 and the evolution of years of schooling completed by race among adults from 1992 to 2009, respectively. It can be seen that Blacks consistently accumulate less human capital in the form of formal education (lower quantity). They are overrepresented on the lower levels of achievement (less-than-primary and primary) and underrepresented among holders of high school and college degrees. Despite an overall increase in educational attainment, in the 20 years since 1992, there is a constant difference of 2 years of completed schooling between Blacks and Whites born between 1957 and 1974.

Can these differences in completed years of schooling explain the disparities in earnings potential we observed above? In order to address this question, we employ simple regression analysis. We compute both unemployment and log-hourly-wage differentials before and after controlling for years of formal education in a sample of males aged 30–35 during the 1992–2009 period. Our findings indicate that accounting for educational disparities accounts for roughly 50% of the differences between Blacks and Whites. Whereas differences in unemployment rates are reduced from 2 to 1 percentage point, those in hourly wages drop from 0.53 to 0.24 log points. Racial differences remain significant, nonetheless.

In fact, in Fig. 7.6 we reproduce estimates for log-hourly-wage density functions (males only) stratified in four education groups: no schooling or preschool

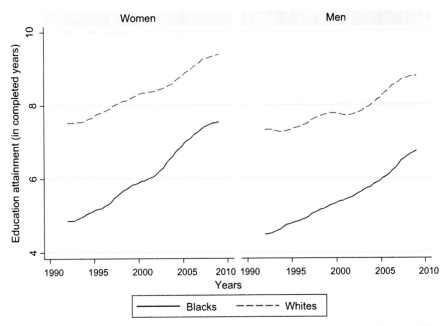

Fig. 7.5 Education attainment by race over time (completed years) for adults age 35, Brazil 1992–2009. Data source: Brazilian Household Survey (PNAD), IBGE.

only (0–1 year), elementary education only (5 years), primary education only (9 years), and high school education only (12 years). It can be seen that in all cases, but particularly for groups with more education, the differences between Blacks and Whites (favoring the latter) are still sizable. Estimates for differences in mean hourly wages in these groups are 0.20, 0.20, 0.25, and 0.26 log points, respectively. In any case, such stylized facts indicate that gaps in the quantity of education seem to be one of the central pieces for the understanding of differences in socioeconomic outcomes between Black and White Brazilians.

What is clearly left out of this picture is that despite having the same years of education, there is no guarantee that these Black and White adult Brazilians were exposed to education of the same quality. In other words, treating years of completed education as a homogeneous set of skills within the Brazilian population is likely no more than wishful thinking. Unfortunately, data that could further aid the understanding of racial gaps in wages are not available in Brazil. For the adult population described in the figures above, there is no data collection on the amount of skills accumulated or even the type of education (private versus public) acquired, as is common in North-American data that include information like the Armed Forces Qualification Test (AFQT) scores.[11] Even if only speculative, yet based on the North-American literature, we are left with the conclusion that the reduction on

---

[11] See O'Neil (1990), Maxwell (1994), Neal and Johnson (1996), Heckman (1998), and Carneiro et al. (2005).

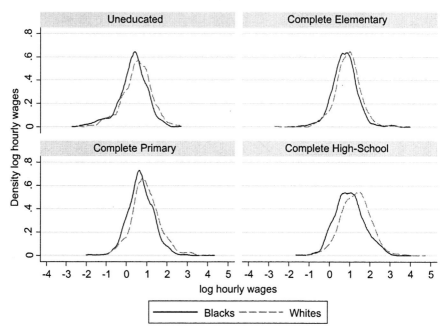

Fig. 7.6 Log wage distributions for adults aged 30 to 35, Brazil 2001. Data source: Brazilian Household Survey (PNAD), IBGE

Brazilian racial socioeconomic gaps in the years to come will be proportional to the differences in both the quantity and the quality of education acquired by Blacks and Whites. With this in mind, we turn to the description and assessment of recent policies that either directly or indirectly changed (or are likely to change) trends in human capital accumulation across racial groups.

Recent Trends in Attainment Gaps Based on Aggregate Data

The 1990s marked a decade of change in Brazil. After years of struggle with chronic inflation and economic turmoil, the country started experiencing stability in 1995. The control of inflation represented a particularly dramatic improvement in the life of the unbanked (and nonindexed) poor, representing a maintenance of purchasing power previously unthinkable. In that sense, planning and investment in the education of children became more attractive to poorer parents than they previously were, increasing the demand for schooling.

Most importantly, once macroeconomic instability ceased to be the main focus of the Brazilian government, new policies started being designed and implemented in different spheres. In terms of education policy, there was a significant regulatory wave. We point to three new policies: First, initial steps were taken in the establishment of a system of accountability based on national examination of students (Sistema Nacional de Avaliacao da Educacao Basica, SAEB) that led to the implementation of national targets for improvements in 2007. Second, the federal government created the so-called Bolsa Escola Program, a conditional cash transfer mechanism that paid families to enroll and keep their kids in school.[12] Finally, there was a sharp change in the distribution of the federal budget for education affecting both amounts and regional distribution of resources for school construction, maintenance, and improvement (Fundo de Manutencao e Desenvolvimento do Ensino Fundamental, FUNDEF).

Together, these systemic changes produced progress in standard educational policy targets. There was, for example, an increase in the rates of enrollment of school-aged children. This "democratization" process has had a major impact on the composition of the student body and has increased the representation of a deprived portion of the population within classrooms.

From the perspective of the central discussion of this chapter, the democratization has brought to the classroom students with darker skin tones, who would have been out of school otherwise. This pattern can be seen in Fig. 7.7, where we reproduce patterns of school enrollment at age 7 (elementary school entry age) from 1989 to 2009. The reduction in the racial gaps regarding the access to education at this age is truly remarkable. Even if not directly resulting from policies that target racial differences, this increase in access to education has enormous potential in reducing gaps in socioeconomic outcomes among future generations of Brazilian Blacks and Whites.

The fulfillment of such potential would require that children of disadvantaged backgrounds not only entered but also stayed in school, however. This does not seem to be the case. When we examine the evolution of enrollment at age 15 during the same period (Fig. 7.8), we conclude that there was no reduction in dropout rates that followed differential patterns across races. In other words, over time Black children became more likely to enter school but not more likely to finish primary education relative to Whites.

However, some educational policies could help transform this massive entry of students in the system into accumulated years of education. We explore the fact that in Brazil the education policy is decentralized to investigate a specific student-retention initiative. In particular, starting in 1996, the state of Sao Paulo's public school system adopted an automatic promotion scheme. This policy grouped contiguous grades into cycles, with retention occurring only at the end of each cycle. In the case of Sao Paulo, two cycles were created. Cycle 1 encompasses grades 1–5 and cycle 2 covers grades 6–9. High schools were not included in the automatic promotion scheme. Under such rules, a student is promoted to the next grade if she attends more than 75 % of the classes, irrespective of her mastery of the material that was covered during the academic year. Insufficient proficiency can result in

---

[12] This was later phased into the current Bolsa Familia Program, the largest conditional cash transfer program in the world.

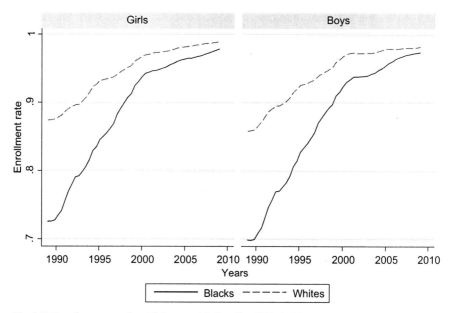

Fig. 7.7 Enrollment rates for children aged 7, Brazil—1989–2009. Data source: Brazilian Household Survey (PNAD), IBGE

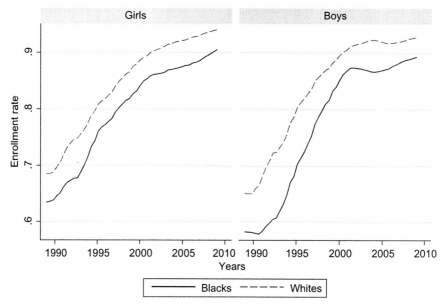

Fig. 7.8 Enrollment rates for children aged 15, Brazil—1989–2009. Data source: Brazilian Household Survey (PNAD), IBGE

grade retention only at the end of each cycle, nonetheless. In this case, the pupil must repeat the last grade within that cycle.

Several international organizations, including the World Bank, support this policy as an effective way to curb low-grade completion and to decrease dropout rates. The general lines of the argument are that grade retention could adversely affect noncognitive skills (like confidence and self-esteem), increasing anxiety levels and harming learning capacity. In this scenario, a better alternative would be the promotion to the next grade despite the insufficient performance.[13]

In any case, the results from this policy (coupled with the democratization) in terms of racial gaps can be observed in Figs. 7.9 and 7.10. It is remarkable to note that in Sao Paulo, convergence in attainment (for 10- and 15-year-olds, respectively) between Blacks and Whites is much more pronounced than it is in other parts of the country. The timing of convergence coincides with the adoption of automatic promotion. Even if not aiming directly at racial issues, by benefiting students at the bottom of the skill distribution, automatic promotion has a disproportional impact in enrollment and dropout rates among Blacks. These findings indicate that most of the differences in the rate of primary education are bound to become irrelevant for the understanding of Black–White socioeconomic outcomes in the near future, at least in Sao Paulo.

However, when we take a closer look at the enrollment in high school and college, the picture that emerges is less optimistic. As can be seen in Fig. 7.11, there is no reduction in the gap in high school graduation rates. In fact, we also detect that high school enrollment racial gaps in Sao Paulo have even been growing since 2003 (not shown). Differences in access to college are also pronounced. Figure 7.12 shows that since 1992 there has been no sign of reduction in gaps. Policies directly or indirectly aimed at closing racial differentials in both high school and college levels seem ineffective so far in terms of enrollment.

Moreover, even if Black and White individuals are more likely to have closer levels of schooling (measured in years of formal education), it is still an open question whether the quality of education received by each group can be considered comparable. In order to discuss these challenges further, we attempt to extract information by following students' trajectory within Sao Paulo's education system.

Measuring Education Gaps in Sao Paulo: Lessons from Longitudinal Microdata

In this section, we investigate the racial gap in education in two main dimensions: (1) student progression in the school system and (2) student performance on standardized tests. In both cases, we draw our conclusions exploring unique longitudinal data that we were able to construct from administrative information. Table 7.1

---

[13] See King et al. (2008).

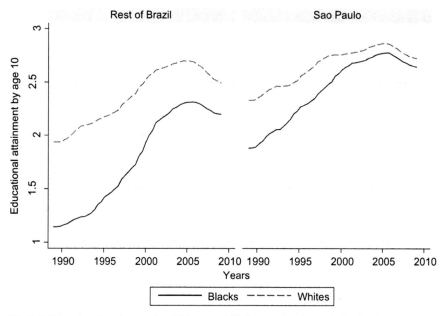

Fig. 7.9 Educational attainment for children aged 10 (in completed years), Sao Paulo versus Rest of Brazil—1989–2009. Data source: Brazilian Household Survey (PNAD), IBGE

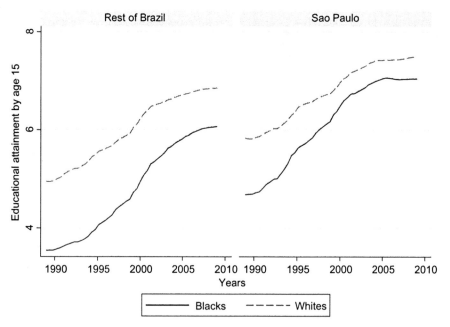

Fig. 7.10 Educational attainment for children aged 15 (in completed years), Sao Paulo versus Rest of Brazil—1989–2009. Data source: Brazilian Household Survey (PNAD), IBGE

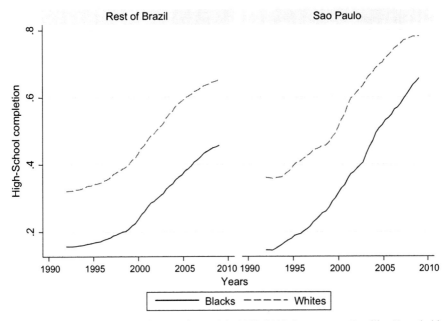

Fig. 7.11 High-school completion rate (by age 24), 1992–2009. Data source: Brazilian Household Survey (PNAD), IBGE

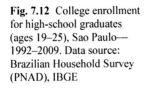

Fig. 7.12 College enrollment for high-school graduates (ages 19–25), Sao Paulo— 1992–2009. Data source: Brazilian Household Survey (PNAD), IBGE

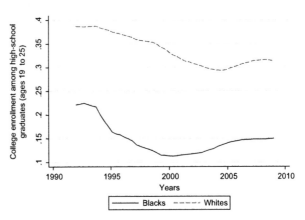

focuses on the progress of the White students through the Sao Paulo educational system (private and public schools included). The figures tell us that in the year 2011, 281,500 students out of the 346,000 who were enrolled in the first year of elementary school (second grade) in 2007 reached the sixth grade without interruption; that is, 81.4 % of the second graders of the 2007 cohort did not repeat a grade or leave school throughout these four schooling years. Table 7.2 reproduces the same analysis for Black students. We detect a difference across racial groups.

Table 7.1 Attrition rates for White students, all types of schools

	2nd grade	3nd grade	4rd grade	5th grade	6th grade	Total
2007	345,838					345,838
	100.00%					100.00%
2008	13,763	323,050				336,813
	3.98%	93.41%				97.39%
2009	1,924	25,650	306,152			333,726
	0.56%	7.42%	88.52%			96.50%
2010	400	5,044	26,820	298,699		330,963
	0.12%	1.46%	7.76%	86.37%		95.70%
2011	131	1,245	6,012	33,506	281,517	322,411
	0.04%	0.36%	1.74%	9.69%	81.40%	93.23%

Table 7.2 Attrition rates for Black students, all types of schools

	2nd Grade	3nd Grade	4rd Grade	5th Grade	6th Grade	Total
2007	186,135					186,135
	100.00%					100.00%
2008	9,977	169,970				179,947
	5.36%	91.32%				96.68%
2009	1,664	19,184	157,237			178,085
	0.89%	10.31%	84.47%			95.68%
2010	356	4,530	19,292	152,112		176,290
	0.19%	2.43%	10.36%	81.72%		94.71%
2011	117	1,136	5,252	24,896	139,044	170,445
	0.06%	0.61%	2.82%	13.38%	74.70%	91.57%

Only 74.7% of the Black second graders (in 2007) reached the sixth grade in 2011. The racial dropout/failure gap is larger for Blacks in each adjacent pair of schooling years. The difference across groups is particularly large after the first year of elementary education and in the transition from elementary to middle school (fifth to sixth grade). During this latter transition, about 5% of the White students (enrolled in the second grade in 2007) fail or abandon the school, whereas 7% of the Black students in the same cohort do not make it to middle school in an appropriate time.

Tables 7.3 and 7.4 reproduce the analysis presenting the school progress figures for the students of the 2007 cohort between grades 8 and 12. Again, they show important differences in school progress across racial groups. Only 51% of the Black students in the eighth grade reach the last year of high school (grade 12), whereas 62% of the White students do so. The pattern observed for younger students also shows up among older ones. Racial gaps are pertinent over all school years investigated and are again particularly relevant at the transition from middle to high school (ninth to tenth grade). During this transition, about 11% of the White students (enrolled in the eighth grade) fail or drop out, whereas 15% of the Black students do not make it to high school at the time they should.

The measurement of school-years transition probabilities also allows a more careful investigation into the automatic promotion scheme adopted in schools directly managed by the Sao Paulo school authority. In Fig. 7.13, we compare the

Table 7.3 Attrition rates for White students, all types of schools

	8th Grade	9th Grade	10th Grade	11th Grade	12th Grade	Total
2007	250,896					250,896
	100.00%					100.00%
2008	11,176	229,145				240,321
	4.45%	91.33%				95.79%
2009	2,329	22,153	201,168			225,650
	0.93%	8.83%	80.18%			89.94%
2010	576	5,750	36,859	173,259		216,444
	0.23%	2.29%	14.69%	69.06%		86.27%
2011	60	893	10,648	27,231	156,705	195,537
	0.02%	0.36%	4.24%	10.85%	62.46%	77.94%

Table 7.4 Attrition ratesfor Black students, all types of schools

	8th Grade	9th Grade	10th Grade	11th Grade	12th Grade	Total
2007	142,758					142,758
	100.00%					100.00%
2008	9,219	125,078				134,297
	6.46%	87.62%				94.07%
2009	2,076	17,440	103,898			123,414
	1.45%	12.22%	72.78%			86.45%
2010	538	4,906	26,345	84,799		116,588
	0.38%	3.44%	18.45%	59.40%		81.67%
2011	67	752	8,355	17,847	73,852	100,873
	0.05%	0.53%	5.85%	12.50%	51.73%	70.66%

transition probabilities of students in state-run and municipality-run schools in the system. The former all adopted automatic promotion, whereas only a minority of the latter has done so in this period. We find that racial differences in attrition rates are indeed virtually nonexistent in schools that adopt automatic promotion.

To what extent do differences in attrition between school levels result from students' own learning experiences? We investigate this after observing sizable differences in fifth- and ninth-grade SARESP mathematics test performances for Blacks and Whites in schools directly managed by the Sao Paulo school authority, which are reproduced in Fig. 7.14. Tests of difference in means indicate gaps of 0.34 and 0.29 standard deviations, respectively, favoring Whites.

Then, in Fig. 7.15, we cross performance in standardized tests in mathematics at the end of elementary education (x-axis) with attrition rates in terms of entry into middle school (y-axis). The dashed line illustrates attrition levels for White students, whereas the solid line represents the difference in attrition between Blacks and Whites (dotted lines indicate 95% confidence intervals). Attrition rates decrease rapidly as test scores increase. Importantly, once test scores in mathematics are accounted for, there is no detectable gap in attrition rates between Black and White students. This evidence suggests that all the relevant differences in 1-year attrition rates between the races at this schooling level come from underlying differences in proficiency.

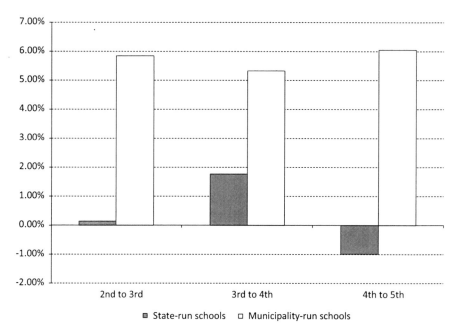

Fig. 7.13 Difference in attrition rates (Black versus White), by school system Sao Paulo—2007–2010

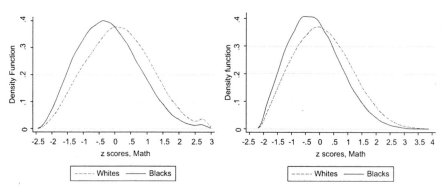

Fig. 7.14 5th and 8th Grade Math Scores, Sao Paulo Public Schools 2008. Data source: SARESP

Despite similarities in patterns, the difference in attrition between Blacks and Whites in the transition between middle and high school is not fully explained by ninth-grade mathematics tests scores, as we show in Fig. 7.16. There is an indication, therefore, that decisions to enroll in high school are more elaborate, and focus on other dimensions is not directly captured by standardized test material.

Since proficiency gaps are responsible for a large share of racial differences in educational attainment decisions, we turn to a more careful investigation of their

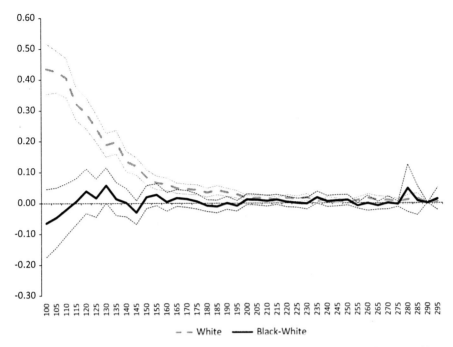

Fig. 7.15 Attrition Rates by Proficiency Level in Math elementary to middle school transition. Data source: SARESP and matriculation records

prevalence and persistence. Because the standardized test scores for initial grades (grades 2 and 3) are not computed using IRT, they cannot be directly compared to scores of the other grades. Therefore, we report results using a z-score transformation of the percentage of correct answers by each student. The computed gaps for grades 2, 3, and 5 were obtained using the cohort of students who were tested in the second grade in 2007 and who did not fail or drop out of the school system at least until fifth grade. Gaps for grades 7 and 9 were calculated using data on the cohort of students who were in the fifth grade in 2007 and did not fail or drop out of the school system at least until the ninth grade. That is, these students were tested in grade 7 in 2007 and in grade 9 in 2009. Lastly, gaps for grade 11 were obtained using data on the cohort of students who were in grade 9 in 2007 and who did not fail or drop out of the school system at least until twelfth grade. Therefore, these students were tested in grade 9 in 2007 and in grade 12 in 2010.

By selecting our sample in this way, we avoid mixing attrition issues with longitudinal evolution of proficiency. We calculate the racial gaps employing three statistical models. The first model delivers the raw differences between Black and White students, without accounting for potential differences in the school environment and students' socioeconomic characteristics. The second model accounts for differences in observable socioeconomic characteristics. The third model compares students conditional on their attending the same school and having similar socioeconomic

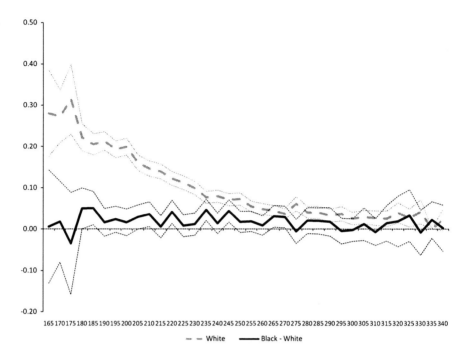

Fig. 7.16 Attrition Rates by Proficiency Level in Math middle to high school transition. Data source: SARESP and matriculation records

characteristics. Figure 7.17 presents the results for the four versions (results for the nontested grades—4, 6, 8, 10, and 11—were obtained through linear interpolation).

As expected, differences in socioeconomic characteristics and the school environment account for about 55% of raw racial gap, especially in initial grades; that is, the calculated gaps using model 3 (gray line) are roughly 55% of the raw gap (black solid line). However, even after controlling for the school environment and students' socioeconomic background a gap remains for all grades. The racial gap slightly increases during elementary school (from 0.09 to 0.13 standard deviations) and returns to its initial level during middle and high school years. Notwithstanding, the evidence is consistent with a constant racial gap over time. In particular, it reveals the existence of a gap that children bring to school at the time of entry, which is neither explained away by socioeconomic differences captured by parental education and ownership of durables nor eliminated by the training offered in these public schools.

Figure 7.18 reproduces the same exercise using IRT scores (therefore, grades 2 and 3 are not used). This time we display the evolution of the IRT scale for Black and White students across grades. The same pattern obtained with the standardized percentage of correct answers is observed for IRT scales; that is, the observed racial gap in proficiency seems to be constant across grades.

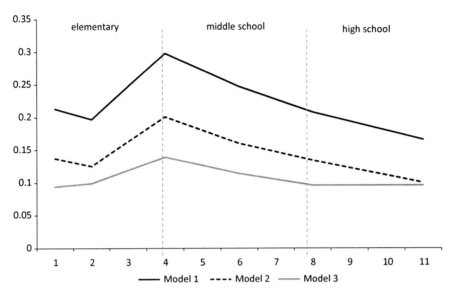

Fig. 7.17 Math Proficiency Gaps (z-scores % of correct answers) over time in school. Data source: SARESP

Therefore, our results indicate that in spite of the recent trend of reduction in the racial gap in years of education, led by the democratization in school access, the proficiency gap does not follow the same trend. These findings suggest that even if the democratization process eventually closes the secular racial gap in years of education, Blacks will stilllag behind Whites in proficiency. A remarkable message of our exercise is that the usual explanations for the existing racial gap in proficiency, such as differences in school quality, school environment, and socioeconomic background between Black and White students, explain only about 55 % of the gap. Blacks still underperform Whites of identical background by 10 % of a standard deviation in mathematics tests.

These findings are compatible with US evidence suggesting that differences in skills between Blacks and Whites emerge during infancy, affecting both cognitive and noncognitive aspects of child development and becoming more prominent while children attend elementary school.[14] We interpret the early appearance and the dynamics of such racial gaps as a call for a better understanding of the role that a child's race plays in the school and in classroom settings. Based on evidence uncovered in this chapter, one can arguably say that among the greatest challenges of the Brazilian basic educational system is designing and adopting policies capable of closing these gaps. In order to achieve this goal, it is necessary to identify the main causes of the proficiency gap that go beyond the usual explanations related to differences in school quality and socioeconomic background.

---

[14] See Fryer and Levitt (2004b).

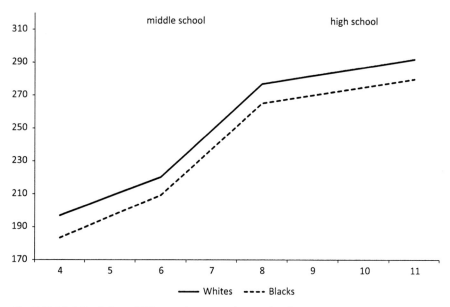

Fig. 7.18 Math Proficiency IRT scores by race and over schooling years. Data source: SARESP

A possible alternative explanation would be if teachers treat Black and White students differently, unfavoring the closing of preexisting gaps. We combine student-level data on standardized test scores with data on students' report cards in order to tackle this issue. We explore the fact that SARESP's grading is color-blind (relative to a child's phenotype), because it is done electronically, and that the state schools in Sao Paulo adopted a uniform criterion-referenced rule (grades must be an integer number between 0 and 10). The rationale for the empirical exercises performed here is to see whether White and Black students with the same blindly graded mathematics score (SARESP) receive different grades.

We perform these empirical exercises in two steps. At first, we identify the existence and robustness of the impact of race over the differences between nonblind (assigned by teachers) and blind measures of proficiency. Second, we go about investigating whether there is any detectable sign of such measures over observed student behavior. We investigate several alternative measures of proficiency and look at both cardinal and ordinal measures. The cardinal measure used is teachers' direct assessment (0–10 scale). The ordinal measures encompass an indicator for best performance in the classroom (achieving the maximum score within the classroom) and the percentile rank within the classroom.

The exercises for both cardinal and ordinal measures are conducted in a logical sequence, with gradual inclusion of controls that aim first at making the teachers' assessments and scores more similar (because teachers can assess writing ability in evaluations, whereas the test is multiple choice). We use scores in the SARESP writing sample to control for these effects. We then add demographic characteristics that may explain racial differences (Blacks tend to be older). Finally, socio-economic characteristics that may be correlated with race and may explain differences (e.g., personal appearance or interactions between parents and teachers) are added. Table 7.5 presents the results for teachers' assessment in mathematics. The introduction of controls reduces observed racial differences about 70 to 80%, but a significant difference remains, suggesting that teachers do evaluate White students more favorably than their Black counterparts.

A possible explanation for this finding is that students' behavioral indicators are available to teachers during classroom interactions and may influence their assessment of a child's aptitude. Therefore, racial differences in class behavior could explain away the gap in Table 7.5. To address this issue, we consider alternative proxies for behavior in an attempt to check the sensibility of results. We explore information correlated with behavior from different sources, such as (1) teacher attendance records, assuming the students who miss more classes are the worst behaved even when attending (we used attendance in the first six school months), (2) physical education records of attendance and grades in the first six months of classes (because PE grades are under the responsibility of a different teacher and should basically reflect obedience in group activities, we consider this to be a strong predictor of behavioral problems), (3) self-reported absence in classes, and (4) history of school transfers and failures (going 2 years back), which should mostly reflect behavioral problems (considering the lenient rules for grade approval).

Table 7.6 presents our results. It reveals that the introduction of behavior controls seems to have no effect over the estimated racial differences, indicating that disparities in behavior are not driving the results. Our reading of these results is that there are still differences in assessments that are not explained by the controls included. This is, loosely speaking, an indication of discrimination within schools or that students are different in dimensions (observable by teachers) well beyond the ones we are capable of measuring.

Tables 7.7 and 7.8 present the patterns of discrimination we encounter when investigating ordinal measures. The ordinal effects are larger than the cardinal ones, representing 10–30% of the original differences. The bottom line is that Blacks are less likely to be best in class and more likely to be underranked relative to Whites, reinforcing the evidence that teachers do treat Black and White students differently.

These results are particularly worrisome in a scenario where parents and children themselves update investment and effort decisions after extracting signals regarding scholastic abilities from report cards. Intraclassroom differentials would then feed back into the parental/individual decision process (Lundberg and Startz 1983; Coate and Loury 1993). In other words, if children's perceived ability increases the returns or reduces the costs of investments, or if a teacher's assessment influences key non-cognitive aspects of a child's life (such as self-esteem, confidence, and motivation), this mechanism could reinforce racial gaps in the accumulation of human capital. Its impact would also depend on how labor markets are structured (based on the presence or absence of tournament-like contracts).

We conclude our empirical analysis on this matter with a glance at the impacts of biased grading on parental and child decisions related to the accumulation of

Table 7.5 Teachers' assessments gap in mathematics

		(1)	(2)	(3)	(4)	(5)	(6)	(7)
5th grade	White	7.025,146						
		(0.011456)						
	Black	672,483	-0.300317	-0.235375	-0.092498	-0.075833	-0.069773	-0.062495
		(0.010329)	(0.009900)	(0.008684)	(0.006155)	(0.005563)	(0.005533)	(0.005511)
	Obs	206,799	206,799	206,799	206,799	206,799	206,799	206,799
7th grade	White	6.119520						
		(0.009405)						
	Black	5.733836	-0.385684	-0.322063	-0.163994	-0.133855	-0.120165	-0.104906
		(0.008554)	(0.008000)	(0.006995)	(0.005816)	(0.005630)	(0.005551)	(0.005528)
	Obs	302,687	302,687	302,687	302,687	302,687	302,687	302,687
9th grade	White	6,045.883						
		(0.009919)						
	Black	5,671.388	-0.374496	-0.313657	-0.186931	-0.124394	-0.099902	-0.087054
		(0.009259)	(0.008969)	(0.007857)	(0.007051)	(0.006778)	(0.006659)	(0.006652)
	Obs	266,816	266,816	266,816	266,816	266,816	266,816	266,816
Controls								
Classroom fixed-effects (FE)		–	–	Yes	Yes	Yes	Yes	Yes
Proficiency scores		–	–	–	Yes	Yes	Yes	Yes
Writing ability		–	–	–	–	Yes	Yes	Yes
Demographics		–	–	–	–	–	Yes	Yes
Socio-economic status (Ses)		–	–	–	–	–	–	Yes
Teacher Accounts		–	–	–	–	–	–	–
Physical Education		–	–	–	–	–	–	–
Self-report absence		–	–	–	–	–	–	–
History		–	–	–	–	–	–	–

Obs: Robust standard errors clustered at the classroom level are presented in parentheses

Table 7.6 Teachers' assessments gap in mathematics—behavior controls

		(1)	(2)	(3)	(4)	(5)	(6)	(7)
5th grade	Gap	-0.062495	-0.069843	-0.062608	-0.062975	-0.062895	-0.065559	-0.062821
		(0.005511)	(0.005432)	(0.005356)	(0.005478)	(0.005508)	(0.005317)	(0.005356)
	Obs	206,799	206,799	206,799	206,799	206,799	206,799	206,799
7th grade	Gap	-0.104906	-0.110643	-0.110877	-0.107540	-0.105787	-0.110146	-0.110076
		(0.005528)	(0.005217)	(0.005179)	(0.005392)	(0.005427)	(0.005015)	(0.005131)
	Obs	302,687	302,687	302,687	302,687	302,687	302,687	302,687
9th grade	Gap	-0.087054	-0.096063	-0.109268	-0.092605	-0.069890	-0.089534	-0.089547
		(0.006652)	(0.006187)	(0.006109)	(0.006439)	(0.005957)	(0.005505)	(0.005615)
	Obs	266,816	266,816	266,816	266,816	266,816	266,816	266,816
12th grade	Gap	-0.078740	-0.083853	-0.100538	-0.079506	-0.072856	-0.090911	-0.093563
		(0.011102)	(0.010381)	(0.010525)	(0.010774)	(0.010777)	(0.009935)	(0.010260)
	Obs	73,785	73,785	73,785	73,785	73,785	73,785	73,785
Controls								
Classroom fixed-effects (FE)		Yes	Yes	Yes	Yes	Yes	Yes	FE
Proficiency scores		Yes	Yes	Yes	Yes	Yes	Yes	Proficiency scores
Writing ability		Yes	Yes	Yes	Yes	Yes	Yes	Writing ability
Demographics		Yes	Yes	Yes	Yes	Yes	Yes	Demographics
Socio-economic status (Ses)		Yes	Yes	Yes	Yes	Yes	Yes	Ses
Teacher Accounts		Yes	–	–	–	Yes	–	Teacher Accounts
Physical Education		–	Yes	–	–	Yes	Yes	Physical Education
Self-report absence		–	–	Yes	–	Yes	–	Self-report absence
History		–	–	–	Yes	Yes	Yes	History

Obs: Robust standard errors clustered at the classroom level are presented in parentheses

Table 7.7 Gap in the probability of being best-in-class in mathematics

		(1)	(2)	(3)	(4)
5th grade	White	0.18482			
		(0.002041)			
	Black	0.140972	−0.043848	−0.018976	−0.018614
		(0.001747)	(0.001920)	(0.001530)	(0.001535)
7th grade	White	0.103831			
		(0.001207)			
	Black	0.073270	−0.030561	−0.013719	−0.013617
		(0.001160)	(0.001166)	(0.001006)	(0.001011)
9th grade	White	0.102309			
		(0.001233)			
	Black	0.077895	−0.024414	−0.010302	−0.010365
		(0.001336)	(0.001265)	(0.001090)	(0.001097)
12th grade	White	0.113466			
		(0.002271)			
	Black	0.092004	−0.021462	−0.010385	−0.010688
		(0.002331)	(0.002398)	(0.002268)	(0.002280)
Controls					
Classroom fixed-effects (FE)		−	−	Yes	Yes
Proficiency scores		−	−	Yes	Yes
Writing ability		−	−	Yes	Yes
Demographics		−	−	Yes	Yes
Socio-economic status (Ses)		−	−	Yes	Yes
Teacher account		−	−	Yes	−
Physical Education		−	−	Yes	Yes
Self-report absences		−	−	Yes	−
History		−	−	Yes	Yes

Obs: Robust standard errors clustered at the classrom level are presented in parentheses

skills. We regress dropout rates, migration to private schools (search for quality), student satisfaction with the learning process, and motivation to learn mathematics against the cardinal and ordinal biases discussed above for fifth and ninth graders (end-of-cycle students). Grade inflation has implicit impacts on the automatic promotion policy. Yet, their impacts go beyond the promotion effect. Children who have their grade inflated above their actual proficiency are not affected in terms of dropout (even after conditioning on being promoted). The children are also not more satisfied with the learning process. Yet, underranking and underscoring seem to induce students to be less likely to migrate to a private school. In that sense, the overall impact of differential grading by teachers seems to have negative impacts over children's motivation and over the probability of investment in higher-quality private education.

Even though discriminatory behavior by teachers driven by taste cannot be ruled out as a source of explanation for our findings, statistical discrimination is also consistent with the results. In a school environment similar to the one suggested by

Table 7.8 Gap in within-class percentile rankings in math

		(1)	(2)	(3)	(4)
5th grade	White	0.509169			
		(0.000731)			
	Black	0.473094	−0.036075	−0.011097	−0.010652
		(0.000623)	(0.001293)	(0.000891)	(0.000898)
7th grade					
	White	0.512904			
		(0.000580)			
	Black	0.464474	−0.048430	−0.017631	−0.017609
		(0.000503)	(0.001040)	(0.000847)	(0.000867)
9th grade					
	White	0.508350			
		(0.000583)			
	Black	0.463386	−0.044963	−0.014336	−0.014336
		(0.000562)	(0.001106)	(0.000899)	(0.000918)
12th grade					
	White	0.492637			
		(0.000821)			
	Black	0.460244	−0.032392	−0.017562	−0.018021
		(0.001260)	(0.002021)	(0.001901)	(0.001957)
Controls					
Classroom fixed-effects (FE)		–	–	Yes	Yes
Proficiency scores		–	–	Yes	Yes
Writing ability		–	–	Yes	Yes
Demographics		–	–	Yes	Yes
Socio-economic status (Ses)		–	–	Yes	Yes
Teacher account		–	–	Yes	–
Physical Education		–	–	Yes	Yes
Self-report absences		–	–	Yes	–
History		–	–	Yes	Yes

Obs: Robust standard errors clustered at the classroom level are presented in parentheses

Aigner and Cain (1977), where well-intentioned teachers evaluate competence of their students based on proficiency examinations and on a catch-all noncognitive-abilities term, differential evaluation may still emerge due to relative imprecision in ability signals. As in Cornell and Welch (1996), this can be the case even when all individuals are rational and believe, correctly, that there are no average differences between people of various races in terms of true competence.

One should refrain from jumping, based on these findings, to the conclusion that teacher evaluations need to be replaced by system-wide standardized evaluations. It is important to consider that society may value schools' role in the formation of human capital that is not directly capitalized in terms of proficiency. Nonetheless, the results presented here indicate that education authorities should work toward improving screening methods. Possible interventions include addition of grading guidelines, teacher training, reduction of teacher turnover, fine-tuning of automatic-promotion schemes, direct (and possibly independent) evaluation of students' non-cognitive traits, and provision of student-level standardized test results to teachers.

Table 7.9 Proportion of Blacks outperforming Whites in ENADE, by pre-existing differences in ENEM scores

	Original gap in ENEM-2006 standard deviation units				
Exam section	0	10.0%	25.0%	35.0%	50.0%
Major-specific	0.490	0.494	0.494	0.488	0.488
material	(0.464–0.516)	(0.475–0.513)	(0.479–0.510)	(0.474–0.502)	(0.475–0.501)
General	0.477	0.495	0.497	0.499	0.498
examination	(0.451–0.504)	(0.476–0.514)	(0.481–0.512)	(0.485–0.513)	(0.486–0.511)
ENADE	0.496	0.502	0.504	0.498	0.498
2007—Total	(0.470–0.522)	(0.483–0.521)	(0.488–0.520)	(0.484–0.512)	(0.485–0.510)

In essence, we find that Black and White students get to the end of their basic education experience not only with different levels of proficiency in mathematics but also receiving different signals (from their teachers) about their scholastic ability. A question remains: Can color-sighted policies undo what public schools seem to be doing to the generation and maintenance of racial gaps in educational attainment among their students?

Experimenting with Color-Sighted Policies

Racial inequality has been recently placed on the forefront of the Brazilian policy agenda. This led to current experimentation with affirmative action policies in tertiary education admissions or financing and in public-sector hiring. In 2002, the federal government created an official affirmative action program in the hiring for the public administration sector (Decreto Lei 4228/2002), and in 2003 (via Lei 10678/2003) it established a special secretary for the promotion of racial equality (Secretaria Especial de Politicas de Promocao da Igualdade Racial).

Despite the oddness of policies based on racial identification in a country with such a blurred concept of race, a growing number of colleges have been adopting a quota system in admissions. Brandao (2007) provided a detailed account of early (starting in 2001) adoption of a quota system by state and federal institutions in the states of Rio de Janeiro, Mato Grosso do Sul, Bahia, Parana, Mato Grosso, and Alagoas. Mostly, these are also combined with social quotas. Finally, in 2004, the federal government created the ProUni (College for All) program, which awards full or partial scholarships to low-income students who studied in tuition-free high schools (public or private) as its main college-attendance incentivization policy. Since its conception, ProUni has also reserved scholarships for Black students in a proportion corresponding to the Black population in each state of residence, as long as the socioeconomic requirements have also been fulfilled.

Given the dimension of the ProUni program, we see this as an opportunity to examine the characteristics of such affirmative action policies in Brazil. We focus attention on two main aspects in terms of efficiency: (1) static efficiency, the empirical existence of efficiency loss in terms of college performance among those who had already finished high school, and (2) dynamic efficiency, the extent to which college admission policies influence high school enrollment, performance, and graduation rates of future generations. Whereas we tackle the first with existent data, we can only speculate about the second.

We know that Black and White Brazilians graduate from high schools with different levels of mastery of the school material. This happens even after we control for both family- and school-level characteristics, as we showed above. In principle, then, quotas in college admission would operate to give Blacks a similar chance of entry despite their underperformance in high school exit examinations. However, to what extent is their college education affected by the unfavorable background? Can the college operate to compensate Blacks and make them competitive at the labor market level (compared with their White counterparts)?

To answer these questions, we simulated the impact of quotas over individual college performance by employing longitudinal data on students who took the ENEM (high school material) and ENADE (college material) exams. Therefore, we studied a selected subsample of Blacks and Whites who entered college. We then compared the relative college performance of Whites and Blacks at different levels of high school proficiency gaps.

Table 7.9 presents the results of this exercise. We studied the percentage of pairs of students in which the Black student scored above the White student in the ENADE exam (with the 90% confidence interval presented underneath). This percentage is computed in different columns for pairs with different original difference high school performance (ENEM exam). That is to say, as we move from left to right, we see pairs of students in which the Black student was further behind his White counterpart. Differences in ENEM scores are set at 10, 25, 35, and 50% of one standard deviation in performance (equivalent to a variation of 13 percentage points in percentage of correct answers). Different lines in the table examine the performance in different sections of the ENADE exam, either with coverage of major-specific material (college content) or with coverage of high school material (general fundamentals).

We find that when interpreting quotas by their score-subsidy counterparts, no evidence of loss of efficiency in college performance is observed. This indicates that affirmative action in Brazil, at least considering this simple exercise, has the potential of compensating for the unfavorable background of Black students without compromising their performance in the first year of college activities. This is true in the two exam components that we examined and is prevalent even when considering large subsidies. In particular, bonus points equivalent to 0.25 standard deviations in the ENEM exam (which represents the same relative size of racial gaps in the SARESP exam at the end of high school) would not represent any detectable loss of efficiency in college performance.

How about the dynamic incentives set in motion by the of the quota policy? We see this as a clear challenge. If, on the one hand, by reducing costs of admission the policy can encourage Black students to not drop out of high school, on the other hand, it can discourage effort in the learning process. Of course, its net result will depend on the fine-tuning of the quota system in place or on the amount of ENEM-score's subsidy they correspond to. More research on these dynamic impacts needs to be performed before any final conclusion about the success of such recent Brazilian affirmative action initiatives is reached.

Conclusions

In this chapter, we document the prevalence and extent of socioeconomic differentials between Black and White Brazilians. We then relate these outcomes to differences in the accumulation of human capital across races. Findings indicate that differences in both quantity and quality of formal education are pervasive. We uncover that recent trends in enrollment rates and in attrition reduction observed in some Brazilian states can generate reduced socioeconomic differences among future cohorts. Nonetheless, we also find worrisome evidence on the persistence of gaps in the quality of education provided to Blacks and Whites, as well as the possible gap-reinforcing role played by public elementary, middle, and high schools.

The recent Brazilian experience with color-sighted policies and affirmative action in college admissions is also discussed. We find no reason to believe that such policies lead to immediate loss of efficiency in college training, quite on the contrary. According to a simulation exercise, we have reasons to believe that even if awarded large admission test score subsidies, Blacks would not fall behind Whites while in college. Nonetheless, we still believe that more research is needed before reaching conclusions regarding incentives created by such policies for future generations of Black students.

Dynamic incentives may in fact be at the center of differences in proficiency observed when children start school. Are parents somehow feeding negative expectations regarding returns to investments in education of their children in their decisions? This would render a perverse self-fulfilling equilibrium that can be really hard to dismantle. Recent research on the investment decisions of mixed-race parents who have White-looking and Black-looking children in Brazil suggests that this can be the case.[15]

Finally, we have way less to say about informal mechanisms of discrimination and social segregation that may operate within Brazilian schools and reinforce differences in performance. Some version of the "acting White" phenomenon is likely at play in Brazil, but it is only by gathering more data on peer networks that we will be able to evaluate these hypotheses. This can be a fruitful avenue of research on Brazilian racial relations and achievement gaps' dynamics.

Acknowledgment The authors thank CAPES/Brazilian Ministry of Education (Observatorio da Educacao, Project 3313) for funding. Rangel also acknowledges the support of CNPq/Brazilian Ministry of Science and Technology (Research Productivity Fellowship). We benefited from

---

[15] See Rangel (2008).

extensive discussions with Fernando B. Botelho. Fernando Carnauba provided invaluable research assistance. The opinions expressed here represent the views of the authors and not those of any of the funding agencies. All remaining errors are the responsibility of the authors. Contacts for comments: rmadeira@usp.br and rangelm@usp.br.

References

Aigner, D., & Glen, C. (1977, Jan). Statistical theories of discrimination in the labor market. *Industrial and Labor Relations Review, 30,* 175–187.

Alexander, N., Guimarães, A., Hamilton, C., Huntley, L., & James, W. (2001). *Beyond racism: Race and inequality in Brazil, South Africa, and United States.* Boulder: Rienner Publisher.

Arias, O., Yamada, G., & Tejerina, L. (2004). *Education, family background and racial earnings inequality in Brazil.* Washington, DC: manuscript, Inter-American Development Bank.

Brandão, A. A. (2007). Cotas Raciais no Brasil: A primeira avaliação. Livros: Sindicato Nacional dos Editores de.

Campante, F., Crespo, A., & Leite, P. (2004). Wage inequality across races in Brazilian Urban Labor Markets: Regional Aspects. *Revista Brasileira de Economia, 58*(2), 185–210.

Carneiro, P., Heckman, J., & Masterov, D. (2005). Labor market discrimination and racial differences in premarket factors. *Journal of Law and Economics, 48*(1), 1–40.

Coate, S., & Loury, G. C. (1993a, December) Will affirmative-action policies eliminate negative stereotypes? *American Economic Review, American Economic Association, 83*(5), 1220–1240.

Coate, S., & Loury, G. (1993b, May). Antidiscrimination enforcement and the problem of patronization. *American Economic Review, American Economic Association, 83*(2), 92–98.

Cornell, B., & Welch, I. (1996). Culture, information, and screening discrimination. *Journal of Political Economy, University of Chicago Press, 104* (3), 542-571.

Fryer, R., & Levitt, S. (2004a). Causes and consequences of distinctive black names. *The Quarterly Journal of Economics, MIT Press, 119*(3), 767–805.

Fryer, R., & Levitt, S. (2004b). Understanding the Black-White test score gap in the first two years of school. *The Review of Economics and Statistics, MIT Press, 86*(2), 447–464.

Fryer, R. (2010, November). The importance of segregation, discrimination, peer dynamics and identity in explaining trends in the racial achievement gap, in handbook of social economics, Volume 1B. (Ed. Benhabib, J., Bisin, A. and Jackson, M. O.)

Hasenbalg, C., Valle Silva, N. D., & Lima, M. (1999). Cor e estratificação social. Contra Capa: Rio de Janeiro.

Heckman, J. (1998). Detecting discrimination. *The Journal of Economic Perspectives, 12*(2), 101–116 (Spring).

Herring, C., Keith, V. M., & Horton H. D. (2004). *Skin Deep: How race and complexion matter in the "Color-Blind" era,* Institute for Research on Race & Public Policy.

King, J. (1971). The Biology of Race. Harcourt Brace Jovanovich Publisher

King, E. M., Orazem, P., & Paterno, E. M. (2008). Promotion with and without learning: Effects on student enrollment and dropout behavior. Staff General Research Paper 12968, Iowa State University, Department of Economics.

Lundberg, S. J., & Startz, R. (1983) Private discrimination and social intervention in competitive labor markets. *American Economic Review, LXXIII,* 340–347.

Massey, D., & Denton, N. (1988) The dimensions of residential segregation. *Social Forces, LXVII,* 281–315.

Maxwell (1994, January). The effect on Black-White wage differences of differences in the quantity and quality of education. *Industrial and Labor Relations Review, 47*(2).

Neal, D., & Johnson, W. (1996, October). The role of premarket factors in Black-White wage differences. *Journal of Political Economy, University of Chicago Press, 104*(5), 869–895.

O'Neil (1990). The role of human capital in earnings differences between Black and White men. *The Journal of Economic Perspectives, 4*(Autumn), 25–45.

Parra, E. J., Kittles, R. A., & Shriver, M. D. (2004). Implications of correlations between skin color and genetic ancestry for biomedical research. *Nature Genetics Supplement, 36*(11).

Perry, G., Arias, O., López, J. H., Maloney, W., & Servén, L. (2006). *Poverty reduction and growth: Virtuous and vicious circles.* The World Bank.

Pierson, D. (1945) Brancos e pretos na Bahia: estudo de contato racial. São Paulo: Companhia Editora Nacional (Coleção Brasiliana, v. 241).

Rangel, M. (2008). Is parental love colorblind: Allocation of resources within mixed-families. BREAD Working Paper 167, March.

Reichmann, R. (1999). *Race in contemporary Brazil: from indifference to inequality.* USA: Pennsylvania State University Press.

Telles, E. E. (2004). *Race in another America: The significance of skin color in Brazil.* United States: Princeton University Press.

Part IV
Europe

Chapter 8
Narrowing the Achievement Gap: Policy and Practice in England, 1997–2010

Geoff Whitty and Jake Anders

Introduction

The coalition government that was elected in May 2010 set out to 'close' the achievement gap. This ambition went even beyond the previous New Labour Government's ambition to 'narrow' that gap. Given that England does not score particularly well on 'equity' measures within international achievement surveys like Programme for International Student Assessment (PISA), even the lesser of these aspirations was ambitious, to say the least. This chapter explores the progress made under New Labour and assesses future prospects under the current Conservative-led Coalition of Conservative and Liberal Democrat parties.

Many years ago, Bernstein (1970) pointed out that 'education cannot compensate for society', whereas an early critic of New Labour's attainment targets argued that a serious programme to alleviate child poverty might do far more for boosting attainment and literacy than would any modest intervention in schooling (Robinson 1997). Nevertheless, given that the achievement gap is narrower and social mobility greater in some countries that are socioeconomically and culturally similar to England (Sutton Trust 2011), it is reasonable for politicians to believe that education and other social policies can make a difference in regard to the achievement gap(s).

For many years in the last century, there were major concerns about the underachievement of girls. That gender gap has been largely reversed, although not yet in the hard sciences or at the very highest levels in some other subjects. Minority ethnic achievement has also been a concern, although there are stark differences in the performance of different minority groups. However, the key focus in English work on the gap at present is 'social class' differences in educational achievement, even though this term itself is often expressed as 'poverty', 'disadvantage', 'depriva-

---

G. Whitty (✉) · J. Anders
Institute of Education, University of London, London, UK
e-mail: g.whitty@ioe.ac.uk

J. Anders
e-mail: jake@jakeanders.co.uk

J. V. Clark (ed.), *Closing the Achievement Gap from an International Perspective*,   163
DOI 10.1007/978-94-007-4357-1_8, © Springer Science+Business Media B.V. 2014

tion' or 'social exclusion' and is usually measured in terms of socioeconomic status (SES) or eligibility for free school meals (FSMs; Whitty 2001).

The emphasis in this chapter on gaps between social groups as identified through cognitive measures and the achievement of academic qualifications is not intended to suggest that the only purpose of schooling is to achieve such qualifications or that those who fail to do so are deficient, either absolutely or relatively, in other important respects. Indeed, during the period under consideration here, there was, for example, considerable emphasis on the role of education in fostering 'well-being'. Nevertheless, there is a great deal of evidence that life chances in English society are closely linked to school attainment in a myriad of ways and that personal fulfilment and social justice could both be enhanced by narrowing or closing long-standing academic achievement and participation gaps (Schuller et al. 2004). We therefore agree with Kerr and West (2010) that 'despite the dangers of narrowing our view of what education is about', a focus on attainment is justifiable because 'attainment undeniably has important consequences for life chances' (p. 16).

One specific reason why it is important to address this attainment gap in schools is that it has implications for access to higher education. There has been a considerable and persistent gap in England in the rates of participation in higher education between those from higher and lower socioeconomic groups (Kelly and Cook 2007). There has also been a strong and enduring tendency for students at the leading universities to be drawn from more affluent families and from those schools that cater mainly to such families (Boliver 2011).

Although there are undoubtedly still financial and aspirational barriers to widening participation and ensuring fair access in higher education (Whitty 2010a), it is now clear that the major impediment to students proceeding to higher education is low prior attainment. Research by the Institute of Education, the London School of Economics and the Institute of Fiscal Studies found that, although there is a considerable gap in higher education participation between those from different backgrounds, this gap is actually small once prior attainment has been fully taken into account (Chowdry et al. 2010a; Vignoles and Crawford 2010; Anders 2012). It is worth noting, however, that work by Jackson et al. (2007) has argued that a significant proportion of the gaps in prior attainment may be due to non-academic 'secondary effects'.

Prior attainment and choices made in terms of future study at ages 14 and 16 can then have huge consequences for future employment prospects. Low attainment and inappropriate subject choices can be particularly restrictive on opportunities for entry into the professions (Milburn 2009) and science, technology, engineering and mathematics (STEM)-related employment (Coyne and Goodfellow 2008).

The remainder of this chapter will focus largely, though not exclusively, on socioeconomic differences in educational attainment. It will begin by looking briefly at the evidence on the performance of different social groups in the preschool period and then concentrate on the compulsory phase of schooling before touching on differential levels of participation in higher education at the end of mainstream schooling. In so doing, it will demonstrate the potential impact of early failure on later achievement throughout the life course, as well as identifying the sorts of inter-

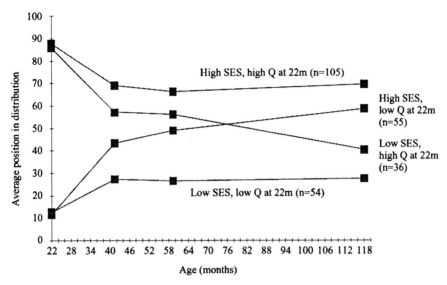

Fig. 8.1 Average rank of test scores at 22, 42, 60 and 120 months, by SES of parents and early rank position. The definition of categories with sample observations are as follows: *high SES*—father in professional/managerial occupation and mother similar or registered housewife (307 observations); *low SES*—father in semiskilled or unskilled manual occupation and mother similar or housewife (171 observations); *medium SES*—those omitted from the high- and low-SES categories (814 observations). (Source: Feinstein 2003)

ventions that the evidence suggests might begin to break the enduring link between social background and educational achievement.

The Preschool Attainment Gap in England

Politicians of all three major parties have made use of a graph produced by Feinstein (2003) that purports to show that, even before starting school, children with high cognitive test scores from disadvantaged backgrounds are falling behind less able children from more advantaged backgrounds (Fig. 8.1).

Although some doubt has been raised regarding this analysis on account of the potential for regression to the mean to exaggerate the phenomenon (Jerrim and Vignoles 2011), it is highly unlikely that this would overturn the core finding that high-SES, lower-ability children catch up with their low-SES, higher-ability peers even if they do not necessarily overtake them. In any case, these figures have undoubtedly informed the government policies we discuss below.

More recent analysis of the Millennium Cohort Study (MCS) confirms the existence of socioeconomic differences in attainment by the age of 3. These seem to reflect more than just differences in the distribution of innate ability across the socioeconomic spectrum: the gaps widen between the ages of 3 and 5 with, for example, children in the top quintile of household income showing the fastest progress between these two ages (Goodman et al. 2009). These results hold for other measures of status such as father's occupational class, mother's education and housing tenure. To put these into further context, the Sutton Trust (2011) compared measures of school readiness across different countries, finding that England had larger socioeconomic gradients than do most other Anglophone countries, even though the gaps were smaller than for the USA.

Sure Start was a multifaceted early-years intervention introduced by the New Labour government elected in 1997 and was designed to improve the life chances of those growing up in disadvantaged areas (NESS 2010, 2012). Unlike more narrowly targeted interventions, it was introduced in areas of assessed need rather than targeted on specific individuals wherever they lived. The evaluation failed to identify any positive impact of Sure Start on 'school readiness', as measured by a Foundation Stage Profile score. However, it did identify positive impacts on various aspects of parenting style and child's body mass index (BMI), and the initiative may still be shown to have longer-term effects on educational outcomes. It has been suggested that the lack of an identifiable impact on differences in school readiness is due to 'the introduction of universal free early education for all children whether in Sure Start areas or not' (NESS 2012, p. 12).

Whatever the explanation, there is evidence of a socioeconomic gradient in attainment as early as the Early Years Foundation Stage (EYFS), that is, before the beginning of compulsory schooling at age 5. Because no individual-level data are available, ONS (2006) has to use an area-based proxy for low SES. As Sure Start centres were explicitly located in deprived areas, children in these areas are on average from families with lower SES. The analysis shows that the percentage of 5-year-olds achieving a 'good' level of development by the end of the EYFS is lower in schools in Sure Start areas than it is elsewhere. This holds for personal, social and emotional development; communication; language; and literacy; but it is more marked in the latter.

The Attainment Gap in English Schools

With some local exceptions, the English school system is divided into two main phases. Primary (including infant and junior) schooling runs from age 4 or 5 to age 11, followed by secondary schooling continuing up to the compulsory minimum school leaving age of 16 and in most cases extends on to 18 or 19. Most children change schools at age 11. Since 1988, the system has been divided into Key Stages. Following the EYFS referred to earlier, Key Stage 1 (KS1) runs from ages 5 to 7, Key Stage 2 (KS2) from ages 7 to 11 and Key Stage 3 (KS3) from ages 11 to 14. Key Stage 4 (KS4), which ends with most pupils sitting for General Certificate of Secondary Education (GCSE) examinations or equivalent qualifications, lasts from ages 14 to 16. Now that most students remain in some sort of education or training until age 18, and will soon be required to do so, many schools refer to education in the 16–19 age range as Key Stage 5 (KS5). Much of this is carried out in 11–19 schools, but in some cases, it is carried out in separate post-16 institutions—called *sixth form colleges, tertiary colleges* or *further education colleges*. Post-16 (KS5) education is diverse in character with large numbers of vocational qualifications available in addition to the academic track leading to Advanced Level (A-Level) examinations, which have traditionally been the threshold qualification for entry to higher education.

Rather than having separately elected school boards as in the USA, most publicly funded schools in England (a category that includes most church schools) have traditionally been maintained by multifunctional local authorities. However, as with charter schools in the USA, there have been attempts since the 1980s to improve failing state schools by giving them autonomy from local authorities and involving private sponsors in their governance. Half of all publicly funded secondary schools and a few primary schools now have autonomous 'academy' status or are in the process of acquiring it. A private fee-paying sector caters to only about 7% of the whole school population but educates nearly 20% of those in KS5 (DCSF 2008). Although such schools are not exclusively for higher-SES students, highly selective elite independent schools are overwhelmingly populated by such students, and students from such schools dominate entry into the leading universities. Most state schools are comprehensive (non-selective), but there remain a small number of academically selective schools at the secondary level. Such 'grammar schools' have often been regarded as a route to social mobility for able working-class children, but the intakes of the few that remain are heavily weighted toward the higher end of the SES scale.

Most data seem to show that there is a socioeconomic gradient in attainment throughout the English schooling system. Using data from attainment in 2005 (DfES 2006), although there is always a gap between FSM-eligible and non-FSM eligible students in terms of relative performance, this does not grow inexorably through the different stages of schooling up to KS3. At each stage, the performance of children eligible for FSM is always around 85–90% of that of the rest of the cohort. However, this widens at KS4, where FSM-eligible young people achieve roughly three quarters the average point score of the rest of the cohort (DfES 2006).

Survey data analysed by Goodman et al. (2009) present a slightly different picture when more detailed measures of SES, rather than simply FSM eligibility, are used. These authors find a widening gap in attainment through children's educational careers up until KS3 (age 14), but find that it narrows somewhat for KS4 results. The difference between these two analyses is accounted for by the fact that FSM eligibility splits the population into a deprived group and the rest, whereas this analysis generally compares a broader (compared with FSM) lower group with a smaller (compared with non-FSM) higher group. Nevertheless, taken together, these studies do point to a widening of the socioeconomic gap during English children's educational careers.

Other changes in inequality through the educational career are also presented in these studies. For example, Goodman et al. (2009) show that, though in earlier years of education the gender gap in attainment comes and goes (but with girls always ahead where a gap is observed), it widens more consistently through the secondary school years (pp. 27–28). Perhaps more surprisingly, they also suggest that 'wide ethnic differences amongst young preschool children appear to narrow over time, and are quite small by the time young people reach GCSE' (p. 28).

Even the evidence on a socioeconomic gradient itself is not without its dissenters, however. Saunders (2012) questions the basis of much of the evidence on social mobility of which the underperformance of children from poorer backgrounds is a major part. Whatever one's view of this critique there is little doubt that concerns surrounding social mobility impact government policy over this period (Feinstein 2003).

As mentioned earlier, the New Labour government considered it a key part of its educational policy to narrow the attainment gap between children from different socioeconomic backgrounds. Given this goal, it is perhaps surprising that data on the trends for this gap are patchy. Although there are figures on the gap at a particular point in time, they often use different measures of attainment and/or different comparator groups, making it difficult to assess the trends. Also, in the initial period of New Labour government, apart from Education Action Zones, an ill-fated area-based initiative, the major emphasis was on driving up standards overall. It was only the failure of this to impact social differences in attainment that led to specific policies after 2001 to address the attainment gap, with a thrust in this direction after 2005. Although there were increases in average levels of attainment in the first period of New Labour government, some have argued that even these increases were at least partly achieved through grade inflation (Tymms 2004). This paper does not look into this matter in depth, except in so far as it affects our attempt to isolate the change in the socioeconomic attainment gap.

We might initially think that neither a general rise in standards, nor possible grade inflation, would impact the trends in attainment gaps as measured by, for example, the proportion gaining five or more top (A*–C) grades at GCSE. However, there is no guarantee that this will be the case, and thus caution is urged in interpreting changes in gaps. This is because even if grades were to rise uniformly across the board, different numbers of individuals from different parts of the socioeconomic spectrum may be pushed across the threshold. The trends are nevertheless likely to be indicative of the direction of travel, but it may be important to check them against other research.

Figure 8.2 shows trends in the attainment gap up until 2003 and suggests a slight narrowing of the gap between students from manual and non-manual families.

ONS (2006) provides further data on a wide range of changes in attainment gaps, although individual-level data are provided only between 2002 and 2005. These figures show a reduction in the attainment gap between pupils eligible and those not eligible for FSM in terms of those obtaining no GCSEs (or equivalents) and the proportion obtaining five or more A*–C GCSEs (or equivalents). However, there was a slight increase in the same gap where it was a requirement that the set of GCSEs included English and mathematics. These figures show a stronger trend toward narrowing when Index Deprivation Affective Children and Infants (IDACI, an area-based indicator of deprivation) is used instead of FSM eligibility. This is be-

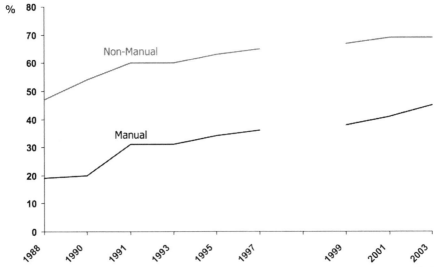

Fig. 8.2 Percentage of cohort achieving 5+ A*–C GCSEs by parents' social class: 1988–2003 (%). (Note: Discontinuity exists between 1997 and 1999 because of a change in the classification of social class from Socio-economic group (SEG) to National Statistics Socio-economic Classification (NS-SEC). *Manual* and *non-manual* categories have been constructed by grouping more detailed breakdown of social class groups. The 'other' group has been excluded from the analysis. (Source: DfES 2006, analysis of Youth Cohort Study cohorts 4–12, sweep 1)

cause this measure compares the most deprived with the least deprived, rather than the most deprived with the rest, and there is evidence of generalised catching up between the bottom three deprivation quartiles and the top. In the FSM measures, this catching up by pupils in the middle reduced the relative gains of the bottom compared with the top.

This has been updated covering a longer period by the more recent data of Hansard (2012), which is summarised in Table 8.1. These show a mixed picture, but there is a broad trend toward small reductions in attainment gaps in the official figures. The exception below is the measure that excludes GCSE equivalents (such as vocational qualifications). This would seem to reflect the trend toward the use of such alternative qualification by schools for lower-performing pupils. Over this period, the equivalence between these qualifications was favourable toward the alternative qualifications and did not necessarily reflect their value to the individual in the labour market or in seeking to continue their studies. Totally excluding equivalents probably goes too far the other way, as it seems unlikely that these qualifications had no value. However, it indicates one of the routes through which the recorded attainment gap was narrowed and indicates that this may not have reflected a genuine reduction in inequality (de Waal 2008).

Other research has attempted to get around the problems of changing standards in a variety of ways. Jerrim (2012) uses data from the PISA, a study of 15-year-olds' achievement conducted across the Organisation for Economic Co-operation and

Table 8.1 Change in percentage of pupils who have achieved various attainment benchmarks between 2005/06 and 2010/11 by free school meal (FSM) eligibility. (Our calculations based on Hansard (2012), in turn based on National Pupil Database)

	FSM	All others	Gap
Percentage not achieving a GCSE or equivalent	−2.8	−1.1	−1.7
Percentage achieving five A*–C grades at GCSE (including English and mathematics), including equivalents	15	14.3	−0.7
Percentage achieving five A*–C grades at GCSE (including English and mathematics), excluding equivalents	8.9	9.9	1

In some cases, our data include only pupils who have achieved vocational qualifications rather than those entered for the qualification. For GCSEs, all pupils who are entered are included. Figures for 2005/2006 to 2010/2011 are based on final data. Figures for 2010/2011 include AS levels, full and short GCSEs, double-award GCSEs, applied GCSEs and the accredited iGCSEs and their legacy qualifications. Figures for 2005/2006 to 2009/2010 include only full GCSEs, double awards and accredited iGCSEs and their legacy qualifications

Development (OECD) nations every 3 years, as part of a cross-national comparison. The analysis primarily focuses on reading skills because of the data available from PISA. It should, however, be noted that there are caveats associated with the comparability of PISA data from different years, and Jerrim (2011) advises caution in the interpretation of his results. Outcomes are based on PISA test scores (where 40 points are roughly comparable to a year of schooling), while SES is measured using quintile groups derived from occupational status.

Jerrim's analysis suggests an overall reduction from 2000 to 2009 in the attainment gap between those in the top and bottom quintile groups of 15 points (roughly equivalent to catching up by a term of schooling). However, this result sits on the edge of statistical significance. He also considers the changes taking place at different points of the attainment distribution. This analysis suggests that just looking at the average hides a more complex story. At the top end of the attainment distribution there is barely any change, whereas, at the bottom, a larger and statistically significant reduction in attainment gap of 25 points (roughly equivalent to two terms of schooling) is observed. Figure 8.3, reproduced from Jerrim (2012), shows these changes in attainment gaps over time at different levels of attainment.

Jerrim discusses the potential policies such changes could be associated with. 'Anecdotally, much of the investment made in disadvantaged children in England is designed to help this group reach a basic level of skill (i.e. to push up the lower tail). Indeed, academics, policymakers and the media frequently discuss England's "long tail of low achievement" and the need to increase the proportion of disadvantaged children (for example, those receiving free school meals) reaching a certain floor target (for example, five GCSEs at grades A*–C). Although this is clearly important, much less attention seems to be paid to helping disadvantaged children who are already doing reasonably well to push on and reach the top grades' (Jerrim 2012, p. 176).

Sullivan et al. (2011) take an alternative approach to dealing with the potential problem of rising attainment overall. They treat educational qualifications as a po-

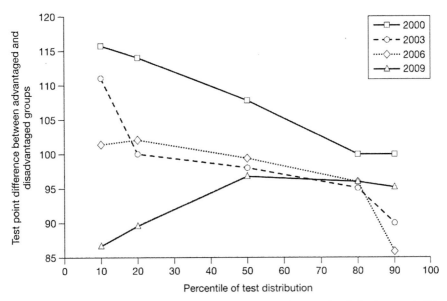

Fig. 8.3 Comparison of PISA test point difference between advantaged and disadvantaged children at different points of the attainment distribution. Running along the horizontal axis are the percentiles of the national PISA reading test distribution. Figures on the vertical axis refer to the estimated difference in test scores between children from different socioeconomic backgrounds, measured by Highest International Social and Economic Index (HISEI) of occupational status. comparing the most advantaged (top national HISEI quintile) with the least advantaged (bottom national HISEI quintile) backgrounds. (Source: Jerrim 2012, predictions from quantile regression estimates based upon the PISA data sets)

sitional good. As such, the absolute level of attainment is not regarded as important. However, the paper points out that their 'relative measure deals with overall credential expansion/inflation, but cannot deal with differential credential inflation, whereby credentials which are designed for lower achieving students are given a disproportionately high face-value in relation to their actual intellectual, educational and labour-market value' (Sullivan et al. 2011, p. 221). This is important in that it implies that some of the apparent increase in attainment at the lower end may be illusory or of little value in the employment market.

Nevertheless, the paper finds broadly similar results to those above, suggesting 'social class inequalities persist…they tend to be greater at higher levels of attainment [and]…class inequalities at all levels have been declining' (Sullivan et al. 2011, pp. 234–235). They argue these results are robust and that their use of a positional measure of attainment still shows 'clear, albeit much more modest, trends towards class equalisation' (p. 235).

Cook (2011) also presents evidence of a reduction in the attainment gap between 2006 and 2010. It uses performance relative to the mean in sciences, modern languages, mathematics, English, history and geography, generally regarded as the core subjects. In this case, the size of the reduction looks relatively modest and

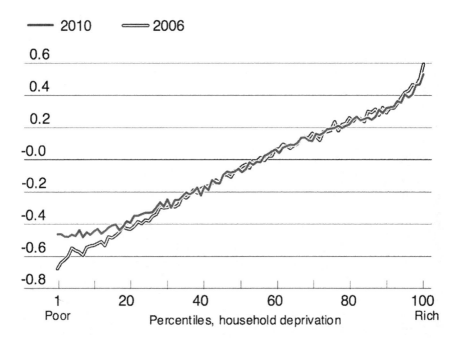

Fig. 8.4 Graph showing relation between household deprivation and relative performance in GCSE point score in core subjects. *Vertical axis* shows standard deviation from mean GCSE point score performance in the core subjects of sciences, modern languages, mathematics, English, history and geography. Percentiles of household deprivation derived using Index Deprivation Affective Children and Infants (IDACI). (Source: Cook 2011, analysis of National Pupil Database)

concentrated among those in the bottom fifth of households ranked by deprivation (Fig. 8.4).

The analysis shows a steady weakening of the overall correlation between the two factors in the years between 2006 and 2010. Interestingly, this is the case particularly for KS4 attainment overall, where performance on some vocational courses is included. Again, it could be argued that this lends support to the charge that part of the decline in the socioeconomic attainment gap is due to individuals switching to alternative courses. However, as the core measure still shows a decline, not all of the reduction in the gap can be dismissed as illusory, even if one were to accept the argument that the alternative courses are somehow less rigorous or marketable.

Government data provide evidence of the narrowing of gaps in terms of other student characteristics. ONS (2006) indicates that, using top (5+A*–C) GCSE scores, the main low-performing minority ethnic groups all closed the attainment gap relative to White pupils. For example, in 2003, 52.4% of White students achieved 5+A*–C scores, increasing to 55.9% in 2005, whereas the figure for Black Caribbean students increased from 33.9 to 42.0%. At KS2, Pakistani and Bangladeshi pupils narrowed the gap compared with White pupils, although the gap between White and Black African pupils widened slightly (ONS 2006, p. 6).

The relative performance of Looked After Children (those in local authority or foster care) paints a less positive picture (DfE 2011a). Whereas at KS2 the attainment gap (measured by the difference in proportion of children achieving high scores—at least Level 4—in both English and mathematics) has reduced from 35 percentage points to 31 percentage points, the movement at KS4 is in the opposite direction. The gap between Looked After Children achieving top grades in both English and mathematics at GCSE has increased from around 37 percentage points to almost 45 percentage points. The positive news here is that the proportion achieving the benchmark did rise for Looked After Children, but the improvement was faster among the rest of the cohort.

Thus, although by most measures there was a small reduction in the attainment gap under the New Labour government of 1997–2010, it must be regarded as a disappointing achievement when compared with the aspirations of successive Prime Ministers and Secretaries of State for Education. Not surprisingly, the Coalition government has tended to dismiss even the limited narrowing of the gap that was achieved under New Labour, regarding it as a poor return on the public resources invested. This picture is summarised and restated in the Coalition government's Social Mobility Strategy (HM Government 2010b) and is presented graphically in Fig. 8.5.

What Contributed to the Narrowing of the Gap?

'High quality education for the many rather than excellence for the few' was New Labour's slogan immediately following the 1997 election. This was symbolised in the first instance by the abolition of the Thatcher government's Assisted Places Scheme, which provided publicly funded means-tested scholarships to enable academically able children from poor families to attend elite private schools. Though ostensibly targeted at working class children ill-served by failing inner-city comprehensive schools, early take-up of the scheme was actually dominated by middle-class families who otherwise might have sent their children to good suburban schools, but whose income was low enough to qualify for the scheme (Edwards et al. 1989). The resources freed by the abolition of this scheme were diverted to the state sector to reduce class sizes in infant schools. This was presented as a socially redistributive measure, but it did not actually have that effect. Most large classes were in marginal suburban electoral districts, not in disadvantaged areas, suggesting that the policy was driven at least in part by the findings of election opinion polling rather than educational research (Whitty 2006).

There were considerable numbers of educational initiatives during the period of New Labour government, reflecting a variety of different understandings about how best to close the gap. They ranged from area-based interventions such as Education Action Zones, Excellence in Cities and the London Challenge, through National Strategies for Literacy and Numeracy, remodelling the school workforce including the use of more teaching assistants, improving school leadership training,

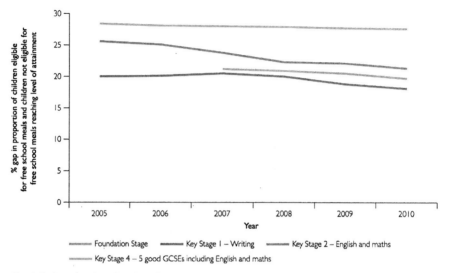

Fig. 8.5 Gaps in educational performance have narrowed only slightly despite significant investment. (HM Government (2010b, p. 20) drawing on data from the following sources: Department for Education, various Statistical First Releases such as Foundation Stage Profile Attainment by Pupil Characteristics in England 2009/10 (2011)—gap in % reaching a good level of development; Key Stage 1 Attainment by Pupil Characteristics in England 2009/10 (2011)—gap in % reaching expected level in reading; Key Stage 2 Attainment by Pupil Characteristics in England 2009/10 (provisional) (2011)—gap in % reaching expected level in English and mathematics; GCSE and Equivalent Attainment by Pupil Characteristics in England 2009/10 (2010)—gap in % achieving 5 GCSEs A*–C including English and mathematics; and Level 2 and 3 Attainment by Young People in England Measured Using Matched Administrative Data: Attainment by Age 19 in 2009 (provisional) (2010)—gap in % achieving a Level 3 qualification by age 19)

enhancing teacher quality, creating a network of specialist schools and founding academy schools outside the local authority system, to the 'personalisation' of education through individually targeted interventions such as Reading Recovery. In addition, there was Every Child Matters, a multiagency policy that addressed a wider 'children's agenda'.

Oddly, New Labour seemed to recognise the importance of wider structural and cultural influences in its broader policies, especially in the Sure Start initiative and around the wider children's agenda, but it did not always apply such insights to its understanding of differential performance in schools. Instead, many of New Labour's key school policies seemed to be founded 'on the belief that quality differences between schools are primarily the responsibility of schools themselves and can therefore be tackled by initiatives at the school level' (Thrupp and Lupton 2006, p. 315). This was unfortunate in that it sometimes led to a failure to 'join up' policies.

Furthermore, the vast numbers of educational policies introduced by New Labour led to charges of 'initiative-itis', while the tendency to alter them even before they had been properly evaluated has meant that it is virtually impossible to determine across the system as whole which policies were effective in narrowing the gap. This was despite the fact that New Labour politicians avowedly adopted an 'evidence-based' approach to policy and employed the rhetoric of 'what works' with the same enthusiasm as their North American peers (see Ofsted 2010b; Whitty 2012).

For some policies, Education Action Zones, Excellence in Cities and the employment of teaching assistants, the evidence is equivocal or suggests little or even negative impact (Power et al. 2004; Machin et al. 2007; Blatchford et al 2012). We shall therefore focus here on some of the policies for which there does seem to be credible evidence that they did make an impact on the attainment gap.

The National Strategies

The National Strategies for Literacy (from September 1998) and Numeracy (from September 1999) were a key early policy enacted by Labour to attempt to raise standards overall. An evaluation of a major plank of the National Strategy for Literacy, namely the 'Literacy Hour', was conducted by Machin and McNally at the London School of Economics. This identified a significant impact of the Literacy Hour in its piloted form as part of the earlier National Literacy Programme (NLP). It found that 'reading and English Key Stage 2 levels rose by more in NLP schools between 1996 and 1998' than it did in the comparator schools, which had not yet introduced the policy (Machin and McNally 2004, p. 27).

A more critical view has been taken by a series of reports by Tymms and colleagues (Tymms 2004; Tymms et al. 2005; Tymms and Merrell 2007). These question the extent to which standards have truly increased by using secondary data on pupil performance that are argued to be more comparable over time. Although it does seem likely that some of the increase in apparent performance has been due to grade inflation it should not detract from quasi-experimental evidence, such as that used by Machin and McNally, because there is no particular reason to think inflation would affect the pilot schools more than comparator schools.

However, the results found by Machin and McNally relate to early impacts of the intervention. It seems plausible that part of these effects is simply due to the increased focus generated by the introduction of these strategies. Indeed, the evaluation of the National Literacy and Numeracy Strategies commissioned by the Department of Education and Skills (DfES) suggests that 'the initial gains in the 1999 national tests were likely due largely to higher motivation on the part of teachers and others at the local level' (Earl et al. 2001, p. 5). This would also explain the tailing off in improvements observed in general performance over the period.

More generally, Earl et al. (2001) were positive about the impact the strategies were having in terms of implementation, suggesting that they brought about large shifts in priorities within almost all schools in the country. They describe the strategies as 'successful' at more than one point in their report. However, in a critique similar to the one later developed by Tymms, Goldstein (2002) suggests the report relied too much on test performance at KS2 to justify extrapolating from successful implementation to success in raising standards.

Machin and McNally (2004) also noted particularly strong effects at lower levels of attainment (but still positive effects for those already achieving above the target level) and an increased impact for boys (who were otherwise lagging) compared to girls. The results on differential impacts at varying levels of ability fit well with the suggestion by Jerrim (2012) of a reduction in the attainment gap at the bottom of the ability distribution and suggest that the strategies may have been more effective in this respect than their critics claim.

Evaluation of the National Strategies is a difficult task for several reasons. Elements such as the Literacy and Numeracy strategies were rolled out rapidly and comprehensively, quickly becoming a pervasive part of the education system. The strategies also had many elements, reaching across EYFS, primary, secondary, behaviour and attendance, and school improvement programmes. Many evaluations point only to overall improvements in attainment over the period (DfE 2011b), implicitly treating almost all New Labour education policies as part of the National Strategies. They also tend to provide only descriptive evidence, and we have no indication of what would have happened in the presence of different or unchanged policies. Indeed, the schools' inspectorate (Office for Standards in Education, Children's Services and Skills—Ofsted) has pointed to the failure to evaluate which elements of the National Strategies were successful as a serious shortcoming, partly stemming from the sheer number of initiatives introduced in a relatively short period of time. Its report does, however, praise the impact the National Strategies have had on increased debate around pedagogy, suggesting almost all schools feel they have led to an improvement in teaching and learning and the use of assessment (Ofsted 2010b, p. 5).

Specific evaluation of the Narrowing the Gaps element of the National Strategies was carried out by York Consulting (Starks 2011). This evaluation focussed on support and resources for both children eligible for FSM and Gypsy, Roma, Traveller (GRT) children. It reports finding evidence of increased use of the practices that appear effective in improving pupil attendance, motivation, confidence and attainment. These included capacity building by local authorities to support schools in achieving goals, improved engagement with parents and intelligent tracking of pupil attainment. For the reasons referred to above, there is little specific quantitative evidence of how this feeds through into outcomes beyond the national trends in attainment gaps identified earlier. The limited case study evidence on the reduction of gaps is not particularly encouraging, with only three out of the eight case study schools reducing the attainment gap. However, it is not clear how representative these case studies were, and the conclusion appears to relate to a limited time frame, although it is not entirely clear exactly what this is. The report suggests that the strategies were anyway not fully implemented by the end of the period, and it argues that with continued support we may see further positive results.

Ultimately, the National Strategies seem to have had a limited impact on the attainment gap, although their overall impact plateaued in later years. By then, and well before it lost the 2010 election, the New Labour government had decided that such large-scale national initiatives were no longer appropriate. Its Children's Plan envisaged much greater local and professional autonomy in driving improvement in the future (DCSF 2007). This was consistent with a wider trend toward handing more responsibility to schools and federations of schools, including autonomous academies and chains of such academies (Curtis 2009).

Academies

Academies were based on an expectation that giving greater autonomy to schools with dynamic leadership teams and private sponsorship would improve their performance. Some of these academies were new schools in disadvantaged areas, whereas others were existing schools deemed to be failing under local authority supervision. An official evaluation conducted by PricewaterhouseCoopers on behalf of the DfES (PWC 2008) notes an increased level of performance in these schools relative to the national average. However, this methodology has been criticised (Machin and Vernoit 2011) on two main counts. Firstly, new academies during the period of evaluation had a more disadvantaged intake relative to the national average. Secondly, changes in the SES of the intake frequently accompanied the opening of an academy, and these have the potential to further undermine the validity of the comparison.

An evaluation by the National Audit Office (NAO) used a more select group of comparator schools, based on their intake and performance relative to the academies prior to conversion. This found increases in performance, but the analysis suggests that this was largely driven by the 'substantial improvements by the less disadvantaged pupils' (NAO 2007, p. 27). Although improvements are of course to be welcomed, this does not seem particularly promising for reducing attainment gaps between students from higher socioeconomic backgrounds unless there are substantial peer effects. On the other hand, as Maden (2002) once put it, successful schools tend to have 'a "critical n" of more engaged, broadly "pro-school" children to start with' (p. 336), so a longer-term perspective may be helpful here.

In their own study, Machin and Vernoit (2011) went further to try and overcome the potential for selection bias in the choice of comparator schools. They used maintained schools that went on to become academies after their data collection period. Their analysis yielded preliminary results suggesting that in the academies an extra three percentage points of pupils achieved top grades (five A*–C) at GCSE (or equivalents). However, they only identified this effect in academies that had been open for more than 2 years at the time of their evaluation. Interestingly, their results suggested that despite the same increase in the SES of the school's intake noted above (and the consequent reduction for neighbouring schools) there were also increases in performance in these neighbouring schools, perhaps due to increased competition. This finding runs counter to the claims made by most critics of academies, who regard their success as coming at the expense of other local schools. Unfortunately, the incoming Coalition government cancelled an evaluation of academies commissioned by the previous government, which might have resolved this issue amongst others.

There is no doubt that some of the academies founded under New Labour proved successful in improving the attainment of disadvantaged students. However, not all academies have performed so well in this and indeed other respects. As Curtis et al. (2008a) argued, 'Academies are in danger of being regarded by politicians as a panacea for a broad range of education problems'. They pointed out that given the variable performance of academies to date, 'conversion to an Academy may not always be the best route to improvement' and that care needed to be taken 'to ensure they are the "best fit" solution to the problem at hand' (p. 10).

The London Challenge

There are also other New Labour programmes and initiatives that have been evaluated in sufficient depth to give an indication of the sort of interventions that can be effective in narrowing the gap. The transformation of schooling in London in this period is worthy of particular attention. Wyness (2011) notes that, although the demographic character of London would lead one to expect that educational outcomes in London would be inferior to those in the rest of the country, London students actually perform better than those from the rest of the country at most ages and levels of attainment. Performing as well as the rest of the country at KS1, London students 'pull away from their non-London counterparts at Key Stage 2, with the gap remaining constant, or increasing at Key Stage 4' (Wyness 2011, p. 47). It has even been claimed that London is the only capital city in the developed world whose schools perform better than those in the rest of their nations (Stewart 2011).

One of the possible explanations Wyness offers for this is the London Challenge, a policy introduced in 2003 at a time when there was something of a 'moral panic' about the performance of London's schools. Its overall brief was ambitious and extensive (DfES 2005). Although it included market-based elements, others seemed to respond to the potentially negative effects of such policies. It was consistent with the New Labour emphasis on standards, and recognised the importance of concerted collective efforts to raise achievement among those schools and children who had been languishing under existing policies. The first Commissioner for London Schools, Tim Brighouse, describes London as trying to be the first place to show that schools could contribute to 'cracking the cycle of disadvantage' (Brighouse 2007, p. 79).

The London Challenge was initially a 5-year partnership between central government, schools and boroughs (districts within London) to raise standards in London's secondary school system. Provision included transforming failing schools into academies, making pan-London resources and programmes available to all schools, giving individualised support for the most disadvantaged students and intensive work with five of the 33 London boroughs and more particularly with 'Keys to Success' schools within them. These were the schools in London facing the biggest challenges and in greatest need of additional support. Each school received bespoke solutions through diagnostic work and ongoing support (Brighouse 2007).

Provision was extended in 2006 to include work with primary schools and activities in relation to students' progression to further and higher education. There has been additional continuing professional development for teachers through the Chartered London Teacher scheme and for head teachers through the London Leadership Strategy.

Some politicians have privileged particular policies in their accounts of the success of the London Challenge. For example, the present Secretary of State for Education, Michael Gove, recently claimed that the three most important elements were sponsored academies, the use of outstanding schools to mentor others and a focus on improving the quality of teaching—especially through Teach First (the English equivalent of Teach for America; Gove 2012). This emphasis is perhaps not surprising given the centrality of these particular policies to his own party's preferred reforms, which are discussed briefly at the end of this chapter.

Even so, there is certainly evidence that each of these particular policies had a positive impact on schools in their own right (Machin and Vernoit 2011; Earley and Weindling 2006; Muijs et al. 2010). However, we are not aware of any research that shows that they were necessarily the most important elements in the success of the London Challenge or in narrowing the attainment gap in London. In reality, New Labour's London Challenge programme, whose success Gove was praising, was a multifaceted policy, and it included elements that seem to be out of step with the present government's approach. It involved a range of interventions at the level of 'the London teacher, the London leader, the London school and the London student' (Brighouse 2007, p. 80 ff).

This means that unfortunately, as had national policies, it is difficult to identify which parts of the intervention had the positive effect. Nevertheless, the overall approach of London Challenge does seem to have had a tangible impact, although there may have been other factors at work in London at that time (Wyness 2011; Allen 2012). National performance data show that between 2003 and 2006, the national rate of improvement in the number of students achieving five or more GCSE passes with grades A*–C at age 16 was 6.7%, whereas in London it was 8.4%, and in the 'Keys to Success' schools in London it was 12.9% (DfES 2007a).

Toward the end of its existence, the London Challenge was extended to other English cities as the City Challenge (DfES 2007b). Hutchings et al. (2012) present evidence that these programmes had impacts on reducing the number of underperforming schools and increasing the performance of those eligible for FSM faster than the national average. However, only in London (and in Greater Manchester in the primary phase) has this been translated into a closing of the attainment gap over the period 2008–2011.

Even in London, it was initially suggested that the improvement in the overall performance of London schools noted above derived largely from an increase in attainment among the more advantaged students in the schools that were receiving the most intensive interventions. However, subsequently it was found that not only were the 'Keys to Success' schools improving at a faster rate than the norm, but also the attainment gap for disadvantaged children in London was itself narrowing faster than elsewhere and narrowing fastest in these particular schools. Using FSM entitlement as a proxy for economic disadvantage, data provided to us by the DfES showed that attainment at age 16 for this group of pupils within 'Keys to Success' schools rose by a larger amount than for the non-FSM pupils (13.1 points compared to 12.3 points for the latter between 2003 and 2006). Michael Gove drew attention to this particular success for poorer children in London when he noted that whereas in England more generally '35 per cent of children on free school meals achieve five good GCSEs with English and Maths … in inner-London 52 per cent meet [this benchmark]' (Gove 2012). He also noted that this is not far off the national average for pupils, regardless of their background.

An Ofsted report on the impact of London Challenge described continuing positive impacts beyond the initial period. It noted that the primary schools that joined the London Challenge 'are improving faster than those in the rest of England', partly attributing this to schools continuing to participate in development programmes for teachers after the support given as part of London Challenge had ended (Ofsted 2010a). The report was positive about the possibilities for maintaining the gains from London Challenge due to changes it has engendered in practices (such as increased use of performance data to track progress) and ethos (such as motivating staff to share good practice with other schools). Such collaboration may have countered the more negative effects of school choice mechanisms, so it will be important to monitor what happens in London now that the initiative as a whole has finally come to an end but market-oriented policies remain in place. On this issue, Hutchings et al. (2012) found encouraging evidence that schools that were part of the initial London Challenge scheme, but no longer funded as Keys to Success schools after 2008, continued to improve at a faster rate than did the national average despite the extra support ending.

Extended Schools

There were also other promising developments in London and elsewhere in England. Extended schools and full service extended schools (similar to full service schools or 'wrap-around schooling' in the USA) were introduced to provide an extended day and/or additional services on school sites. The evaluation of New Labour's pilot programme of full service extended schools found that the number of students reaching the national benchmark at age 16 (five good GCSEs) in such schools rose faster than the national average and that it brought particularly positive outcomes for poorer families by providing stability and improving their children's engagement in learning. Encouragingly in terms of the concerns of this chapter, the final report indicated that the achievement gap between advantaged and disadvantaged students, based on FSM eligibility, had narrowed in these schools (Cummings et al. 2007, p. 126).

Reading Recovery

Support for Reading Recovery was an example of a policy targeted directly at individuals rather than at schools or areas and was part of a broader personalisation agenda that developed in the later years of the New Labour administration. Reading Recovery originated in New Zealand but was introduced in England by the Institute of Education and was given government funding, and it eventually became a key component of the national Every Child a Reader programme. It aims to provide one-on-one support to children falling behind their peers in the first few years of school. As such, it aims to break the cycle of low self-esteem and lack of confidence resulting from falling behind, itself hampering further progress. A Reading Recovery evaluation (NatCen 2011) saw improvements in reading ability and reading-related attitudes and behaviours of children receiving help from the programme. It is worth noting, however, that this is a purely descriptive analysis; no comparator group can be identified because pupils who should receive the Reading Recovery are selected in schools only where it is being implemented. As such, we cannot say what progress these children would have made in the absence of Reading Recovery. It could be the case that some would have caught up by themselves or through pre-existing support mechanism, or alternatively that they would have fallen further behind. The same evaluation also used a quasi-experimental method to estimate a wider impact of Every Child a Reader. This found an encouraging impact on school-level reading and writing attainment of between two and six percentage points in the later years of the intervention.

Teach First

There has been an increasing recognition 'that getting the right people to become teachers is critical to high performance' (Barber and Mourshed 2007, p. 16). Teach First, like Teach for America, was an initiative to recruit highly qualified graduates into teaching in particularly disadvantaged schools. It began work in London in 2002. An evaluation by Muijs et al. (2010) that schools with Teach First teachers achieve higher attainment for their students than do comparable schools (as matched by type of school, gender intake, performance levels, student intake characteristics, location and school size). As with any quasi-experimental method, we cannot be sure the results are causal, because the matching will not ensure that the schools are truly comparable. Indeed, because schools can choose to partner with Teach First, there seems considerable scope for those with more proactive leadership or more capacity to benefit from Teach First teachers to be driving these results. The evaluation attempts to assess this possibility by also comparing Ofsted evaluations of Teach First and comparator schools, finding little significant difference. It also finds evidence of a mild, but significant, correlation between the number of Teach First teachers in a school and its student outcomes, a pattern we would expect where such teachers are making a real difference to the pupils' attainment. Although this does not give us specific evidence on closing the attainment gap, because all Teach First schools have disadvantaged intakes, it seems plausible that this initiative can help to reduce between-school attainment gaps.

Beyond Competition?

Apart from the case of academies, these gains have been derived from initiatives that, to some extent at least, run counter to the central thrust of recent policies in England and elsewhere that see school improvement as coming through market competition and choice between autonomous schools. London Challenge recognised the particular challenges facing schools in the capital and the need for them to work together, while one of the key features of extended schools was multi-agency co-operation and schools providing services for local communities. Reading Recovery required substantial resources to be devoted to the needs of a small number of disadvantaged children, arguably at the expense of investment in the needs of more affluent students whose parents are often seen as calling the tune in our current education system (Ball 2003). Teach First teachers made a collective contribution to improvement across the system as a whole, as well as serving in the individual schools to which they were allocated.

Thus, in their different ways, these initiatives have recognised the importance of countering wider influences on educational performance to a greater extent than is evident in the dominant market rhetoric adopted by recent governments (Whitty 2008). Taken together, they also provide support for the warning made by Ravitch (2010) in the USA 'that, in education, there are no shortcuts, no utopias, and no silver bullets' (p. 3).

Another recent review of the evidence on whether schools can narrow the gap, carried out at the University of Manchester, suggests that, though the ability of schooling to lessen the impact of deprivation on children's progress is limited by factors beyond the control of the school system, 'carefully designed school improvement interventions…can help schools to narrow the gap in attainment' (Kerr and West 2010, pp. 8–9). However, the authors also argue that '[n]either general nor targeted interventions have, thus far, demonstrated substantial sustained improvements that can be spread widely' (p. 37). They conclude that structural and 'beyond-the-school strategies' are necessary, arguing that 20 years of competition between schools has done little to improve the lot of disadvantaged students but that 'collaboration between schools has shown some promising results'. They also advocate an overhaul of school governance and management structures and suggest that 'radical changes across children's services [will be] needed to support sustained improvements in children's outcomes' (Kerr and West 2010, p. 45).

Access to Higher Education

The New Labour government introduced a series of policies designed to narrow the participation gap between traditional and non-traditional entrants to higher education, the latter meaning those from lower socioeconomic groups and some particular ethnic minorities. These policies included new student financing arrangements to offset increased fees, the establishment of an Office for Fair Access (OFFA) to ensure that universities took their responsibilities in this area seriously, and Aim-Higher, an outreach initiative that helped universities to work closely with schools to increase aspiration, achievement and enrolments.

A report from OFFA (Harris 2010) found that, though these widening participation efforts had had a positive impact overall, the picture was different if the group of what the report calls 'highly selective' institutions was considered separately. These institutions included Oxford, Cambridge and other research-intensive universities. Although the overall higher education participation rate of the least-advantaged 40 % of students had increased since the mid-1990s, the participation rate of the same group at the most selective third of universities had stayed constant. Furthermore, the gap between the most and least advantaged had actually increased in these universities as those from the most advantaged backgrounds (the top 20 %) were now more likely to attend these institutions than they were in the mid-1990s.

In relation to the concerns of this chapter, it is important to note that these figures seem to be influenced not so much by bias in selection by universities, but through a lack of qualified applicants from disadvantaged backgrounds. Indeed, Anders (2012) finds little evidence of different success rates among university applicants with similar attainment at the end of KS2 (age 11). As such, most of the overall participation gap is driven at or before the decision to apply to university, with factors such as lower prior attainment or lower educational expectations by young people from more disadvantaged backgrounds (Chowdry et. al. 2010b) potentially meaning that they do not apply in the first place (see also Sutton Trust 2004).

For those who do apply, the pattern of subjects they study is also socially skewed (Whitty 2010a). For students wanting to study STEM subjects, it is not just attainment that counts but specifically attainment in the right subjects. Even having the potential to study STEM subjects at university requires decisions to be taken relatively early in a student's school career, as STEM subjects usually have more specific requirements with regard to entry qualifications than with regard to many other subjects (Coyne and Goodfellow 2008). Harris (2010) observed that one has only to recognise that 'the range of sciences offered in independent and selective schools is often wider (than in non-selective state schools), and that science-based subjects such as medicine are disproportionately offered by selective universities and at least some of the reasons for a skewed application pool are immediately very clear (p. 73). Thus, although the main imperative in terms of further widening of participation and fair access must be to enhance attainment in school, improved information, advice and guidance is also important—particularly for some STEM subjects, such as engineering and medicine, where the combination of prior qualifications needed is especially tightly specified.

Toward the end of the New Labour government, a study by the Institute of Education identified a need to develop the AimHigher initiative through more work with younger children, involving parents where possible, more sustained interventions engaging all students and not just a select group, as well as doing more work on subject-specific issues (Tough et al. 2008). Another project recommended that schools should ensure that students know about the full spectrum of universities, that school staff should be open with students regarding the nature and standing of different universities and that there should be a change in the university recruitment timetable to benefit 'first-generation' applicants who generally have lower predicted test scores and are also likely to apply to the more selective universities only if high scores are predicted in their A-level examinations (Curtis et al. 2008b).

Postscript: Policies and Prospects Post-2010

As indicated at the start of this chapter, the Coalition government that was elected to replace New Labour in May 2010 has made a commitment to 'closing' the achievement gap as part of a wider commitment to increasing social mobility, which it claims had stalled under New Labour (HM Government 2010a). This government is led by the Conservative Party under Prime Minister David Cameron and the Secretary of State for Education, Michael Gove, is one of his closest allies. The general thrust of their policies is to continue and accelerate the emphasis on seeking improvement through school autonomy, competition and choice that was pioneered by Margaret Thatcher's Conservative government but continued by New Labour under Tony Blair, along with a reassertion of traditional approaches to schooling (Whitty 1989, 2008).

Whereas the academies policy of the Blair government discussed above sought to use academy status mainly to prioritise the replacement or improvement of failing schools in disadvantaged areas, the Conservative-led Coalition has potentially extended this status to virtually all schools. Schools highly rated by Ofsted, a disproportionate number of which are in more affluent areas, can be granted academy status automatically if they so desire. Meanwhile, parents, teachers and others are being encouraged to open publicly funded 'free schools', which, like academies, will be outside the jurisdiction of local authority. It remains an open question whether such policies will help to 'close' the gap or effectively 'open' it up again.

However, the nature of the new government's educational policy is to some extent influenced by the social justice agenda of the Liberal Democrat party, whose votes give the Coalition its majority in parliament. Among the policies that are directly linked to the commitment to close the attainment gap is a 'pupil premium' to be paid on top of the normal grant for every school-age student in receipt of FSMs in state schools. This is consistent with the earlier trend of linking resources to individuals in need regardless of the neighbourhood in which they are receiving their schooling. Unfortunately, welcome as this payment is, the level of it is below that envisaged by the Liberal Democrats prior to the election, and it replaces other targeted benefits that were paid under New Labour. Most seriously, the fact that it is being introduced at a time of major expenditure cuts in other areas means that some schools will barely notice its impact. Furthermore, the money is not ring-fenced or mandated for particular purposes, and retrospective monitoring of its use by Ofsted will be the main mechanism for ensuring that it is actually used to benefit the education of the disadvantaged. An early survey of teachers for the Sutton Trust (2012) was not encouraging. It suggested that little of the £1.25 billion allocated through the pupil premium for disadvantaged children in 2012–2013 would be spent on activities that are known to boost attainment.

Another initiative, which may help in this respect in the future, is the creation by the government of an Education Endowment Foundation, a grant-making charity dedicated to raising the attainment of disadvantaged pupils in English primary and secondary schools by challenging educational disadvantage, sharing evidence and finding out what works. One of the ways in which it will do this is by providing independent and accessible information through a Learning and Teaching Toolkit (EEF 2012) that provides guidance to schools on how best to use the pupil premium to improve the attainment of their pupils by summarising educational research from the UK and elsewhere. This has so far identified effective feedback, metacognition and peer tutoring as three strategies that have been shown to have high impact at low cost, on the basis of strong evidence. In the case of peer tutoring, it suggests that children from disadvantaged background may derive particularly large benefits from this strategy. It also identifies the high impact of early years' intervention, but notes the high costs involved in this.

There is currently considerable controversy about whether the government's curriculum policies will help to close the gap. There is, for example, a commendable emphasis on early literacy but an undue commitment to 'synthetic phonics' as the only way to teach reading, despite evidence that, though it can indeed be an effective strategy with disadvantaged children, it is not a panacea and that a more mixed approach is desirable (Wyse and Parker 2012).

Another policy announced by Michael Gove was the 'English Baccalaureate', an award to students but also effectively a new performance measure for secondary schools based on the percentage of students achieving high grades in specified subjects, i.e. English, mathematics, science, history or geography and a foreign language. This may initially affect disadvantaged students adversely, as they are more likely to have been exposed to alternative curricula than are more advantaged students on a university entrance track.

A linked policy has been to reduce the number of 'equivalent' qualifications that are permitted to be used in school performance tables as alternatives to the GCSE qualifications at age 16. This will have an impact of the number of vocational qualifications taught in schools and places a further emphasis on a return to conventional academic qualifications. Ironically, in view of the Coalition government's enthusiastic embrace of the academies programme, some of the New Labour academies that moved sharply up the performance tables in recent years did so partly by introducing these alternative qualifications (de Waal 2009).

The government's response to concerns about its traditionalist curriculum policy has been that social justice requires equal access to high-status knowledge and that there is little point in students succeeding in courses that are deemed to have little value by universities, employers and the wider society. However, though there may well be a good argument for ensuring that all students should have the opportunity to gain access to 'powerful knowledge' (Young 2010), if indeed that is what is the traditional curriculum provides, the government will need to give more attention than it has done hitherto to reforming the pedagogy through which these subjects are taught (Whitty 2010b). Exley and Ball (2011) argue that some current policies involve a return to the nineteenth century and that we need to remember that few disadvantaged children and families benefited from the type of schooling that predominated in those days. So the jury remains out on how far current policies will contribute positively to continuing narrowing the gap in school attainment—let alone closing it.

A further issue is the Coalition government's policies for further and higher education. They have removed Education Maintenance Allowances that supported disadvantaged students to stay in full-time education beyond age 16 and replaced it with a much less expensive and extensive scheme. In universities, they have introduced higher fees alongside income-contingent loans to be repaid by graduates while earning. Although this means that no families will have to pay the fees upfront, there is a concern that some students will be unwilling to take on the levels of debt envisaged. The early evidence gives limited credence to those who anticipated a reduction in the rate of applications when the increase was introduced. However, this decline has been fairly even across SES, but has been particularly evident among older students (UCAS 2012). It will be years before we know the extent and nature of the changes' impact on patterns of recruitment to higher education and the professions.

A decision to bring to an end the work of AimHigher has led to controversy, but the government is pledged to secure the strengthening of universities' widening participation strategies and to hold universities to account for them. It has also called for better information, advice and guidance in schools and has proposed that higher education entry statistics, including entry to elite universities, should be a new performance indicator for secondary schools.

Finally, with a view to longer-term strategies for closing the gap, the Coalition government commissioned two important reports. Both these reports were written by Labour Members of Parliament, demonstrating that closing the gap is a key cross-party priority. The first of these, the Field report (Field 2010), was the product of a review of the evidence on poverty and life chances. Although part of its remit was to consider how to reduce poverty across the life cycle, it developed a particular focus on the importance of children's development in their first 5 years to their future life chances. It recommended a much greater focus on the EYFS, with some of the funding from other phases of education gradually being shifted to these early years. However, it also recommended spending this reallocated funding in much more targeted ways, through programmes such as support for parenting skills. Particularly important for the concerns of this chapter, it recommended that schools be held accountable through the inspection system for reducing attainment gaps, not just increasing attainment (Field 2010, p. 64). This recommendation has effectively been implemented through the latest school inspection framework and through an addition to the annual school performance tables that will show how effective they are in achieving progress with students at three different levels of ability.

Two other recommendations that received considerable attention were to introduce new 'life chance indicators' to supplement financial indicators of poverty and to divert future increases in child-related social benefits to increase early-years provision. Taken together, these represent a shift in focus away from poverty as a lack of money and toward non-monetary 'factors in young children which we know to be predictive of children's future outcomes' (Field 2010, p. 9). This has the potential to be highly significant for the future direction of policy aimed at reducing achievement gaps.

The Allen report (Allen 2011) specifically considered how the government should take forward its early intervention strategy. It surveyed literature on the particularly rapid pace of cognitive development before the age of 3, concluding that if the child does not get the best start in life, it can seriously hamper their social and economic outcomes later in life. It painted a positive picture of the potential for all children, regardless of their socioeconomic background, when it stated that 'what parents do is more important than who they are' (Allen 2011, p. 23). In particular, it recommended targeting school readiness during the EYFS, attempting to ensure that socioeconomic gaps between children entering school described earlier in this chapter are closed. Again, there are clear implications here for future policy directions. For example, it seems likely that following these two reports, we will see an increased focus on developing parenting skills.

If the present government did move decisively in that direction, it would signal an acceptance of the conclusion of Kerr and West (2010) that 'efforts to improve schools must be accompanied by efforts to support disadvantaged families' (p. 41). As Mortimore and Whitty (1997) argued under a previous government, 'society needs to be clearer about what schools can and cannot be expected to do' (p. 12). This does not mean that schools cannot make a difference, or that they do not have a particularly important role in helping to narrow the attainment gap and thereby enhancing the life chances of disadvantaged children. It does mean that they cannot do it alone.

Acknowledgements We would like to thank Rebecca Allen, John Jerrim, Sandra Leaton Gray, Sarah Tang, Anna Vignoles and Emma Wisby at the Institute of Education, University of London, for their help in preparing this chapter.

References

Allen, G. (2011). *Early Intervention: The next steps. An independent report to her Majesty's Government*. London: Cabinet Office.

Allen, R. (2012). How can London schools be so good, given the high cost of living for teachers? IoE Blog 22/05/2012. London: Institute of Education. http://ioelondonblog.wordpress.com/2012/05/22/how-can-london-schools-be-so-good-given-the-high-cost-of-living-for-teachers/.

Anders, J. (2012). The link between household income, university applications and university Attendance. *Fiscal Studies, 33*(2), 185–210.

Ball, S. J. (2003). *Class strategies and the education market: The middle classes and social advantage*. London: Routledge.

Barber, M., & Mourshed, M. (2007). *How the world's best-performing school systems come out on top*. London: McKinsey & Co.

Bernstein, B. (1970). Education cannot compensate for society. *New Society*, 26 February 1970, 344–347.

Blatchford, P., Russell, A., & Webster, R. (2012). *Reassessing the impact of teaching assistants: How research challenges practice and policy*. London: Routledge.

Boliver, V. (2011). Expansion, differentiation, and the persistence of social class inequalities in British higher education *Higher Education, 61*(3), 229–242.

Brighouse, T. (2007). 'The London Challenge–a personal view'. In T. Brighouse & L. Fullick (Eds.), *Education in a global city: Essays from London*. London: Institute of Education.

Chowdry, H., Crawford, C., Dearden, L., Goodman, A., & Vignoles, A. (2010a). Widening participation in higher education: Analysis using linked administrative data. *IFS Working Paper W10/04*. London: Institute for Fiscal Studies.

Chowdry, H., Crawford, C., & Goodman, A. (2010b). The role of attitudes and behaviours in explaining socio-economic difference in attainment at age 16. *IFS Working Paper 10/15*. London: Institute for Fiscal Studies.

Cook, C. (2011). Poorer children close education gap. *Financial Times*. London: Pearson. http://www.ft.com/cms/s/0/d82fc3cc-eab3-11e0-aeca-00144feab49a.html.

Coyne, J., & Goodfellow, J. M. (2008). *Report to the Secretary of State, DIUS, on universities' links with schools in STEM subjects*. http://www.bis.gov.uk/policies/higher-education-debate. Accessed 2 Nov 2009.

Cummings, C., Dyson, A., Muijs, D., Papps, I., Pearson, D., Raffo, C., Tiplady, L., Todd, L., & Crowther, D. (2007). Evaluation of the full service extended schools initiative: Final Report. *DfES Research Report RR852*. London: Department for Education and Skills.

Curtis, A., Exley, S., Sasia, A., Tough, S. & Whitty, G. (2008a) *The academies programme: Progress, problems and possibilities*. London: Sutton Trust.

Curtis, A., Power, S., Whitty, G., Exley, S., & Sasia, A. (2008b). *Primed for success? The characteristics and practices of state schools with good track records of entry into prestigious UK universities*. London: Sutton Trust.

Curtis, A. (2009). Academies and school diversity. *Management in Education, 23*(3), 113–117.

DCSF (Department for Children, Schools and Families) (2007). *The Children's Plan: Building brighter futures*. London: Stationery Office.

DCSF (Department for Children, Schools and Families) (2008). Schools and Pupils in England: January 2007 (Final). *Statistical First Release 30/2007*. http://www.dcsf.gov.uk/rsgateway/DB/SFR/s000744/index.shtml. Accessed 9 March 2009.

DfES (Department for Education and Skills) (2005). *London Schools: Rising to the Challenge*. London: Department for Education and Skills.

DfES (Department for Education and Skills) (2006). Social Mobility: Narrowing Social Class Educational Attainment Gaps. *Supporting Materials to a speech by the Rt Hon Ruth Kelly MP Secretary of State for Education and Skills to the Institute for Public Policy Research*. London: Department for Education and Skills.

DfES (Department for Education and Skills) (2007a). Secretary of State Announces Extension of the London Challenge Programme. Archived press release. http://webarchive.nationalarchives.gov.uk/20070603200104/http://dfes.gov.uk/londonchallenge/.

DfES (Department for Education and Skills) (2007b). *City Challenge for World Class Education*. London: Department of Education.

DfE (Department for Education) (2011a). Outcomes for children looked after by local authorities in England, as at 31 March 2011. *Statistical First Release SFR30/2011*. London: Department for Education.

DfE (Department for Education) (2011b). The National Strategies 1997–2011. A brief summary of the impact and effectiveness of the National Strategies. *DfE Reference: 00032–2011*. London: Department for Education.

de Waal, A. (2008). *School Improvement-or the 'Equivalent'*. London: Civitas.

de Waal, A. (2009). *The Secrets of Academies' Success*. London: Civitas.

Earl, L., Levin, B., Leithwood, K., Fullan, M., & Watson, N. (2001). OISE/UT Evaluation of the implementation of the National Literacy and Numeracy Strategies. Second Annual Report. Watching & Learning 2. *DfES Report DfES-0617/2001*. London: Department for Education and Skills.

Earley, P., & Weindling, D. (2006). Consultant leadership—a new role for head teachers?, *School Leadership & Management, 26*(1), 37–53.

Edwards, T., Fitz, J., & Whitty, G. (1989). *The state and private education: An evaluation of the assisted places scheme*. Basingstoke: Falmer Press.

EEF (Education Endowment Foundation) (2012). Teaching and Learning Toolkit. www.educationendowmentfoundation.org.uk

Exley, S., & Ball, S. J. (2011). Something old, something new…understanding Conservative education policy. In H. Bochel, (Ed.), *The Conservative Party and social policy*. Bristol: Policy Press.

Feinstein, L. (2003). Inequality in the early cognitive development of British children in the 1970 Cohort. *Economica, 70*(177), 73–97.

Field, F. (2010). *The foundation years: preventing poor children becoming poor adults*. The report of the Independent Review on Poverty and Life Chances. London: Cabinet Office.

Goldstein (2002). The national literacy and numeracy strategy (NLNS) evaluation: Part 2. Bristol: Centre for Multilevel Modelling, University of Bristol. http://www.bristol.ac.uk/cmm/team/hg/oisereport2.html.

Goodman, A., Sibieta, L., & Washbrook, E. (2009). Inequalities in educational outcomes among children aged 3 to 16. *Final report for the National Equality Panel*. London: National Equality Panel.

Gove, M. (2012). *Michael Gove at the National College annual conference*. http://www.education.gov.uk/inthenews/speeches/a00210308/michael-gove-at-the-national-college-annual-conference. Accessed 12 July 2012.

Hansard (2012). *House of Commons Hansard Written Answers for 12 July 2012*. London: Houses of Parliament. http://www.publications.parliament.uk/pa/cm201213/cmhansrd/cm120712/text/120712w0003.htm.

Harris, M. (2010). *What more can be done to widen access to highly selective universities?: A Report from Sir Martin Harris, Director of Fair Access*. Bristol: OFFA. http://www.offa.org.uk/wp-content/uploads/2010/05/Sir-Martin-Harris-Fair-Access-report-web-version.pdf. Accessed 20 August 2010).

HM Government (2010a). *The coalition: our programme for government*. London: Cabinet Office. http://programmeforgovernment.hmg.gov.uk/. Accessed 1 July 2010.

HM Government (2010b). *Opening doors, breaking barriers: A strategy for social mobility*. London: Cabinet Office. https://www.gov.uk/government/uploads/system/uploads/attachment_data/file/61964/opening-doors-breaking-barriers.pdf. Accessed 1 July 2010.

Hutchings, M., Greenwood, C., Hollingworth, S., Mansaray, A., Rose, A., Minty, S., & Glass, K. (2012). Evaluation of the City Challenge programme. *DfE Research Report DFE-RR215*. London: Institute for Policy Studies in Education, London Metropolitan University.

Jackson, M., Erikson, R., Goldthorpe, J., & Yaish, M. (2007). Primary and secondary effects in class differentials in educational attainment. *Acta Sociologica, 50*(3), 211–229.

Jerrim, J. (2011). England's "plummeting" PISA test scores between 2000 and 2009: Is the performance of our secondary school pupils really in relative decline? *DoQSS Working Paper 11-09*. London: Department of Quantitative Social Science, Institute of Education.

Jerrim, J. (2012). The socio-economic gradient in teenagers' reading skills: How does England compare with other countries? *Fiscal Studies, 53*(2), 159–184.

Jerrim, J., & Vignoles, A. (2011). The use (and misuse) of statistics in understanding social mobility: regression to the mean and the cognitive development of high ability children from disadvantaged homes. *DoQSS Working Paper 11-01*. London: Department of Quantitative Social Science, Institute of Education.

Kelly, K., & Cook, S. (2007). Full-time young participation by socioeconomic class: A new widening participation measure in higher education. *DfES Research Report RR806*. http://www.dcsf. gov.uk/research/data/uploadfiles/RR806.pdf. Accessed 21 Oct 2008.

Kerr, K., & West, M. (Eds.). (2010). *Social inequality: Can schools narrow the gap?* Insight 2. Chester: British Educational Research Association.

Machin, S., & McNally, S. (2004). The Literacy Hour. *CEE Discussion Papers 0043*, London: Centre for the Economics of Education, London School of Economics.

Machin, S., McNally, S., & Meghir, C. (2007). *Resources and Standards in Urban Schools*. London: Centre for the Economics of Education.

Machin, S., & Vernoit, J. (2011). Changing school autonomy: Academy schools and their introduction to England's education. *Centre for the Economics of Education. CEE DP 123*. http://cee. lse.ac.uk/ceedps/ceedp123.pdf.

Maden, M. (Ed.). (2002). *Success against the odds five years on*. London: Routledge.

Milburn, A. (2009). *Unleashing aspiration: The final report of the Panel on Fair Access to the Professions*. HMSO. http://www.cabinetoffice.gov.uk/strategy/work_areas/accessprofessions. aspx. Accessed 12 Jan 2010.

Mortimore, P., & Whitty, G. (1997). *Can school improvement overcome the effects of disadvantage?* London: Institute of Education.

Muijs, D., Chapman, C., & Armstrong, P. (2010). *Maximum impact evaluation: The impact of Teach First teachers in schools*. Final Report. Manchester: University of Manchester.

NAO (National Audit Office) (2007). *The Academies Programme*. Report by the Comptroller and Auditor General HC 254 Session 2005–2007. London: Stationery Office.

NatCen (National Centre for Social Research) (2011). Evaluation of every child a reader. *DfE Research Report DFE-RR114*. London: Department for Education.

NESS (National Evaluation of Sure Start Team) (2010). The impact of sure start local programmes on five year olds and their families. *DfE Research Report DFE-RR067*. London: Department for Education.

NESS (National Evaluation of Sure Start Team) (2012). The impact of sure start local programmes on seven year olds and their families. *DfE Research Report DFE-RR220*. London: Department for Education.

ONS (Office for National Statistics) (2006). *Statistics of Education: Trends in Attainment Gaps: 2005*. London: Office for National Statistics.

Ofsted (2010a). London Challenge. *Report number 100192*. London: Ofsted.

Ofsted (2010b). The national strategies: A review of impact. *Report number 080272*. London: Ofsted.

Power, S., Whitty, G., Gewirtz, S., Halpin, D., & Dickson, M. (2004). Paving a 'third way'? A policy trajectory analysis of education action zones. *Research Papers in Education, 19*(4), 453–475.

PWC (PricewaterhouseCoopers) (2008). *Academies evaluation: Fifth annual report*. London: Department for Education and Skills.

Ravitch, D. (2010). *The death and life of the great American school system: How testing and choice are undermining education*. New York: Basic Books.

Robinson, P. (1997). *Literacy, numeracy and economic performance*. London: Centre for Economic Performance, London School of Economics.

Saunders, P. (2012). *Social Mobility Delusions*. London: Civitas.

Schuller, T., Preston, J., Hammond, C., Brassett-Grundy, A., & Bynner, J. (2004). *The benefits of learning: The impact of education on health, family life and social capital*. London: Routledge.

Starks, L. (2011). *The national strategies-evaluation of the support and resources for narrowing the gap*. Leeds: York Consulting LLP.

Stewart, W. (25 Mar 2011). The challenge now is to hang on to this success, *Times Educational Supplement*. http://www.tes.co.uk/article.aspx?storycode=6074357.

Sullivan, A., Heath, A., & Rothon, A. (2011). Equalisation or inflation? Social class and gender differentials in England and Wales. *Oxford Review of Education, 37*(2), 215–240.

Sutton Trust (2004). *The missing 3000: State school students under-represented at leading universities.* London: Sutton Trust.

Sutton Trust (2011). *What prospects for mobility in the UK? A cross-national study of educational inequalities and their implications for future education and earnings mobility.* London: Sutton Trust.

Sutton Trust (2012). *The use of the Pupil Premium, NFER Teacher Voice Omnibus 2012 Survey.* Slough: National Foundation for Educational Research.

Thrupp, M., & Lupton, R. (2006). Taking school contexts more seriously: The social justice challenge. *British Journal of Educational Studies, 54*(3), 308–328.

Tough, S., Sasia, A., & Whitty, G. (2008). *Productive partnerships? An examination of schools' links with higher education.* London: Sutton Trust.

Tymms, P. (2004). Are standards rising in English primary schools? *British Educational Research Journal, 30*(4), 477–94.

Tymms, P., Coe, R., & Merrell, C. (2005). Standards in English schools: Changes since 1997 and the impact of government policies and initiatives. *A report for the Sunday Times.* Durham: CEM Centre, University of Durham.

Tymms, P., & Merrell, C. (2007). Standards and quality in English primary schools over time: The national evidence. *Primary Review Interim Report Research Survey 4/1.* Cambridge: Primary Review.

UCAS (2012). How have applications for full-time undergraduate higher education in the UK changed in 2012? http://ucas.com/documents/ucas_how_have_applications_changed_in_2012_executive_summary.pdf.

Vignoles, A., & Crawford, C. (2010). Access, participation and diversity questions in relation to different forms of post-compulsory further and higher education (FHEs). In M. David, (Ed.), *Improving learning by widening participation in higher education.* London: Routledge.

Whitty, G. (1989). The new right and the national curriculum: State control or market forces? *Journal of Education Policy, 4*(4), 329–341.

Whitty, G. (2001). Education, social class and social exclusion. *Journal of Education Policy, 16*(4), 287–295.

Whitty, G. (2006). Education(al) research and education policy making: Is conflict inevitable?. *British Educational Research Journal, 32*(2),159–176.

Whitty, G. (2008). Twenty years of progress? English education policy 1988 to the present. *Educational Management Administration & Leadership, 36*(2), 165–184.

Whitty, G. (2010a). Who you know, what you know or knowing the ropes? New evidence in the widening participation debate. Paper presented at the Festival of Education, Wellington College, July 3–4.

Whitty, G. (2010b). Revisiting school knowledge: Some sociological perspectives on new school curricula. *European Journal of Education, 45*(1), 28–45.

Whitty, G. (2012). Policy tourism and policy borrowing in education: A Trans-Atlantic case study. In G. Steiner-Khamsi & F. Waldow (Eds.), *World yearbook of education 2012: Policy borrowing and lending in education.* New York: Routledge.

Wyness, G. (2011). *London schooling: Lessons from the capital.* London: CentreForum.

Wyse, D., & Parker, C. (2012). *The early literacy handbook.* London: MA Education Ltd.

Young, M.F.D. (2010). Alternative educational futures for a knowledge society. *European Educational Research Journal, 9*(1), 1–12.

Chapter 9
The Achievement Gap in Science and Mathematics: A Turkish Perspective

Mustafa Sami Topcu

Introduction

Turkey is a highly developing country with a booming economy and the youngest population ratio in Europe. Approximately a quarter of the nation's population consists of elementary and high school students. Despite this young population rate, the achievement levels of most students in science and mathematics are below expected levels. According to national (e.g., high school entrance examination, HSEE) and international examinations (e.g., the Programme for International Student Assessment, PISA), Turkish students' achievement levels in science and mathematics are quite low compared to those of other Organisation for Economic Cooperation and Development (OECD) countries. Based on HSEE in 2010, eighth-grade students' mathematics mean score was 5 out of 20 and their science mean score was 6.76 out of 20. In terms of general science and mathematics achievement ranking, Turkey was 29th among 30 OECD countries in PISA 2006 and was 32nd among 34 OECD countries in PISA 2009. Although Turkey did show national improvements compared to previous PISA results, these improvements in science and mathematics achievement did not reflect in international rankings. In addition to low achievement levels of Turkish students in the international context, there are also achievement gaps in the national context. As a specific example, although the percentage of the students enrolling in undergraduate programs in science high schools was 90–98%, this percentage decreased to 10–18% in vocational and technical high schools. In general, achievement gaps exist in the Turkish education system because of four reasons: large quality differences in school types, extremely competitive nationwide examinations, highly standardized and teacher-centered science and mathematics teaching (from elementary schools through college), and the effects of large socioeconomic background differences on science and mathematics achievement. In response to these challenges, the Ministry of National Education (MONE) has taken many precautions. Recent science and mathematics curricula reforms have

M. S. Topcu (✉)
Department of Elementary Science Education, Muğla Sıtkı Koçman University, Mugla, Turkey
e-mail: msamitopcu@gmail.com

J. V. Clark (ed.), *Closing the Achievement Gap from an International Perspective,*
DOI 10.1007/978-94-007-4357-1_9, © Springer Science+Business Media B.V. 2014

been built on more student-centered and flexible learning and teaching methods instead of standardized teaching. Parallel with this reformist movement, the number of nationwide examinations, especially at elementary school levels, has decreased. Moreover, MONE continues to decrease the number of school types, especially at high school levels, because the biggest achievement gaps in science and mathematics exist between high school types. As a last effort, in order to decrease the effects of socioeconomic background differences on science and mathematics achievement, studies continue by MONE.

Education System in Turkey

Education starts with preschool education for 3–5-year-olds, which is not compulsory. In 2002, the MONE developed a preschool education policy that aims to spread preschool education across the country. Parallel with this attempt, the enrollment rate in preschools has increased year after year. In addition, the first preschool curriculum for 3- to-6-year-olds was developed in 2002. Following the preschool education period, compulsory elementary education starts with children whose ages range from 6 to 14 years; it is free at public schools. The elementary education period takes 8 years without a break, and students are awarded a certificate of elementary education after their graduation. Following the elementary education period, high school education takes 4 years and is not compulsory for students graduating from elementary schools. Starting in 1997, radical changes in the Turkish education system have started, and, as a first attempt, the compulsory elementary education period increased from 5 years to 8 years. Until 1997, vocational and technical high schools had included both middle and high school parts. In 1997, the middle parts of vocational and technical high schools were closed, and elementary education was extended to 8 years without a break. In 2005, the high school education period was extended from 3 years to 4 years. With the recent (April 2012) law amendment that comprised radical changes in the current education system, it was decreed that high school education is going to be compulsory for all citizens starting from fall 2012. Therefore, the compulsory education period extended to 12 years. The most distinctive characteristic of this change is that the middle parts of vocational and technical high schools will be reopened and that compulsory education will be divided into three parts. According to this last law amendment, all children aged 6–7 years must start and complete their elementary education in the first 4-year period. These students must continue and complete their middle school education without a break after their elementary education in the second 4-year period. In the last 4-year period, they must complete their high school education.

Based on the 2011–2012 academic year, there are 1,169,556 preschool students with 55,883 teachers, 10,979,301 elementary school students with 515,852 teachers, and 4,756,286 high school students with 235,814 teachers. According to Turkish Statistical Institute (TurkStat), there are 17 million students at the elementary and high school levels with 800,000 teachers (TurkStat 2012a). Approximately a quarter of the nation's population consists of elementary and high school students. There are 168 universities (103 public and 65 private) in Turkey, and these universities include around 42,124 faculty members (Council of Higher Education (COHE) 2012). While the number of the university students in public universities is 3,135,813, the number of the university students in private universities is 160.560. Turkey's population is 74,724,269. Considering the student population and total population of Turkey, it might be claimed that the percentage of young population in Turkey is rather high, and this young population consists mostly of elementary school, high school, and university students.

There are two types of elementary schools in Turkey: public and private. Public elementary schools are free to all citizens, and the teachers of public elementary schools are assigned by MONE. The price of private schools is high, and teachers of private elementary schools are assigned by the management of the schools. Teachers of public schools are assigned by MONE until their retirement, whereas the teachers of private schools are assigned by the management of the private schools for a 1- or 2-year contract. Depending on private school teachers' performance, their contracts may or may not be renewed. However, this contract system is not active in public elementary schools. Regardless of their performance, most teachers of public elementary schools continue teaching until they retire. In contrast with elementary schools, there are more than 20 types of high schools (e.g., Anatolian high schools, Anatolian teacher high schools, public science high schools, private science high schools, social sciences high schools, and vocational and technical high schools). Although students or their parents select only one of two types of elementary schools (public or private), they have to select one of the high schools whose number is high (more than 20). Similar to the elementary school system, although public high schools are free and teachers of these schools are assigned by MONE, the price of private high schools is high, and teachers of these schools are assigned by the management of the private schools. The contract system for high schools is the same as the elementary school contract system.

In order to enter prestigious and successful high schools in Turkey, students have to take the HSEE just after their graduation from elementary schools. This examination is a nationwide examination and is conducted by MONE once a year on a fixed date across the country. The examination is a standardized examination and includes multiple-choice questions. This examination contains 23 Turkish language skills, 20 mathematics, 20 science, 20 social science, and 17 English language skills questions. In 2009, approximately 1 million eighth-grade students took the HSEE. Mean score in mathematics of eighth-grade students was 2.35 out of 20, with a 4.75 standard deviation; their science mean score was 5.25 out of 20, with a 5.62 standard deviation (MONE 2009b). In 2010, 1,008,302 students took the HSEE. Mathematics mean score of eighth-grade students was 5 out of 20, with a 5.62 standard deviation; their science mean score was 6.76 out of 20, with a 5.83 standard deviation (MONE 2010). The number of elementary school students is increasing year by year, and students' science and mathematics test scores are also increasing. However, this improvement is insufficient because students still answer only about a quarter of the questions correctly. Based on their HSEE scores, students can enroll in different kinds of high schools; for example, whereas students whose HSEE score is high can enroll in science or Anatolian high schools, students whose high school placement test score is low can enroll only in vocational and technical high schools. This condition could create an important inequality problem. After high school education, though the percentage of the students enrolling in undergraduate programs in science high schools can reach 90%, this percentage can decrease to 10% in vocational and technical high schools (Berberoglu and Kalender 2005).

In addition to the high school examination and placement system, it is necessary to discuss the university examination and placement system in Turkey. The only way to enter university programs in Turkey is to take the university entrance examination (UEE). Similar to the HSEE, the UEE is a kind of standardized examination and includes multiple-choice questions. The UEE mainly consists of two stages. Applicants who pass the first stage of the UEE (higher education transition examination) take the second stage. These examinations are conducted nationwide by the Measurement, Selection, and Placement Center (OSYM) once a year on a fixed date across the country. The first-stage examination is an elimination examination, conducted annually to select students for the second-stage examination, and includes Turkish language skills, mathematics, and science and social sciences questions. The second-stage examination is more specific compared to the first-stage examination. The second-stage examination, called the *undergraduate placement examination,* is conducted annually, to place students in the undergraduate programs. This examination consists of five types of examinations, and, depending on students' target undergraduate programs, students apply for some of these examination types. Based on the scores taken from these examinations, students are accepted to different programs of the universities. For example, in order to enter medicine or engineering programs, they should take the examination type that includes mostly science and mathematics questions. In order to enter law or economics programs, they should take the examination type that includes mostly mathematics and Turkish language skills questions. In 2010, 1,587,866 students took the UEE, and 349,579 of them were able to enter an undergraduate program (TurkStat 2012b). In 2011, 1,759,403 students took the UEE, and 350,911 of them were able to enter an undergraduate program (TurkStat 2012b). The number of applicants entering the UEE is increasing. Consistent with this development, the number of students entering undergraduate programs is also increasing. The main reason for this improvement is that the number of universities has increased from 70 to 168 in the last 5 years. There are 103 public and 65 private universities in Turkey (COHE 2012).

Science and Mathematics Curricula Reforms in Turkey

In Turkey, the most important curriculum reform in elementary and middle school science and mathematics of the last 30 years was implemented in 2006. The main idea behind this reform was to provide educational equality for every citizen and to improve students' science and mathematics achievement in national and international contexts (MONE 2006, 2009a). The science and mathematics curricula have been developed by committees consisting of university professors, education specialists, and science and mathematics teachers. In the process of curriculum development, opinions from teachers, education inspectors, students, and their parents were gathered to improve the science and mathematics curricula. In 2004 and 2005, pilot studies were implemented in some selected cities of the country in order to explore whether or not the program is well functioning. In light of the pilot studies, the curricula were revised, and starting from 2006, these revised reformed curricula have been implemented across the country. Parallel with this process, science and mathematics textbooks and instructional materials, consistent with the vision and aims of the science and mathematics curricula, were developed and sent to schools by MONE. The Turkish government has been providing free science and mathematics textbooks to teachers and students since 2003.

It is also necessary to deal with the main philosophy and the aims of both reformed science and mathematics curricula. The main philosophy used in the elementary and middle school science and mathematics curricula was constructivism, and the vision of these curricula was to make students scientifically or mathematically literate, regardless of students' individual differences (MONE 2006, 2009a). These curricula reforms aim to create more constructivist-based and student-centered classrooms instead of traditional-based and teacher-centered classrooms. Moreover, both curricula suggest formative assessment instead of summative assessment. In the reformed science curriculum, the name of "science" was changed to "science and technology." The main aims of the science and technology curriculum are to develop students who can utilize science and technology knowledge to make informed decisions in daily life; analyze scientific knowledge critically; conceptualize the nature of science; analyze interactions between science, technology, and society; and understand the strengths and limitations of scientific and technological developments (MONE 2006). The science and technology curriculum specifically focused on the development of students' science process skills, attitudes towards science, and science citizenship education.

The literacy concept was especially highlighted in the science and technology curriculum. In this curriculum, science and technology literacy consists of seven dimensions: the nature of science and technology, key science concepts, scientific process skills, science–technology–society–environment relationships, scientific and technical psychomotor skills, scientific values, and attitudes and values towards science. The main objectives of the science and technology curriculum are that students will be able to:

- Understand and learn about the natural world
- Develop their curiosity towards scientific and technological developments
- Understand the nature of science and technology and relationships between science, technology, society, and environment
- Have skills about building knowledge through research, reading, and discussion
- Develop their knowledge, experience, and interest towards jobs related to science
- Learn how to learn
- Use science and technology to gain new knowledge and solve problems when they are confronted with unusual situations

- Use scientific process and principles while they make personal decisions
- Be aware of social, economic, ethical, and personal health values, and environmental problems related to science and technology, as well as take responsibility and make informed decisions related to these issues
- Like knowing about and understanding science, appreciate investigating and gaining logical insight, and think about the results of actions
- Improve economic productivity in their professional lives by using knowledge, understanding, and skills (MONE 2006, p. 9)

Specifically, the vision of the mathematic curriculum is built on the principle that every child can learn mathematics. Learning mathematics includes acquisition of basic concepts and skills about mathematics, thinking about mathematics, conceptualizing general problem-solving strategies, and appreciating mathematics as an important vehicle in daily life. The main objectives of the mathematics curriculum are that students will be able to:

- Understand mathematical concepts and systems, form relationships among them, and use these concepts and systems in daily life and other learning areas
- Have necessary basic knowledge and skills to continue with advanced education in mathematics or other related areas
- Make inferences related to logical induction and deduction
- Express their own mathematical thinking and reasoning in the process of solving mathematical problems
- Use mathematical terminology and language in order to explain and share mathematical ideas logically
- Use prediction and mental processing skills effectively
- Develop problem-solving strategies and use these strategies to solve problems in daily life
- Develop models and relate models to verbal and mathematical expressions
- Develop self-confidence and positive attitudes towards mathematics
- Appreciate the power of mathematics and the structure of it which contains a relationships network
- Promote and develop intellectual curiosity
- Conceptualize the historical development of mathematics, and the role and value of mathematics history in the development of human thought
- Develop their characters so as to be systematic, careful, patient, and responsible
- Develop their research and knowledge-gaining skills
- Relate mathematics to art and develop aesthetic views (MONE 2009a, p. 9)

Challenges and Solutions in the Turkish Education System

Despite the young population rate and the booming economy, most students' achievement levels in science and mathematics are not at expected levels in Turkey. According to national (e.g., HSEE) and international (e.g., PISA) examinations,

Turkish students' achievement levels in science and mathematics are quite low in contrast with students in other OECD countries. In terms of general science and mathematics achievement ranking, although Turkey was 29th among 30 OECD countries in PISA 2006, Turkey was 32nd among 34 OECD countries in PISA 2009 (OECD 2007a, 2010a). Although Turkey showed national improvements compared to previous PISA results, these improvements in science and mathematics achievement did not reflect in international rankings.

There have been many challenges in the Turkish education system. One of the most important challenges is achievement gaps in national and international contexts. In this chapter, *achievement gaps* refer to the differences of students' mathematics or science achievement depending on educational factors (e.g., school types or students' socioeconomic backgrounds), especially in the national context. These achievement gaps or differences can be observed on individual, group, school, and/ or regional levels. In this context, this chapter seeks to answer questions like the following: How do school types, nationwide examinations, and students' socioeconomic backgrounds influence students' achievement in science and mathematics and lead to achievement gaps? What are possible solutions to achievement gaps in science and mathematics?

In general, achievement gaps exist in the Turkish education system because of four reasons: large quality differences in school types, extremely competitive nationwide examinations, highly standardized and teacher-centered science and mathematics teaching (from elementary schools through college levels), and socioeconomic background differences impacting science and mathematics achievement.

Challenge 1: Large Quality Differences in School Types

According to studies (e.g., Alacacı and Erbaş 2010; Berberoglu and Kalender 2005; Delen and Bulut 2011), the biggest predictor of students' science or mathematics achievement in the Turkish context seems to be school types. In Turkey, the number of high school types is very high. While elementary school types are at expected levels (public and private schools), there are more than 20 types of secondary schools. In addition to this school type variability, there are big science and mathematics achievement gaps between these schools. Interschool variability is much higher than intraschool variability in terms of Turkish students' science and mathematics achievement (OECD 2004, 2007a, 2010b). For example, in Turkey, "the performance gap between students in urban schools and those in rural schools is more than 45 score points after accounting for differences in socio-economic background" (OECD 2010b, p. 14). A lot of research was conducted to identify the factors related to students' science and mathematics achievement gaps. Delen and Bulut (2011) investigated predictors of science and mathematics achievement of 4,996 Turkish students using PISA 2009 data and found that 62 % of the variation in science scores and 67 % of the variation in mathematics scores can be explained by the variability between schools. They explained this result by the huge achievement gap among schools in Turkey. Although they explored the effect of school types on science and mathematics achievement, they did not specifically explore which school types (e.g., science high schools or vocational and technical high schools) in Turkey influenced students' achievement in science and mathematics. Berberoglu and Kalender (2005) analyzed PISA 2003 and UEE 1999, 2000, 2001, and 2002 results in order to explore the effects of regions, years, and school types on students' achievement in the Turkish context. Both PISA and UEE results showed that whereas public schools were under the OECD average of mathematics scores, private schools were above the OECD average. As an exception, public science high schools were above this average score. Moreover, regional areas in Turkey did not influence students' achievement scores at the expected level. Berberoglu and Kalender (2005) concluded that, more than regional differences, school types are important in determining students' mathematics achievement. According to MONE's education report, among OECD countries, the biggest achievement gap between high schools existed in Turkey (MONE 2004).

Alacacı and Erbaş (2010) explored the factors that explain mathematics achievement based on the data collected from PISA 2006. Results showed that whereas 62% of the variation in mathematics scores was explained by differences between schools, the remaining variation was explained by variability in individual student characteristics. Whereas the OECD average of intraschool variance is 63%, the interschool variance is 37% (OECD 2007a). As opposed to OECD countries, interschool variance (55%) was higher than intraschool variance (45%) in Turkey (OECD 2007a). Alacacı and Erbaş (2010) also reported that in PISA 2006, 12% of the Turkish students were from science high schools, Anatolian high schools, and foreign language intensive high schools, and their mathematics literacy levels were above OECD average. However, 77% of the Turkish students were from general and vocational high schools, and their mathematics literacy levels were below OECD average.

The analyses of PISA 2003, 2006, 2009 and UEE 1999, 2000, 2001, 2002 revealed that Turkey has the most heterogeneous schools of any OECD country. Consistent results from these national and international assessments bring attention to this challenge in the Turkish education system. This challenge seems to be the biggest factor widening science and mathematics achievement gap in Turkey. Moreover, this challenge brings to mind an important issue of educational equality among the citizens of Turkey.

Solution 1: Resolving Quality Differences in School Types

It is important to note that although there might not be big science and mathematics achievement gaps between public and private elementary schools, these gaps widen between high school types. Because of big differences in the quality of high school education in Turkey, science and mathematics achievement gaps between high schools become more obvious. For example, science and Anatolian high schools have tougher academic programs in science and mathematics and expect higher levels of achievement from students than other high schools do (OECD 2007b). Science and Anatolian high schools designate much more time for science and mathematics in their curricula, and teacher quality in these schools is better than it is in general high schools and vocational and technical high schools (Alacacı and Erbaş 2010). It is a policy in Turkey that best qualified and experienced teachers are selected and placed in science and Anatolian high schools (Alacacı and Erbaş 2010). In response to these challenges, it might be suggested that strong science and mathematics programs be enacted in all high schools. In addition, teacher quality across high schools could be more homogeneous. The examinations to select qualified teachers could be removed, and highly qualified and experienced teachers could be placed into high schools equally.

As a solution to narrow differences between high schools, MONE has started to decrease the number of school types at high school levels. In 2010, 350 general high schools were converted into Anatolian high schools, and by the end of 2013, all general high schools are going to be Anatolian high schools. In the near future, MONE is planning to convert Anatolian teacher high schools into Anatolian high schools. In the long run, the aim of MONE is to collect similar high schools under one umbrella and to narrow science and mathematics achievement gaps between high schools. In addition to unequal distribution of qualified and experienced teachers in schools, there is an unequal distribution of students in schools. Citizens often categorize schools based on their achievement in nationwide examinations. In other words, there are several prestigious and successful public schools in a typical Turkish city, and most parents want to enroll their children in these schools, but only a few parents can succeed. In order to hinder unequal distribution of the students at elementary schools, MONE enacted a rule that all elementary school students must enroll in an elementary school in their own living region. With this precaution, MONE tries to prevent the collection of all students in the same school who are successful or have high socioeconomic status.

Challenge 2: Competitive Nationwide Examinations

The Turkish education system is highly influenced by nationwide central examinations from elementary schools through doctorate admissions (Topcu 2011). Turkey has been conducting many central examinations annually since the 1980s. These examinations generally line, select, and place students into different kinds of schools. However, these examinations are rarely used to determine students' learning level and to what extent are the curriculum objectives being reached. On the other hand, nationwide examinations are reality for the Turkish education system because of the large young population in elementary and high schools. As stated before, approximately a quarter of the nation's population consists of elementary and high school students. In order to select and place elementary school students into high schools, central examinations seem an indispensable part of the Turkish education system. However, placing students into schools depending on the results of competitive examinations increases achievement gaps between high schools, in turn leading to discrimination at the school level (Berberoglu and Kalender 2005; Koseleci-Blanchy and Sasmaz 2011). Because there are big achievement gaps between high schools, most of the students want to enroll in successful high schools such as science and Anatolian high schools. In order to enter prestigious and successful universities, high schools types are important in Turkey. For example, most of the students from science high schools can enroll in any university program easily because they take high-quality education in these schools. Therefore, the HSEE holds an important place in students and parents' lives. Because this examination is performed by MONE each year, parents and teachers emphasize to students that effort and hard work are the most important contributors to a student's achievement (Topcu 2011). Accordingly, students take private lessons and participate in additional courses, beginning in sixth grade, to be successful in nationwide examinations. Unfortunately, overemphasizing these national examinations (e.g., HSEE and UEE), school teachers, administrations, and parents mostly want students to improve their abstract conceptualization and algorithmic problem solving, with little emphasis on analyzing or utilizing scientific knowledge in daily life.

Because not all students can enroll in high-quality high schools in Turkey, centralized examinations are compulsory in order to rank and select students. This problem also exists in the university entrance process. Approximately 2 million students apply for the UEE every year in Turkey. However, one out of five students gains a right to enroll in an undergraduate program in Turkey. More specifically, in 2011, 1,759,403 students took the UEE, and 350,911 of them were able to enter an undergraduate program (TurkStat 2012b). Similar to centralized HSEE, the number of students taking UEE is high; therefore, centralized UEE is compulsory for ranking and selecting students in the current Turkish education system. Although there are many criticisms of these nationwide examinations, the large number of applicants is a serious problem in placing these students in high schools or universities (Cakiroglu and Cakiroglu 2003). Moreover, the large number of university applicants makes the admission interview process difficult.

Solution 2: Decreasing the Number of Nationwide Examinations

National centralized examinations widen achievement gaps in science and mathematics (Berberoglu and Kalender 2005). Nowadays, MONE has focused on the challenge of the large number of nationwide examinations and has been trying to solve this challenge. The first concrete step was taken to resolve this challenge in 2010: the level placement examinations in sixth and seventh grades were removed. MONE will probably remove the nationwide examinations for eighth-grade students in the next few years. MONE is planning to eventually remove all nationwide examinations at the elementary school levels. Another plan developed by MONE is to use level placement tests only to obtain information about students' learning levels and curriculum effectiveness, not for the placement of the students into high schools. Also, decreasing only centralized examinations does not seem sufficient to overcome this challenge. Other precautions such as narrowing achievement gaps between schools should be taken by MONE. As stated before, MONE is collecting similar high schools under one umbrella. In addition, MONE is trying to provide equality for high schools, with equal distribution of qualified teachers and students in high schools. Decreasing only the number of nationwide examinations without providing equal education opportunities in high schools cannot overcome this challenge.

Similar to high school admission system, the challenge of nationwide examinations continues in the university admission system. As stated before, the only way to enter university programs in Turkey is to take the UEE. These examinations are highly competitive nationwide examinations that are mostly questioned by the Turkish universities. Most universities claimed that these examinations may not measure students' success in university education. In the last 10 years, the format of these examinations was changed by COHE more than four times; however, some problems still exist. Because there are approximately 2 million students entering UEE every year, it seems impossible to select qualified students for top-ranking universities in an effective way. In contrast with the HSSE, removing UEE seems impossible in the near future because there are large numbers of university applications. Yet the number of universities in Turkey was not at its expected level. To resolve this problem, the government and COHE increased the number of universities from 71 to 171 in the last 5 years. Parallel with this development, the quotas in the university programs are greatly increasing each year. In addition to increasing the number of universities, COHE takes care of the quality of Turkish universities. Similar to the efforts showed by MONE in order to narrow achievement gaps between high schools, COHE started to narrow achievement gaps between universities. For example, COHE has started the faculty development program OYP (Ogretim Uyesi Yetistirme Programı) in order to meet the needs of developing universities. In the OYP, academically qualified students graduating from university programs are assigned as research assistants in their faculty, and they are required to complete their master's and doctoral education at a given time. After they complete their master's and doctoral education in developed universities, they become faculty members in developing universities. In addition to human resources support given by COHE, the government gives financial and research supports to developing universities. For example, the government allocates increasing funds for developing universities. In addition, more research and development projects of these developing universities are supported by the scientific and technological research council of Turkey (TUBITAK, Turkiye Bilimsel ve Teknolojik Arastırma Kurumu). TUBITAK also provides common research databases to faculty members in libraries of all Turkish universities.

Challenge 3: Highly Standardized and Teacher-Centered Science and Mathematics Teaching

Despite the recent elementary and middle school science and mathematics curricula reforms, implementations of standardized and teacher-centered science and mathematics teaching in most of Turkish classrooms continue (Akpınar and Ergin 2005a). In standardized science and mathematics teaching, students understand science and mathematics as a body of knowledge discovered by scientists (Yilmaz-Tuzun and Topcu 2008), and teachers' role is to deliver this knowledge to students quickly due to the fixed schedule of a state-mandated curricula and central examinations (Topcu 2011). Teachers must complete state-mandated science and mathematics curricula in a fixed amount of time. Accordingly, teachers mostly encourage their students to improve their scientific or mathematics content knowledge and to perform well in nationwide examinations (Yalvac et al. 2007). Other dimensions of scientific or mathematics literacy such as understanding the nature of science and using prediction and mental processing skills effectively in mathematics are ignored in these classrooms.

Studies have explored the relationship between the structure of learning environments such as student- or teacher-centered environments and achievement in science and mathematics. Interestingly, according to the PISA 2003 results, although student-centered activities were negatively related to Turkish students' mathematics achievement, teacher-centered activities were positively related to Turkish students' mathematics achievement (Yayan and Berberoglu 2004). The Trends in International Mathematics and Science Study (TIMSS-1999) revealed that student-centered activities were negatively related to students' science achievement (Aypay et al. 2007; Berberoglu et al. 2003; Ozdemir 2003). Moreover, according to the national student assessment program results, although student-centered activities are negatively related to Turkish students' science achievement, teacher-centered activities are positively related to Turkish students' science achievement (Kalender and Berberoğlu 2009). The data derived from national and international assessments showed that Turkish students' science and mathematics achievements are positively related to teacher-centered activities in classrooms. In other words, while teacher-centered methods and activities are implemented, Turkish students seem to be more successful in science and mathematics. There is a big dilemma at this point because the ultimate aim of the recent reform movements in science and mathematics is to enhance science and mathematics achievement and to develop scientifically and mathematically literate citizens.

Recent curricula reforms in science and mathematics have been built on student-centered activities and constructivist philosophy. This dilemma can be explained in two different ways. The first is that teachers may not have sufficient information about the recent science and mathematics curricula. Therefore, they might not implement these reform movements in their classrooms at the expected level. Second, there could be an inconsistency or a gap between theory and practice in the Turkish educational system (Topcu 2011). Although theoretically many teachers mostly believe that traditional teaching strategies and memorization are insufficient for teaching students, these are the primary strategies and methods that they use in their classroom practices (Topcu 2011).

Solution 3: Curricula Reforms and In-Service Teacher Education

The recent science and mathematics curricula reforms have been built on constructivist philosophy and student-centered teaching instead of highly standardized and teacher-centered teaching (MONE 2006, 2009a). However, highly standardized and teacher-centered teaching is still implemented by many teachers in the context of science and mathematics curricula (Ozgun-Koca and Sen 2002; Yilmaz-Tuzun and Topcu 2008). There are four main categories of reasons for this continuing problem: completion of state-mandated science and mathematics curricula in a fixed amount of time, competition of nationwide examinations, lack of in-service education, and lack of understanding of the reformed science and mathematics curricula.

Both the reformed science and mathematics curricula are kind of spiral curricula (MONE 2006, 2009a). The same science or mathematics subjects in these curricula are becoming deeper every year. For example, whereas the subject of "congruence and similarity" is placed superficially in the sixth- and seventh-grade mathematics curricula, the same subject takes place in a more detailed way in the eighth-grade curriculum. Although the spiral curriculum could help students remember previous related subjects better, this curriculum could take much time for teaching subjects. Many science and mathematics teachers seem to have difficulty completing state-mandated curricula in a fixed amount of time (Akpınar and Ergin 2005b; Simsek 2004). However, teachers must complete these curricula in a fixed amount of time because teachers are controlled regularly by school administrators, education inspectors, and even by parents. Actually, simultaneously teaching previous and new subjects is difficult (Schmidt et al. 2002). At this point, MONE could revise the reformed science and mathematics curricula. Instead of repeating the subjects in previous years, some necessary prior knowledge might be given to students by teachers, and then new subjects might be taught to students (Schmidt et al. 2002). Additionally, the number of the subjects in the current science and mathematics curricula might be decreased. By teaching fewer subjects, the subjects that are taught will be learned more thoroughly, and teachers will have sufficient time to complete a state-mandated curriculum in a fixed time. Another reason for using highly standardized and teacher-centered teaching by teachers might be competitive nationwide examinations. Many teachers highlight these examinations while they teach subjects in the curricula. Instead of focusing on how deeply students learn science or mathematics concepts and processes or to what extent curricula aims have been reached, teachers mostly focus on nationwide examinations. The solutions to this problem in Turkish education system have been given in the section of "Decreasing the number of nationwide examinations."

Although the science and mathematics curricula reforms have been implemented since 2006, many science and mathematics teachers still do not have sufficient information and experience regarding the reformed curricula (Bozdogan and Altuncekic 2007; Sahin 2007). Science and mathematics teachers need to take in-service education regarding how to implement these curricula in their classrooms (Akpınar and Ergin 2005b; Bozdogan and Altuncekic 2007). In 2004 and 2005, a few cities were selected for a pilot study of these curricular implementations. Only teachers from these cities experienced these curricular reforms. After 1 year, these reformed curricula were implemented across the country. When implementation of the reformed curricula began, many science and mathematics teachers did not have any experience with these curricula. Science and mathematics teachers seem to have a lack of knowledge about how to implement these curricula in classrooms. Knowledge and proficiencies possessed by teachers may not translate into teaching practice (Dincer and Uysal 2010). For example, the reformed curricula suggested that teachers reflect constructivism in their classroom activities and that student-centered activities should be developed by teachers. In addition, instead of using product-based assessments, teachers should use process-based assessments according to the curricula. These suggestions have been given to teachers theoretically, but practically, teachers mostly do not have sufficient information and experience about how they will implement these suggestions in their classrooms. Although textbooks, student workbooks, and teachers' guides were given to science and mathematics teachers, in-service teacher education programs did not adequately explain to teachers how to use these materials.

The PISA 2006 and 2009 results suggested that the reformed science and mathematics curricula were not implemented successfully across the country. Despite the recent reformed science and mathematics curricula, Turkish students' science and mathematics achievement average scores in PISA 2006 and 2009 were quite below the international science and mathematics average (OECD 2007a, 2010a). At this point, it might be suggested that MONE should have given more importance to in-service teacher education programs across the country. In these programs, the current problems in science and mathematics teaching can be discussed by teachers systematically. For example, these programs should discuss why Turkish students' science and mathematics achievements are so low. In addition to explaining the vision and the aims of the reformed science and mathematics curricula, how teachers successfully implement these curricula could be discussed with teachers both theoretically and practically. As a last solution, MONE should reevaluate teacher placement and assessment processes in the Turkish education system. In Turkey, science and mathematics teachers receive the same promotions and salaries regardless of their performance in public schools. In addition, in a teacher contract, no items relate to teaching performance. MONE renews public school teachers' contracts regardless of their success in teaching. Regarding the teacher placement and assessment processes, a performance evaluation system should regularly assess teachers' performance, especially in public schools. Depending on their performance in science or mathematics teaching, they would receive promotions and salary increase. This system works well for medical doctors in the Turkish health system.

Challenge 4: The Effects of Large Socioeconomic Background Differences on Science and Mathematics Achievement

Many studies including national and international examination results (e.g., UEE and PISA) show that Turkish elementary and middle school students' socioeconomic backgrounds were related to their science and mathematics achievement. The PISA index of economic, social and cultural status (ESCS) is a composite measure of socioeconomic status that includes parent education and occupational status, family wealth, home educational resources, and possessions related to "classical" culture in the family home (OECD 2002). The PISA 2009 results found a high correlation between students' socioeconomic background and their science and mathematics achievement, and 19% of the variance between student achievements was explained by the ESCS (OECD 2010b). Similarly, based on the PISA 2006 results, Alacacı and Erbas (2010) found that the ESCS index was related to Turkish students' mathematics performance.

Turkish students in PISA 2006 mostly came from the lowest levels of socioeconomic background compared to students of other OECD countries. The effect of level of ESCS on student achievement was higher in Turkey than in any other OECD country (OECD 2007a). Dincer and Uysal (2010) also studied the PISA 2006 results and found that socioeconomic background has a large effect on science achievement. Similarly, Yayan and Berberoglu (2004) focused on the PISA 2003 results and found that family background characteristics were important determinants of Turkish students' mathematics achievement. Aypay et al. (2007) analyzed another international exam, TIMSS-1999, and concluded that socioeconomic background strongly affects eighth-grade Turkish students' science achievement. Guncer and Kose (1993) studied the UEE and found that family background explained a large part of the variance (40%) of high school students' academic achievement. Engin-Demir (2009) investigated relationships between family background and mathematics and science scores in middle school students. She found that family background characteristics were related to school achievement, and family background characteristics explained 5.4% of the variance in school achievement.

Based on the ESCS index, schools were classified into three groups: advantaged, disadvantaged, and mixed (OECD 2010b). Turkey is one of the few OECD countries in which the number of socioeconomically mixed schools is low (OECD 2010b). Students seem to cluster in schools based on their socioeconomic background (Alacacı and Erbas 2010; Dincer and Uysal 2010). In other words, students coming from similar socioeconomic backgrounds are generally grouped into the same public or private schools (Alacacı and Erbas 2010; Dincer and Uysal 2010). This condition increases achievement gaps in the Turkish education system combined with school inequalities. For example, whereas high-socioeconomic-status parents enroll their children in private elementary schools, middle- or low-socioeconomic-status parents enroll their children in public elementary schools. Moreover, high-socioeconomic-status parents make their children take private lessons and additional courses given by teachers. As a result, differences in socioeconomic status and school types widen achievement gaps.

As stated before, placing students into schools based on the national examinations increases students' achievement gaps (Berberoglu and Kalender 2005) and quality differences between school types (Koseleci-Blanchy and Sasmaz 2011). This situation also increases competitions in examinations and intensifies the role of socioeconomic background on student achievement. The PISA 2009 results show that schools discriminate on the basis of socioeconomic background, which influences the gap in student achievements (OECD 2010b). Based on the average ESCS index, "while 64% of the children from the bottom quartile [of socioeconomic background] attend disadvantaged schools, only seven percent are able to attend advantaged schools" (Koseleci-Blanchy and Sasmaz 2011, p. 131). Consistent with the ESCS index results, the Gini index measuring unequal distribution of income among individuals or households within an economy showed that Turkey is one of the top three OECD countries when it comes to income inequality (OECD 2007a). Income inequalities are large, and the impact of socioeconomic background on learning outcomes is also large (OECD 2010b). The PISA 2009 results for Turkey revealed that "more than half of all students come from a socio-economic background below that of the least-advantaged 15% of students in the OECD countries" (OECD 2010b, p. 62). Moreover, the proportion of socioeconomically disadvantaged children in Turkey is high compared to most OECD countries, and 58% of Turkish students belong to the internationally most disadvantaged group (OECD 2010a). Therefore, socioeconomic segregation in Turkish schools and the effects of socioeconomic background on Turkish students' performance appear large, and the problem of inequality in education among students continues.

Solution 4: Equality in School Types and Preschool education

One of the most important aims of the recent science and mathematics curricula reforms in Turkey is to improve equality and access to education by all citizens (MONE 2006, 2009a). Accordingly, all citizens should benefit equally from educational opportunities and should gain basic skills and competencies through education in order to understand and resolve problems and issues in social life. MONE has the largest responsibility to achieve these aims in the Turkish education system. One of the most important tasks of MONE is to minimize the effects of socioeconomic background and school types on student achievement. As expected, MONE is developing strategies to overcome inequalities among high school types. For example, MONE is working on unifying similar high schools under one umbrella and decreasing the number of high school types. In another attempt, supporting disadvantaged students (e.g., through scholarships and alleviating the factor of coming from a low socioeconomic background), MONE has tried to overcome inequality in education (Koseleci-Blanchy and Sasmaz 2011).

Socioeconomic segregation at the school level is such a reality in Turkey that MONE took precautions to overcome this problem in 2010. To reduce the effect of socioeconomic background on student achievement, high-quality preschool education has been provided for children whose parents volunteer to take this education. According to the PISA 2009 results, "the average scores of students who received a preschool education of one year or less are 42 points higher than the average scores of those who never attended preschool" (Koseleci-Blanchy and Sasmaz 2011, p. 132). However, "pre-primary education is rare in Turkey, where less than 30% of 15-year-olds attended pre-primary education for any period of time" (OECD 2010b, p. 96). Accordingly, MONE has recently accelerated the spread of preschool education across the country in order to resolve the problem of large differences in socioeconomic backgrounds. The number of preschools is continually increasing, and parallel with this development, parents' interest in preschool education is also greatly increasing.

Turkey's Performance in the Programme for International Student Assessment (PISA): Increasing or Decreasing

There is a need to understand the effectiveness of the recent science and mathematics curricula reforms in Turkey. At this point, student assessment programs and PISA could be useful for determining curriculum effectiveness in terms of national and international perspectives. These examinations periodically collect data about students' science and mathematics achievement, school types, and socioeconomic status. As one of the international student achievement programs conducted within OECD countries, PISA not only focuses on targeted objectives and content knowledge in the curricula, but also assesses students' ability for using knowledge in real-life situations (Koseleci-Blanchy and Sasmaz 2011). PISA is implemented with 15-year-old students every 3 years, and this program deals with the way students utilize their knowledge and skills in order to interpret and resolve scientific and mathematical issues closely related to society.

Turkey first participated in PISA in 2003, and since then, Turkey has participated in all PISA. Therefore, it is possible to interpret Turkey's science and mathematics achievement development in light of the PISA results. Turkey showed improvements in PISA 2009 compared with PISA 2006 in terms of science and mathematics achievement. Whereas the science mean score increased from 424 points in 2006 to 456 in 2009, the mathematics mean score increased from 424 in 2006 to 445 in 2009 (OECD 2007a, 2010a). Two factors were explained by Koseleci-Blanchy and Sasmaz (2011) that might be effective in Turkish students' improved performance. The first one is that the recent science and mathematics curricula reforms might have influenced students' science and mathematics performance in a positive way. The second one is that increasing awareness of PISA implementation across the country might have motivated students, parents, and school administrations. In other words, this awareness might have influenced students' science and mathematics achievement in a positive way. Despite this progress, Turkey's overall average scores are still below the OECD average. While Turkey was 29th among 30 OECD countries in PISA 2006, it was 32nd among 34 OECD countries in PISA 2009 (OECD 2007a, 2010a). Although Turkey showed national improvements, these improvements did not reflect in international rankings. These results suggest that most of the OECD countries improved their science and mathematics achievements compared to previous years. It is also necessary to note that in Turkey, "the proportion of students not enrolled at school by the age of 15 is around 35 % while it is below 10 % in all other OECD countries" (OECD 2010b, p. 41). Considering the proportion of students who are not enrolled in school, it might be claimed that Turkey's average performance could be worse than what was reflected in PISA results.

PISA also gives us important data about inequality in the Turkish education system and gaps in science and mathematics achievement. One of the indicators showing inequality in an education system might be the number of students not reaching basic competency level in science and mathematics (Koseleci-Blanchy and Sasmaz 2011). Students who have not reached this level can be assessed as not having gained necessary skills to actively take part in society (Koseleci-Blanchy and Sasmaz 2011). The large number of students not reaching basic competency level suggests inequality in the education system. It is good news for Turkey that the number of students reaching basic competency level increased in PISA 2009 compared with PISA 2006 in both science and mathematics (OECD 2007a, 2010a). According to mathematics literacy test results, the number of students scoring under the basic competency level decreased from 52 % in PISA 2006 to 42 % in PISA 2009 (OECD 2007a, 2010a). Similarly, based on science literacy test results, the number of the students scoring under the basic competency level decreased from 46 % in PISA 2006 to 30 % in PISA 2009 (OECD 2007a, 2010a). Despite these improvements in both science and mathematics achievement, the percentage of Turkish students remaining under OECD average is still high compared to students from other OECD countries.

Concluding Remarks and Suggestions

The aim of this chapter was to present and analyze science and mathematics achievement gaps and their causes in the context of the Turkish education system. In light of both national (e.g., HSEE) and international (e.g., PISA) indicators, it was concluded that large achievement gaps in science and mathematics exist in Turkey. It is important to determine reasons or challenges pioneering achievement gaps and to resolve these challenges in their own context. Although there are many challenges in the area of science and mathematics achievement in the current Turkish education system, the most distinct challenges could be categorized under four titles: quality differences in school types, competitive nationwide examinations, standardized and teacher-centered science and mathematics teaching from elementary school through college, and the effects of socioeconomic background differences on science and mathematics achievement.

These challenges are all connected. One challenge might be the reason or the result of another challenge. Therefore, if we want to resolve these challenges, we should explore relationship patterns between these challenges and resolve these challenges together. MONE tries to resolve these challenges separately, but a systematic approach is necessary to resolve these challenges together. For example, MONE is trying to equalize high schools, collecting similar high schools under one umbrella. This is an important step to overcome the challenge of large differences in school types, but it is not sufficient because the effects of nationwide examinations on segregation of schools still continue. Instead of testing, ranking, and selecting, these examinations should focus on how much students learn and to what extent curricula aims have been reached. If these nationwide examinations continue to rank and select students, only decreasing the number of high school types and collecting similar high schools under one umbrella could be meaningless because students will want to enroll in the same successful schools whether or not the names of these schools change. While MONE tries to equalize school types, the quality of teachers and students should also be considered. If student and teacher quality is not balanced between schools, the effort to provide equality between schools may not work properly.

One of the most important challenges in the Turkish education system seems to be that many Turkish teachers continue traditional teaching methods despite the recent science and mathematics curricula reforms. Regarding this challenge, it might be suggested that quantity and quality of in-service training could be improved by MONE, and science and mathematics teachers could learn more methods of teaching (especially related to the recent science and mathematics curricula) in these trainings. Also, MONE and COHE should reconsider preservice teacher education and teacher selection and placement systems. Teachers should be highly trained in teacher education programs because highly educated and motivated teachers are indispensable parts of successful education systems. During their teacher education, teachers should develop their pedagogical content knowledge in science and mathematics. Also, research-based teacher education could be suggested for teachers. While they improve their pedagogical content knowledge, they can develop their research skills and practices to follow new teaching approaches and curricular trends. For instance, MONE could encourage teachers to earn a master's degree in science or mathematics education. MONE also should develop strategies pioneering teaching as an attractive profession in comparison to other professions. Unfortunately, teachers have low salaries compared to other professions, and with each passing day, the popularity of teaching as a profession among other professions is decreasing.

Although changing the effects of socioeconomic background on students' achievement gaps does not appear to be an easy task to achieve in the near future, MONE can develop strategies to overcome this challenge. For example, more budget and resources could be separated for education. Disadvantaged schools and students could be supported by the Turkish government. It is important to note that in order to be successful in narrowing the achievement gaps in the Turkish education system, resources should be used more effectively. In 2012, only 2.75 % of the gross domestic product (GDP) was spent on education. Although Turkish students have low science and mathematics achievement scores in PISA, the percentage of 2.75 is low compared to other OECD countries. For example, even though Finnish students have high science and mathematics achievement scores, 6 % of their GDP goes to education. Therefore, perhaps the Turkish government should allocate more budget to education and support investments in educational resources. With this increasing budget and investments, teacher quality and school environment could be improved, which would increase students' academic achievement in science and mathematics. In addition to financial and investment issues, there is a complex bureaucracy in MONE. The Turkish educational system is managed by one center—MONE.

Because many important decisions about teachers' placement in schools and other employment rights are managed by this center, the bureaucracy is becoming more complex for teachers. By emphasizing decentralization, MONE could resolve the problem of complex bureaucracy. Therefore, highly motivated, autonomous, and successful teachers might be the key for successful Turkish students in the future.

References

Akpınar, E., & Ergin, O. (2005a). Yapılandırmacı kurama dayalı fen öğretimine yönelik bir uygulama. *Hacettepe Üniversitesi Eğitim Fakültesi Dergisi, 29,* 9–17.

Akpınar, E., & Ergin, O. (2005b). The role of science teacher in constructivist theory. *Elementary Education Online, 4*(2), 55–64.

Alacacı, C., & Erbas, A. K. (2010). Unpacking the inequality among Turkish schools: Findings from PISA 2006. *International Journal of Educational Development, 30*(2), 182–192.

Aypay, A., Erdogan, M., & Sozer, M. A. (2007). Variation among schools on classroom practices in science based on TIMSS-1999 in Turkey. *Journal of Research in Science Teaching, 44*(10), 1417–1435.

Berberoglu, G., Celebi, O., Ozdemir, E., Uysal, E., & Yayan, B. (2003). Factors affecting achievement levels of Turkish students in the Third International Mathematics and Science Study (TIMSS). *Educational Sciences and Practice, 2,* 3–14.

Berberoglu, G., & Kalender, I. (2005). Investigation of student achievement across years, school types and regions: The SSE and PISA analyses. *Educational Sciences and Practice, 4*(7), 21–35.

Bozdogan, A. E., & Altuncekic, A. (2007). Fen Bilgisi Öğretmen Adaylarının 5E Öğretim Modelinin Kullanılabilirliği Hakkındaki Görüşleri. *Kastamonu Eğitim Dergisi, 15*(2), 579–590.

Cakiroglu, E., & Cakiroglu, J. (2003). Reflections on teacher education in Turkey. *European Journal of Teacher Education, 26*(2), 253–264.

Council of Higher Education (COHE). (2012). Universities. July 2012, http://www.yok.gov.tr/en/content/view/527/222/.

Delen, E., & Bulut, O. (2011). The relationship between students' exposure to technology and their achievement in science and math. *Turkish Online Journal of Educational Technology, 10*(3), 311–317.

Dincer, M. A., & Uysal, G. (2010). The determinants of student achievement in Turkey. *International Journal of Educational Development, 30*(6), 592–598.

Engin-Demir, C. (2009). Factors influencing the academic achievement of the Turkish urban poor. *International Journal of Educational Development, 29,* 17–29.

Guncer, B., & Kose, M. R. (1993). Effects of family and school on Turkish students' academic performance. *Education and Society, 11*(1), 51–62.

Kalender, I., & Berberoglu, G. (2009). An assessment of factors related to science achievement of Turkish students. *International Journal of Science Education, 31*(10), 1379–1394.

Koseleci-Blanchy, N., & Sasmaz, A. (2011). PISA 2009: Where does Turkey stand? *Turkish Policy Quarterly, 10*(2), 125–134.

Ministry of National Education of Turkey (MONE). (2004). *National report on PISA 2003.* Directorate of Educational Research and Development, Ankara, Turkey.

Ministry of National Education of Turkey (MONE). (2006). *Science and technology curriculum of elementary schools (6th–8th grades).* Board of Education.

Ministry of National Education of Turkey (MONE). (2009a). *Mathematics curriculum of elementary schools (6th–8th grades).* Board of Education.

Ministry of National Education of Turkey (MONE). (2009b). *Statistics on the level placement examination 2009.* General Directorate of Educational Technologies, Ankara, Turkey.

Ministry of National Education of Turkey (MONE). (2010). *Statistics on the level placement examination 2010*. General Directorate of Educational Technologies, Ankara, Turkey.

Organization for Economic Co-operation and Development (OECD). (2002). *Education at a Glance, Glossary*, Paris.

Organization for Economic Co-operation and Development (OECD). (2004). *Learning for tomorrow's world. First results from PISA 2003*. Paris.

Organization for Economic Co-operation and Development (OECD). (2007a). *PISA 2006: Science competencies for tomorrow's world*. Paris.

Organization for Economic Co-operation and Development (OECD). (2007b). *Reviews of national policies for education: Basic education in Turkey*. Paris.

Organization for Economic Co-operation and Development (OECD). (2010a). *PISA 2009 results: What students know and can do–Student performance in reading, mathematics and science* (Vol. I). Paris.

Organization for Economic Co-operation and Development (OECD). (2010b). *PISA 2009 results: Overcoming Social Background–Equity in Learning Opportunities and Outcomes* (Vol. II). Paris.

Ozdemir, E. (2003). *Modeling of the factors affecting science achievement of eighth grade Turkish students based on the third international mathematics and science study repeat (TIMSS-R) data*. Unpublished master's thesis, Middle East Technical University, Ankara, Turkey.

Ozgun-Koca, A., & Sen, A. I. (2002). Evaluation of the results of the third international mathematics and science study for Turkey. *Hacettepe Üniversitesi Eğitim Fakültesi Dergisi, 43*, 145–154.

Sahin, I. (2007). Assessment of new Turkish curriculum for grade 1 to 5. *Elementary Education Online, 6*(2), 284–304.

Schmidt, W., Houang, R., & Cogan, L. (2002). A coherent curriculum: The case of mathematics. *American Educator, Summer*, 1–17.

Simsek, N. (2004). A critical approach to constructive learning and instruction. *Educational Sciences and Practice, 3*(5), 115–139.

Topcu, M. S. (2011). Turkish elementary student teachers' epistemological beliefs and moral reasoning. *European Journal of Teacher Education, 34*(1), 99–125.

Turkish Statistical Institute (TurkStat). (2012a). *Number of schools, teachers, divisions, students and graduates by type of school and educational year*. Turkish Statistical Institute, Ankara.

Turkish Statistical Institute (TurkStat). (2012b). *Number of applicants and appointed to tertiary education by school type and educational status*. Turkish Statistical Institute, Ankara.

Yalvac, B., Tekkaya, C., Cakiroglu, J., & Kahyaoglu, E. (2007). Turkish pre-service science teachers' views on science-technology-society issues. *International Journal of Science Education, 29*(3), 331–348.

Yayan, B., & Berberoglu, G. (2004). A re-analysis of the TIMSS-1999 mathematics assessment data of the Turkish students. *Studies in Educational Evaluation, 30*, 87–104.

Yilmaz-Tuzun, O., & Topcu, M. S. (2008). Relationships among preservice science teachers' epistemological beliefs, epistemological world views, and self-efficacy beliefs. *International Journal of Science Education, 30*(1), 65–85.

Part V
Asia

Chapter 10
Achievement Gap in China

Gaoming Zhang and Yong Zhao

Achievement gap in China has received little attention from outside observers, who have typically focused on the stunning academic achievement of Chinese students in various international assessments (Asia Society 2006; Organisation for Economic Co-operation and Development (OECD) 2011a; Tucker 2011). For example, in the Programme for International Student Assessment (PISA) in 2009, Chinese students from Shanghai, one of the most developed cities, outperformed all other 60 countries in all three areas—mathematics, science, and reading, with 12.3% of variance in student performance explained by their socioeconomic background. China achieved a better equity score than the USA and the average OECD countries (OECD 2011a).

But this is not a representative picture of reality in China. The fact is that uneven economic development and unfair social policies have created an ever-widening gap in terms of education achievement and opportunity among Chinese students, although the gap takes different forms from that in the USA or other countries. In this chapter, we present data to show the widening achievement and opportunity gaps in China and discuss its causes. We also summarize the strategies the Chinese government has undertaken to reduce educational inequality.

G. Zhang (✉)
Department of Teacher Education, School of Education,
University of Indianapolis, Indianapolis, IN, USA
e-mail: zhangg@uindy.edu

Y. Zhao
College of Education, University of Oregon, Eugene, OR, USA
e-mail: yongzhao.uo@gmail.com

J. V. Clark (ed.), *Closing the Achievement Gap from an International Perspective,*
DOI 10.1007/978-94-007-4357-1_10, © Springer Science+Business Media B.V. 2014

Searching for Indicators of Educational Inequality in China

In the USA, the achievement gap is indicated by a multitude of measures such as grades, course selection, dropout rates, standardized test scores such as the National Assessment of Educational Progress (NAEP), and state-administered tests mandated by No Child Left Behind Act (NCLB). But such data are not generally available on a national level, and there are no longitudinal national-wide or even statewide assessment programs that allow the disaggregation of data for individual students or subgroups. In China, the most reliable national indicator of achievement is perhaps college attendance, in terms of both percentage of students attending college and the types of college attended. The achievement gap is accordingly indicated by the results of the national entrance exam to college (i.e., *Gaokao*) among different groups of students.

The *Gaokao* is a national exam held annually in June[1]. Although the requirements and administration of the *Gaokao* vary across different areas in the country, there are some general requirements. First, usually high school seniors take the exam. Second, the *Gaokao* consists of a set of exams, and it takes 2–3 days to administer the set of exams. The mandatory subjects are Chinese, mathematics, and a foreign language (usually English). Applicants also need to choose between the tracks for their future undergraduate education (i.e., science/engineering and art/humanities) and take additional tests that are made for the track they select. The *Gaokao* is not a national unified examination. Students in different areas in the country may take different sets of exams, and they are graded variously across the country. However, the administration is highly controlled by the Ministry of Education (MOE), and by the education administration department of each relevant province or municipality directly under the Central Government.

Currently, depending on the location, students are required to rank their preferences of higher education institutions and programs at one of the following three points of time: 1) prior to the exam, 2) after taking the exam but before learning their scores, or 3) after taking the exam and learning their scores. The preferences are given in several tiers including early admissions, key universities, ordinary universities, and 2-year technical colleges. For each tier, students can list four to six choices of institutions and programs with ranked priority.

There are historical and cultural factors responsible for the domination of the *Gaokao* (Zhao 2012). The antecedent to the *Gaokao*, the *keju* (meaning the Imperial Exam or Civil Exam), was well developed in the Tang dynasty (AD 618–907) and had been used throughout the history of China until it was abolished in 1905. In its life span of over 1,300 years, the *keju* was the only measure of achievement in the society. Test takers had to take a set of hierarchically organized exams. Only by passing one exam could a candidate go to the next more advanced exam. The exams were highly competitive. The final exam was open to a very limited number of par-

---

[1] The *Gaokao* was held in July prior to 2003 and was rescheduled to June due to the excessive heat in major areas in July.

ticipants and was administered by the emperor in the capital city. Top performers in the *keju*, especially in the final round, were granted a high position in government. For many people, the *keju* was the only path of upward mobility. When the *Gaokao* was adopted in 1952 by the newly founded People's Republic of China, it was very much alike a modern version of the *keju*. These exams are the only widely recognized measure of achievement in the society. In addition, both exams are highly competitive and serve as almost the only route to upward social mobility. As a result, the pressure to prepare for and take the exams is extremely high.

Even when additional measures are used in K–12 education in China, these measures are valued only when they can have some impact on students' scores in the *Gaokao* (maybe in the long run). The greater the impact is, the more valuable the measures are; for example, talents in arts and sports are remotely considered as relevant to achievement since usually these talents would not boost students' scores in the *Gaokao*. However, in some extreme circumstances, such as ranking at a top 6 position in a sports competition at a provincial or higher level (e.g., national and international competitions), students will be granted some bonus points for their scores in the *Gaokao*. Still, these measures are widely considered as extensions of the essential measure of achievement in China (viz., scores in the *Gaokao*), not as independent achievement measures.

China's over 2,000 higher education institutions are hierarchically organized into four groups according to their status, which are determined by academic and political factors. Sitting on the top are the 75 universities under the direct administration of the Chinese MOE (MOE 2012). Within these most prestigious universities are 31 ministerial-level universities, whose top leadership is directly appointed and managed by the central government and accorded the status of vice ministers in the bureaucracy. The second tier consists of about 1,000 universities and colleges administered by provincial-level governments. The third tier includes private higher education institutions that are approved by the MOE to offer bachelor degrees. The last group consists of tertiary-level vocational and technical education institutions that offer the equivalent of associate degrees.

The status of the university not only determines the resources it receives which indicates the quality of education it offers, but also affects the future of its graduates because many employers make it an explicit requirement to hire graduates from certain groups of universities. For example, one of the minimal qualifications to apply for a teaching position in a school in Fujian province is a "masters degree from ordinary universities or bachelors degree from normal universities under the direct administration of the Ministry of Education," which in essence means that a bachelor's degree from a Ministry-administered institution is considered equivalent to a master's degree from other institutions (Fujian Quanzhoushi Jiaoyuju [Education Bureau of Quanzhou City Fujian Province] 2012). In other words, the status of a university influences the value of the degree in China much more than it does in the USA. As a result, students work extremely hard to get into higher-status universities.

Thus ultimately students' academic achievement is measured by whether they gain entrance to a higher education institution and by the type of institution they attend. In other words, because the *Gaokao* is used as the only measure of achievement in China, the achievement gap inevitably revolves around the results of the *Gaokao* (i.e., the acceptance rate and the enrollment rate in higher education for different groups of students).

The Achievement Gap

Because China is an ethnically homogenous nation, with over 90 % of its population being Han, the achievement gap does not generally concern race and ethnicity. It has more to do with socioeconomic background and geographical location. Because systematic socioeconomic data that would allow detailed analysis of the gaps are not available nationally, the rural–urban divide becomes the common measure.

The rural–urban divide is not simply a location of residence in China but a matter of privileges and rights awarded by the government based on heritage using a system known as *hukou* or *huji* (residency management system). All individuals in China are designated an urban (nonagricultural) or rural (agricultural) status after being born depending on the parents' residency status. Rural and urban residents have access to different privileges and rights. Rural residents, for example, cannot hold government positions or teach in a university. Their children are not allowed to attend schools or participate in the *Gaokao* in the city, although they may have lived there for decades as migrant workers. Thus, typically rural residents enjoy much lower socioeconomic status than urban residents.

On January 4, 2009, Chinese Premier Wen Jiabao shared his great concerns over the urban–rural disparity in access to higher education in his speech at a meeting with scientists and educational leaders from various areas across the country: "When I went to college, 80 %, perhaps higher than that, of my classmates were from rural areas. Now it is different. The percentage of students from rural areas in higher education institutions has plunged."

Wen's observation on the rural–urban disparity in access to higher education is supported by a large body of research literature (Gou 2006; Qiao 2010; Xie and Li 2000). For example, Gou (2006) compiled data of acceptance rate for urban examinees and rural examinees (including first-time examinees and non-first-time examinees) and compared the rates with the acceptance rate for all examinees. As shown in Table 10.1, the acceptance rate of current rural students has always been lower than the overall acceptance rate. Furthermore, the number of rural examinees registered for the *Gaokao* and the enrollment rate of rural examinees never attained the proportion of rural residents in the national population during all of these years (Gou, 2006).

The acceptance rate can be deceiving because it is based on the number of students participating in the *Gaokao*, which takes place at the end of high school. Since China's compulsory education ends at 9th grade, not all students continue their education after that. There are many more students in urban than in rural areas continuing their education. According to statistics from the MOE, 35 % of high school freshmen in 2009 were from urban areas, whereas only 7 % were from rural areas.

Table 10.1 The *Gaokao* examinees' acceptance rate by urban–rural area (selected years between 1989 and 2005). (The table is created based on statistics from multiple tables (Gou 2006))

Year	Acceptance rate (%)							Proportion of rural residents in the national population (%)
	Urban	Urban first-time takers	Urban non-first-time takers	Rural	Rural first-time takers	Rural non-first-time takers	All	
1989	30.4	32	25.7	18.7	17.5	21	23.7	n/a
1990	30.2	29.5	31.7	18.4	16.5	21.4	23.1	n/a
1996	46.6	46.6	46.5	33.3	30.9	37	38.9	69.52
1997	43	43.4	41.7	33.5	29	41.3	37.7	68.09
1998	42.1	40.4	47.2	32.1	29.4	37.2	36.6	66.65
1999	53.8	52	58.9	41.3	36	50.3	47.1	65.22
2000	60.1	58.8	65	54.1	48.9	66.1	57.1	63.78
2001	65.3	63.8	71.6	58.8	54.1	72.1	62.1	62.34
2002	69.1	67.3	75.7	62.9	57.5	76.7	66	60.91
2003	72.5	71.1	78.2	65.9	61.6	78.8	69.1	59.47
2004	73.7	72.4	79.4	68.8	64.7	81.4	71	58.24
2005	68.4	66.5	76.8	62.9	58.3	78	65.4	57.01

In the 2009 high school graduating class, 35% were from urban areas, whereas about 8% were from rural areas (Hu and Zhu 2011).

Although it is widely recognized that urban–rural divide may be the biggest factor responsible for the achievement gap in China, there are mixed findings of the impact of higher education expansion on the urban–rural disparity. The expansion of higher education refers to the period between 1999 and 2005, when higher education enrollment increased at an annual rate of 20%. As a result, the enrollment in higher education in 2003 was 4.7 times higher than that in 1998 (Li 2010). The studies on the impact of the expansion on the achievement gap between urban and rural areas, however, have mixed findings. Some studies conclude that the gap between urban and rural access to higher education opportunities consistently shrank after the expansion (Gou 2006; Qiao 2010; Yang 2006; Yuan 2007). Qiao (2010) reported that the disparity in rural and urban enrollment rates for the entrance exam fell from 13.28% to 5.46% and the index for the rural–urban disparity fell from 3.199 to 1.44. In 1999, the number of enrolled rural students (844,700) exceeded that of enrolled urban students (744,000) by 100,700, and the gap seemed to further grow in recent years. In 2005, approximately 2,692,700 urban students were enrolled, whereas 3,038,100 rural students were enrolled. Rural students at that time exceeded urban students by 345,400 (Gou 2006).

However, the measures used in these studies have caveats and may result in an incomplete picture of the recent trends of the urban–rural disparity in access to higher education. Both the enrollment rate and enrollment number do not take into consideration rural residents who have not registered for the exam. In addition, the growth of enrollment does not count the proportion of urban and rural residents in the national population. Li (2010) argued that a better measure for the urban–rural disparity in access to higher education is randomly selected samples from data sets for censuses. Li randomly selected 19,615 samples, 1% of the target population for people who were born between 1975 and 1985. Li then compared access to higher education between the group of people who were born between 1975 and 1979 and the group born between 1980 and 1985. The 1980–1985 cohort was used to represent the population who just started to experience the expansion of higher education, and the 1975–1979 group represented the population before the expansion. Li used logistic regression models to analyze the impact of rural/urban residency on access to higher education. The study presented important findings. First, in general, the odds of urban students receiving higher education were 6.3 times higher than those for rural students. Second, the 1980–1985 group had easier access to higher education, with 30% higher chances than the 1975–1979 group. This means the expansion between 1999 and 2005 had greatly promoted the opportunities for higher education in general. The third and probably most significant finding is that the urban–rural gap of access to higher education has been widening after the expansion of higher education. The comparison of the two groups suggests that the index for urban access to higher education opportunities was 3.4 times greater than the rural index before the expansion and 5.5 times greater than the rural index after the expansion (Li 2010).

The gap becomes wider when the status of the institutions is considered. Although no national data are available, reports suggest a decrease in the percentage of students from rural areas enrolled in prestigious universities under the direct administration of the MOE. For example, in Tsinghua University, one of the most prestigious universities, only 17.6% of its freshman class in the year 2000 were from rural areas, a 4.1% decrease from a decade ago. Similar distribution remained in 2010. In Peking University, another prestigious university, the percentage of rural students was 16.3% in 1999, a decrease of 2.5% from 1991 (Hu and Zhu 2011). Another report shows that only 6.8% of high school graduates in Enshi, a rural prefecture of Hubei Province, received a score above the cut scores of Tier One universities in 2011, whereas the percentage was 20% for Wuhan, the capital city of Hubei Province, and 24% for Beijing (Tian 2012).

Behind the Achievement Gap

The achievement gap in China is the product of economic, political, social, and cultural factors. Some of the factors, such as economic factors, are universal. Other factors, such as the quota system in the *Gaokao*, local-residency requirement (i.e., the *hukou*), and school choice, are unique features to China.

Universal Factors: Economic Factors

Economic factors are immediately associated with school finance and availability of educational resources and have been widely recognized for their impact on the achievement gap, and China is no exception. Economic and financial factors underlie the urban–rural disparity and the gap between the eastern and the western areas in China. The urban–rural income gap in China has been striking. In 2009, China had its record-high rural–urban income gap since it launched its reform and opening-up policy in 1978 (Fu 2010). The urban per capita net income was RMB 17,175 (approximately $ 2,525), 3.33 times higher than that in the rural area (Fu 2010). In 2010, China's Gini coefficient (a measure of wealth distribution in a society, with a value of 0 suggesting total equality and 1 suggesting extreme inequality) stood at 0.47 and raised a red flag for inequality because it was higher than 0.4, the benchmark level for dangerous levels of inequality (Tobin 2011).

Unique Factors

Background of Unique Factors: Selection and Stratification

In addition to economic inequality, the achievement gap in China is to a great degree driven by a hidden mission of education system in China: selection and stratification. This hidden mission is largely attributed to the constraints of a developing country. Despite the amazing gross domestic product (GDP) growth rate, with a per capita gross national income at about US$ 4,260, China is still an "upper middle-income country" (World Bank 2010). The overall gross enrollment ratio[2] in China was 22.69 % in 2008, ranked 103rd among all the 205 countries in the database by the Encyclopedia of the Nations (2011). When adjusted by purchasing power, the GDP per capita in China is only 35 % of the world's average. In addition, poverty remains a fundamental challenge in China. According to the World Factbook (Central Intelligence Agency 2012), more than 128 million people (13.4 % of the population) live below the poverty line in China (approximately US$ 363 annually). Yet this poverty standard of US$ 1 a day is widely viewed as very low (French 2008).

With such constraints of a developing country, educational opportunities are inevitably limited. For instance, the proportion of tertiary-degree holders of 25–64-year-olds is less than 5 % in China, substantially below the average of 30 % of all OECD countries, the EU21 average of 27 %, and the US average of 41 % (OECD 2011b).

Furthermore, education spending in China has been consistently low. China spends less than US$ 4,000 per student from primary through tertiary education in 2009, far lower than US$ 9,860, the average spending of all OECD countries (OECD 2011c). The highest-spending countries, such as Austria, Denmark, Norway, Sweden, Switzerland, and the USA, spend more than US$ 11,000 each year (OECD 2011c). China also lags behind in terms of the percentage of spending of GDP on education. It was 3.66 % in 2010, and China expects the percentage to

---

[2] *Gross enrollment ratio* is the ratio of total enrollment for all ages to the population of the age group that officially corresponds to the level of the education shown. Tertiary education normally requires the successful completion of education at the secondary level.

jump to 4 by the end of 2012. The goal was achieved in January 2013. But it was a 12-year delay, since the original goal was to reach the 4% mark by the end of 2000. Even with the current 4% spending, China still needs to address its education-spending shortfall since the percentage is still a little lower than a 4.1% average in developing countries and a 5.1% average in developed countries (Xiong 2012).

A natural solution to scarcity and limited resources is to compete and to select (i.e., selecting those with access to educational resources). The competition becomes fiercer when considering that China is the most populous country, with 1.3 billion people in 2011. The following section introduces a few endeavors to make the selection process more effective and convenient or to improve competence to survive in the intense competition. However, many of these endeavors eventually become part of the inequality and in some cases perpetuate the existing achievement gap. These are the factors that are quite unique to China.

Key Schools and School Choice

One of the biggest selection and stratification levers is key schools. Compared with regular schools, key schools have better educational facilities, better equipment, more high-quality teachers, more funding, and selected students. The idea was to concentrate limited resources on building a few high-quality schools. In the 1990s, there were key schools for all K–12 levels, and there were different levels for key schools—national, provincial, and city levels. In other words, selection and stratification was in the educational system from early on.

With the introduction to key schools and additional disparities among schools, people would go after key schools or other better-quality schools. However, Chinese residents cannot go to any school of their choice. They have to meet special requirements. The requirements used to be exceptionally high scores in exit exams. The requirements gradually moved to financial and political factors such as high fees and social connections. Local authorities define a catchment that a particular school serves. Residents with the catchment area (as defined by their place of birth, as marked on the *hukou* document) are expected to go to the designated school. School choice occurs when parents break the restriction and decide to send their child to a different school, usually a key school.

Key schools are a clear example of the "man-made" component of the achievement gap. The creation of key schools appears to be a smart solution by using limited resources to create high-quality schools and to meet the needs of intellectuals for economic and social development. In reality, the existence of key schools inevitably widens the disparity among schools. Not only do key schools have richer and better educational resources, but they also have a highly selected group of students who have strong academic performance, who come from affluent families, or who are linked to people with high social powers. The existence of K–12 schools also brings intense competition to schoolchildren at a young age. The blueprint for a schoolchild is to go to a key kindergarten, a key elementary school, a key middle school, a key high school, and then eventually a top university. Key schools are viewed as a more promising track to the final arena of competition (i.e., the *Gaokao*).

Key schools have aroused heated concerns over equality, especially when considering that elementary and secondary schooling are compulsory. The amended Compulsory Education Law (MOE 2006) suggests that key schools or programs are no longer allowed. The terms of key schools and non-key schools quickly vanished and gave way to new labels such as window schools, exemplary schools, and experiment schools. The regulation has had little impact on eliminating tracks of schools or bridging the gap between schools beyond abolishing the old terms (Xinhua News 2007).

Quota System and the hukou

A highly centralized quota system has been used in the *Gaokao*. Every year each higher education institution has to submit to the MOE an institution plan of admission and enrollment. In this plan, the institution has to propose the number of students it plans to admit for that year, as well as the number of admitted students from different provinces and municipalities directly under the central government. The total number of admitted students in an institution for that year has to be approved (or assigned, in the case of disapproval) by the MOE. Each institution can decide the number of admitted students for each province. However, there are no criteria or guidelines about the institution's decision of the quota for each province. In practice, when a university sets a fixed admission quota for each province, it reserves a big number to its home province. As a result, students living in more prosperous areas with more universities and more prestigious universities get a much higher chance for university education and better university education. For example, in Anhui province, the chance of one out of every 7,826 test takers being admitted to Beijing University is 40 times lower than the admission rate for the *Gaokao* examinees in Beijing. The chance of a Shanghai student getting admitted to Fudan University (in Shanghai) is 274 times higher than that of a student in Shandong province (Sina News 2012).

This quota system can easily lead to the achievement gap across different areas. The *hukou* system also plays a powerful role in the *Gaokao*, because the *hukou* system requires that students can register and take the *Gaokao* only in the place marked on their *hukou* document, which usually means their birthplace. Such requirements further perpetuate the achievement gap by region and make it difficult to narrow the gap.

STEM Gap

The patterns and causes of the general achievement gap can also be applied to various aspects of education in China, including STEM gap. STEM gap in China is mainly caused by geographical, economical, and political factors instead of race or ethnicity factors. In general, students in more developed provinces and areas are more likely to pursue their careers in STEM.

Although STEM may be a new term for educators in China, the concept of STEM is nothing new. China has an intensive focus on STEM education (Asia Society 2005). Science, technology, and mathematics education has been identified by the Chinese government as the priority of education development since the country was established in 1949. This decision is largely driven by the country's immediate and pressing demand for scientists and technical talents for its modernization and economic development. Science, technology, and mathematics careers are so valued in China that there is a widely accepted saying, "As long as you are good at mathematics, physics, and chemistry, you will make a good living anywhere in the world."

Future Directions: Policies and Practice

When Communist China was founded in 1949, it had limited educational resources. Education equity in China gave way to competition and selection and stratification because stratification was considered effective in immediately producing high-quality intellectuals for economic and social development since the 1950s (Tengxun News 2011). Five decades later, the widening achievement gap and unequal access to educational opportunities have made educational equity one of the most pressing issues in China. In *The Outline of China's National Plan for Medium and Long-term Education Reform and Development* (MOE 2010), the road map of education development issued by the MOE in 2010, "making equal access to education," is listed as "a basic state policy." "The fundamental way to achieve this is to allocate education resources in a reasonable way, give priority to rural, poor, remote, border areas and ethnic autonomous areas, and to bridge the gap in education development" (MOE 2010). The action plans include

- setting up a mechanism to safeguard balanced compulsory education development;
- narrowing the gap of teachers' quality in high- and low-performing schools;
- giving preference to rural schools in aspects such as fiscal funding, school construction, and teacher training; and
- encouraging developed regions to assist underdeveloped areas.

Although this outline presents a blueprint for narrowing the achievement gap, this is no easy task because there are historical, cultural, and political factors. As shown in the failed endeavor of abolishing key schools, narrowing the achievement gap requires more than making laws or strategic plans.

In practice, the hottest issue of education equity concerns migrant students—whether or not migrant students can take the *Gaokao* in the place they live and study rather than where they register their *hukou*, which is usually their place of birth. According to the current system, everything has to be linked to the *hukou*. In other words, students have to go back to their birthplace to take the *Gaokao*. The quota system makes the issue more complex because admission scores for the same university may be vastly different for test takers in different places across the country. Taking the *Gaokao* in a particular place (usually a more prosperous area such as Beijing or Shanghai) will give test takers a much greater chance for higher education, given the same test score. The issue gets more prominent as more children are studying in Beijing and Shanghai where their parents are working, but the students do not have the local *hukou* there.

Conclusion

In China, achievement has been narrowly defined with the *Gaokao* scores. The unequal educational opportunities are marked between urban and rural areas, between the Eastern and the Central/Western regions, and between more and less prosperous provincial areas. In contrast with the USA, where race/ethnicity is the primary concern for the achievement gap, ethnicity is a much smaller factor in China. Instead, both the general achievement gap and the STEM gap in China are influenced by economic factors. But more importantly, both gaps are the product of social, political, and historical factors such as the *hukou* requirement, the quota system, and policies of school choices. Many of these factors are brought into the system to fulfill the hidden mission of selection and stratification to cope with the constraints of a developing country. Although the government gets more determined to improve the current situation of educational inequality, it is no easy task with these intervening factors.

References

Asia Society. (2005). *Education in China: Lessons for U.S. educators*. New York: Asia Society. http://asiasociety.org/files/ChinaDelegationReport120105b.pdf. Accessed 18 July 2012.

Asia Society. (2006). *Math and science education in a global age: What the U.S. can learn from China*. New York: Asia Society.

Central Intelligence Agency. (2012). *The World Factbook*. https://www.cia.gov/library/publications/the-world-factbook/fields/2046.html. Accessed 18 July 2012.

Encyclopedia of the Nations. (2011). Gross enrollment rate (%), tertiary, total—Educational statistics. http://www.nationsencyclopedia.com/WorldStats/Edu-tertiary-gross-enrollment-rate.html. Accessed 18 July 2012.

Fu, J. (2010, March 2). Urban–rural income gap widest since reform. *China Daily*. http://www.chinadaily.com.cn/china/2010-03/02/content_9521611.htm. Accessed 18 July 2012.

Fujian Quanzhoushi Jiaoyuju (Education Bureau of Quanzhou City Fujian Province). (2012). Fujian Quanzhoushi Jiaoyuju 2012 Quanzhou Wuzhong Jiaoshi Zhaoping 54 Ming Tonggao (Recruitment Announcement of 54 Teachers for Number 5 School in Quanzhou City, Fujian Province). http://www.jiaoshi.com.cn/display/article_82_40910.html. Accessed 30 July 2012.

French, H. (2008, January 13). Grinding poverty defied China's boom. *The New York Times*. http://www.nytimes.com/2008/01/13/world/asia/13iht poverty.1.9172195.html?_r=1 & pagewanted=all. Accessed 18 July 2012.

Gou, R. 2006. "Cong chengxiang gaodeng jiaoyu ruxue jihui kan gaodeng jiaoyu gongping" [Examining Equality in Higher Education from the Perspective of Rural-Urban Access to Higher Education]. *Jiaoyu fazhan yanjiu [Research in Educational Development]*, 5, 29–31.

Hu, D., & Zhu, J. (2011). Shuoxiao Jiaoxue Zhiliang Chaju Yu Chujing Chenxiang Jiaoyu Junheng Fazhang (Reduce Disparity in Teaching Quality and Promote Equal Education Development Between Urban and Rural Areas). *Jiaoyu Yanjiu Yu Shiyan, 142*(5), 40–43.

Li, C. (2010). Higher education expansion and unequal educational opportunities: *Sociology Research*, 3. 1–34. http://www.sociology.cass.cn/pws/lichunling/grwj_lichunling/P020100802372472813113.pdf. Accessed 18 July 2012.

MOE. (2006). Compulsory Education Law. http://www.moe.edu.cn/publicfiles/business/htmlfiles/moe/moe_619/200606/15687.html. Accessed 18 July 2012.

MOE. (2010). The outline of China's ational Plan for Medium and Long-term Education Reform and Development (2010–2020). http://www.gov.cn/jrzg/2010-07/29/content_1667143.htm. Accessed 18 July 2012.

MOE. (2012). Jiaoyubu Zhishu Gaodeng Xuexiao. http://www.moe.gov.cn/publicfiles/business/htmlfiles/moe/moe_2156/200807/36536.html. Accessed 29 July 2012.

OECD. (2011a). *Strong performers and successful reformers in education: Lessons from PISA for the United States*. Paris: OECD.

OECD. (2011b). Education at a glance 2001: Country note—China. http://www.oecd.org/dataoecd/29/0/48677215.pdf. Accessed 18 July 2012.

OECD. (2011c). Education at a glance 2011: Highlights. http://www.oecd-ilibrary.org/docserver/download/fulltext/9611051ec001.pdf?expires=1344126964 & id=id & accname=guest & chec ksum=C4E0C3B76CB114CE979040B4A4344EF1. Accessed 18 July 2012.

Qiao, J. (2010). On the rural–urban disparity in access to higher education opportunities in China. *Chinese Education and Society, 43*(4), 22–31.

Sina News. (2012, June 27). Question the *gaokao*. *Sina News*. http://ah.sina.com.cn/news/shms/2012-06-27/17447.html. Accessed 7 July 2012.

Tengxun News. (2011). School choice policies. *Tengxun News*. http://news.qq.com/zt2011/ghgcd/49.htm. Accessed 18 July 2012.

Tian, D. (2012, July 26). Ruhe ran nongcunwa bu shuzai qipaoxianshang (How not to let rural children lose at the starting line). http://csstoday.net/Item/18218.aspx. Accessed 30 July 2012.

Tobin, D. (2011, June 29). Inequality in China: Rural poverty persists as urban wealth balloons. *BBC News Business*. http://www.bbc.co.uk/news/business-13945072. Accessed 18 July 2012.

Tucker, M. (Ed.). (2011). *Surpassing Shanghai: An agenda for American education built on the world's leading systems*. Boston: Harvard Education Press.

World Bank. (2000). *China Overview*. http://www.worldbank.org/en/country/china/overview. Accessed 18 July 2012.

Xie, W., & Li, X. (2000). "Gaodeng jiaoyu gongpingxing de diaocha yu yanjiu baogao" [Report on a survey study of equality in higher education]. *Jiaoyu Zhengce de Jinji Fenxi [An Economic Analysis of Education Policy]*, 257–274. Beijing: Renmin Jiaoyu.

Xinhua News. (2007, January 25). Fees for primary and secondary schools may be waived and fees for high school choices will be lowered. *Xinhua News*. http://news.qq.com/a/20070125/000818.htm. Accessed 18 July 2012.

Xiong, B. (2012, March 9). Time for China to address education spending shortfall. *China.org.cn*. http://www.china.org.cn/china/NPC_CPPCC_2012/2012-03/09/content_24853989.htm. Accessed 18 July 2012.

Yang, D. (2006). *Dreams and reality of education equity in China*. Beijing: Beijing University Press.

Yuan, C. (2007, May 29). Priority for students in the western region—more rural students than urban students admitted to colleges in 2006. *China Youth*. http://zqb.cyol.com/content/2007-05/29/content_1775346.htm. Accessed 18 July 2012.

Zhao, Y. (2012). Reforming Chinese education: What China is trying to learn from America. *Solutions: For a Sustainable and Desirable Future*. 2(2), 38–43.

Chapter 11
Employing a Sociohistorical Perspective for Understanding the Impact of Ideology and Policy on Educational Achievement in the Republic of Korea

Sonya N. Martin, Seung-Urn Choe, Chan-Jong Kim and Youngsun Kwak

Introduction

For the last two decades, much has been written about the academic achievement of students from the Republic of Korea[1] in international exams, because Korean students have routinely performed well above the mean in literacy, mathematics, and science assessments (Kang and Hong 2008; OECD 2011; Choi et al. 2011). In fact, an analysis of assessment records from the Trends in International Mathematics and Science Study (TIMSS[2]) from 1995 to 2007 (NCES 2009) indicates that Korean students, on average, score in the top 10% of all participating Organisation for Economic Co-operation and Development (OECD)[3] countries and that Korea is routinely one of the top five countries for the highest achievement in mathematics and science. Results from TIMSS (NCES 2009) show that not only do Korea's students excel in the top percentiles, but even Korea's lowest-performing students (those scoring in the bottom 10th percentile) outperform the lowest-performing

---

[1] Hereafter referred to as *Korea*, and citizens are referred to as *Korean*. The Republic of Korea is commonly referred to as *South Korea*.

[2] The Trends in International Mathematics and Science Study (TIMSS) is an international assessment of the mathematics and science knowledge of the fourth- and eighth-grade students from different countries to enable international comparisons of students' educational achievement (NCES 2009). Data are available only for eighth grade because Korean students have not regularly participated in the fourth-grade assessments since 1995.

[3] The Organisation for Economic Co-operation and Development (OECD) is an international economic organization (currently of 34 countries) founded in 1961 to stimulate economic progress and world trade (OECD n.d.). Comparisons of wealth, education, health, policies, etc., are routinely reported as a basis for understanding growth and development of the OECD member nations (OECD n.d.).

S. N. Martin (✉) · S.-U. Choe · C.-J. Kim
Seoul National University, Seoul, Republic of Korea
e-mail: sm655@snu.ac.kr

Y. Kwak
Korean Institute for Curriculum and Evaluation, Seoul, Republic of Korea

J. V. Clark (ed.), *Closing the Achievement Gap from an International Perspective,*
DOI 10.1007/978-94-007-4357-1_11, © Springer Science+Business Media B.V. 2014

students of all other countries of the OECD in mathematics (see Table 9, p. 17). In science, Korea's lowest-performing students outperform students of all nations with the exception of Japan (see Table 17, p. 42). As a result, many educational researchers are interested in learning more about the Korean education system. The comparative achievement of Korean students in mathematics and science, at both the lowest and highest levels, becomes even more remarkable when we consider that in the preceding 67-year period, the Korean people have faced many hardships and challenges—including the fall of the Korean empire, a 35-year forced occupation and colonization by neighboring Japan, and a civil war.

We begin this chapter with a focus on the sociohistorical context for our analysis of Korean student achievement. In the sections that follow, we discuss (1) the historical impact of Confucian philosophy in shaping societal norms regarding the importance of education and (2) governmental policies and practices that have contributed to Korean students' achievement in mathematics and science since the 1960s. We draw attention to aspects of student achievement as disaggregated by gender and class to shed light on challenges science educators and researchers are currently facing in relation to equality in science achievement at both national and international levels. Specifically, we discuss the impact of high-stakes assessment on tertiary enrollment, growing inequities in achievement due to economic disparity, and the ways in which globalization is fueling rural-to-urban migration practices and increases in immigrant populations in Korea. In doing so, we introduce new policies and research that seek to promote Korea's continued advancement in political, economic, and educational arenas in both regional and international contexts. We conclude our chapter by raising questions and offering suggestions regarding current policies, research initiatives, and innovations in science education in Korea that have implications for researchers in other nations facing similar achievement disparities.

Korea and Education: The Importance of Taking a Historical Perspective

Though educational researchers often cite Korean students' results in international assessments in comparative analysis studies, they rarely offer a sociohistorical context for their achievement. We believe this context is necessary to our discussion of achievement and education in modern-day Korean society because the Korean education system, including science teaching and learning practices of teachers and students, is influenced by historical events. To understand achievement and gaps in achievement among students in Korea, it is necessary to gain an appreciation for the ways in which history informs cultural practices and beliefs. As Korea is an ancient nation, it is impossible in this short chapter to provide an exhaustive account of the events that have undoubtedly helped form modern-day educational beliefs and practices. Thus, we have narrowed our attention to three historical events that have influenced Korean education today: (1) the introduction of Confucianism to Korea, (2) anticolonial feelings cementing education as a gateway for success, and (3) the government-led education expansion policies enacted after the Korean War.

Confucianism and the Introduction of the Civil Service Exam

The Korean people have a long history, which has been traced across several thousand years through archeological and written documentation. Ancient Korea was organized as clan communities that combined to form small city-states with complex political structures, which eventually grew into kingdoms. Modern-day Korea was born from the unification of these kingdoms under a single ruler in 918, marking the beginning of the Goryeo Dynasty (918–1392), from which modern Korea's name is derived (Seth 2006). During the Goryeo Dynasty, an authoritative, aristocratic ruling society with strong civil and military reigning structures flourished. The emergence of a ruling class ushered in a new social class system made primarily of civil servants, military officers, court officers, soldiers, artisans, and a vast population of peasants and slaves (Eckert et al. 1990). As this class system developed and grew, Confucian philosophy became increasingly integrated into Korean society and co-existed with the dominant Buddhist religion that was widely practiced by members of the monarchy, the aristocratic ruling class, and common citizens alike (Connor 2009). Confucianism placed a great virtue on social hierarchies and promoted the ideal of meritocracy through the belief that public-office officials should be selected based on their performance in competitive examinations. As a result, in 958, a Civil Service Examination system, already widely established in China in the sixth century, was instituted as a rigorous, competitive process for selecting officials based on academic excellence in Korea (Connor 2009).

Initially, males from all positions in society were granted the right to sit for annual written exams on topics in history, philosophy, poetry, Chinese language, and other subject areas in order to compete for official posts in society (Connor 2009; Lankov 2012). Because the examination process was used for selecting men for all levels of posts in government, even the sons of poor families could compete for some of these positions. However, to perform well in these exams, men needed to be literate and needed to have opportunities to study, as the exam focused on rote memorization. So although this method did allow some opportunities for upward social mobility, it was largely unavailable to those born outside of the aristocracy, since only members of the aristocratic ruling levels could support their sons to study for the exams (Elman 2000).

Today, Confucianism as an ideological system continues to shape Korean society by structuring social interactions between members of society, dictating morality through the legal system, and shaping curriculum traditions and classroom practices. In addition, as in ancient times, Korean schools are organized around a culture of high-pressure, high-stakes competitive examinations (including college entrance exams, which take place only once a year) (Lee 2006). Addressing the historical impact of this ideology on educational opportunities for Koreans across the social class strata remains a challenge for educators in Korea. In the subsequent sections, we will discuss the ways in which policy reform in teacher education has helped to mitigate some of these challenges. We also raise questions about the impact of globalization on social class and status in Korean society that are emerging as new challenges for educators and policy makers in the twenty-first century.

The Influence of Neo-Confucianism on Education and Society

During the Joseon Dynasty (1392–1897), neo-Confucianism became more popular, and, as a result, a more rigidly hierarchical society developed (Eckert et al. 1990). The rise of neo-Confucian beliefs is tied to the strengthening of a more hierarchical ordering of societal relationships with the belief that harmonious relationships between superiors and inferiors should be cultivated through reciprocation of benevolence and obedience between rulers and subordinates. As a result, relationships became increasingly ordered along the line of king–subject, parent–child, husband–wife, or older sibling–younger sibling. Relationships were also subject to gender subordination (e.g., brother–sister) and among friends, deference was subject to age. This principle served as a foundation upon which interactions among citizens within society were based with the underlying assumption that rulers would be just and benevolent.

Neo-Confucianism profoundly shaped the practices and beliefs of the educated class, specifically with regard to the development of the government and educational system. Over time, the system became strongly influenced not by meritocracy, but rather by heredity, which led to the development of a "hereditary ruling class," known as the *yangban* (Sorenson 1994). The *yangban* ruling class consisted of both land-owning and non-land-owning nobles who served as scholarly officials overseeing administrative duties for the ruling monarchy (Eckert et al. 1990). All descendants of a member of the *yangban* class received special privileges in that they were able to influence local politics and administrative decisions. Arranged marriages enabled *yangban* families to consolidate and maintain class privileges over many generations. However, a family could lose *yangban* status if a male member of the family failed to pass the civil service exam over a three-generation span. Thus, passing the exam became essential if families were to maintain their power and wealth. This requirement meant that educating boys was of particular importance for each family (MEST 2004) and, as a result, for hundreds of years in Korean society, families have placed considerable resources and energy into educating boys as a means to advance the entire family.

Over time, such high-stakes practices fostered the development of a complex set of qualifying examinations for appointment to both civil offices and military positions for which many private and local schools were created to support the sons of the elite to prepare for the exams (Connor 2009). These schools are the earliest form of a systemized education in the Korean history. However, the vast majority of men, largely belonging to a hereditary underclass of peasants and slaves, were not engaged in any form of systematic, state-supported education during the five centuries of the last dynasty (Eckert et al. 1990; Lee 1988). In addition, no women were encouraged to engage in scholarly activities, and a woman's place in society was based solely on marriage and familial connections. As a result, poor families and women were largely disenfranchised from opportunities to participate in government or education in any way. Such exclusionary practices, both stemming from Confucianism, continue to impact the equality of educational opportunities for both women and students from lower socioeconomic status in Korean society today (Lee 2006). We examine these issues in greater detail in later sections of this chapter.

A notable exception in the disenfranchisement of the lower classes during this period of Korean history occurred during King Sejong's reign (second ruler of the Joseon Dynasty, 1418–1450). In 1446, King Sejong published *Hunmin Jeongeum*, which was a description of a new alphabet (later known as *Hangeul*) for Korean, which he had developed with the support of court scholars. King Sejong wanted non-educated citizens to be able to express themselves in their native language (Eckert et al. 1990, pp. 124–25). Although civil servants' examinations and official documents continued to use Chinese characters over the next 400 years, *Hangeul* eventually became the national system for written language and is in use today. Today, the ease of learning the *Hangeul* system is widely attributed to Korea's achievement of near universal literacy rates.[4] The concept of universal literacy was a radical idea which was widely opposed by members of the ruling *yangban* society who made up the King's court (Seth 2006). As is true in many nations, education was reserved for elite members of society, effectively disenfranchising the vast majority of people and preventing any opportunities for advancement in social status through education. At that time, all educational instruction took place in Chinese rather than Korean. Non-educated members of society did not commonly use Chinese. However, with the introduction of a written system for the Korean language, reading and writing would be accessible for those people who were not members of the *yangban* class. In fact, the ease with which *Hangeul* could be learned by members of the underclass was perceived as a threat to the literati class, who were concerned that even women could learn to read and write using this new system (Connor 2009).

The realization of King Sejong's goal to increase literacy rates has relevance to our discussion about the development of the Korean education system because the fall of the Korean Empire has been attributed, in part, to the spread of *Hangeul* among all citizens, regardless of class and social status. As a growing population of skilled tradesmen and clerks in an emerging middle class began to jockey for greater access to the wealth and power of the noble class, seeds of discontent slowly

---

[4] Korea has achieved near universal literacy rates for people aged 15–24; however, these numbers stand in stark contrast to the educational attainment levels of Koreans who were adults before the institution of compulsory education in the 1950s. Adult literacy rates are difficult to calculate, but for people aged 55–64, rates for secondary- and tertiary-education attainment are below the OECD average (MEST 2009).

took root over the next three centuries. Although it took more than 500 years from the time of King Sejong's decree until all Korean citizens (including all class levels and both genders) were afforded access to free, compulsory education, his intention represented an early shift in thinking by some members of the ruling neo-Confucian elite who envisioned a government and education system that was more egalitarian in nature (Eckert et al. 1990). We highlight King Sejong's work to provide an example of the historical tensions existing in Korean society regarding the right to be educated.

The Influence of External Forces on Education in Korea

From the early eighteenth century through the 1870s, Korea saw the emergence of scholars who espoused the need to modernize the nation. These scholars believed that scientific technology, industry, and the reform of social and political institutions were necessary for Korea to develop as a modern nation. Some members of each class level began to question the ability of the nation to advance without an educated public, which consisted of a large population of indentured servants and slave laborers. In response to calls for rule by constitutional monarchy, reform efforts were undertaken by the ruling family to strengthen the Joseon Dynasty. The aim of these reforms was to prevent internal rebellion (Eckert et al. 1990, p. 193), much of which stemmed from unrest by newly educated members of the underclass and progressive members of the ruling *yangban* society.

Part of the government's strategy to control this unrest included secluding Korea from external influences by the adoption of an Isolationist Policy. This policy aimed at repelling foreign intruders, especially Westerners, who might introduce conflicting ideologies that could undermine the monarchy and ruling social class. However, by the late 1870s, pressures from various groups (both internal and external) forced the government to allow sanctioned contact with Western missionaries and government officials from neighboring countries. Formerly referred to as the "Hermit Kingdom" (Eckert et al. 1990), Korea lost its veil of seclusion and in 1883 saw the development of the first modern schools (set up by Christian missionaries) for members of the non-aristocratic ruling class (Seth 2010). As members of the educated elite began to be inspired by Western ideas, changes in ideology regarding the role of women in society also developed. In 1886, an American missionary, Mary Scranton, founded Ewha University, which was the first school for girls in Korea (Connor 2009) and which today is one of the largest universities for women in the world and one of the most prestigious universities in Asia. During the ensuing decade, hundreds of schools were established for elementary education and vocational training, and men and women of all class levels were encouraged to be educated. This time period was short-lived due to growing tensions among international communities that had a vested interest in controlling access to Korean ports and waterways as strategic points for military power in the early 1900s. However, these more egalitarian philosophies regarding education for women and the poor would prove influential in the development of the modern education system in postwar Korea, as future government officials viewed compulsory education as a social investment necessary to produce valuable economic returns.

The Birth of the Modern Education System in Korea

We believe that any discussion of Korea's education system would be remiss if we did not also address the lasting impacts of the Japanese occupation and the Korean War on the initial development and subsequent rebirth of the Korea's modern-day formal education system. The events leading to the fall of the Korean Empire at the end of the nineteenth century, the colonization of Korea by Japan, and the Korean War in the first half of the twentieth century are too complicated to discuss in this chapter; however, from a sociocultural perspective, these events have had a profound impact on how the Korean people have viewed education. Thus, in the sections that follow, we offer the reader a selective historical account of events we feel are pertinent to our examination of present-day Korean student success and achievement. In the remainder of this chapter, we build on these historical events to investigate the effects of policy and practice on achievement equality and to offer directions for future science education research in Korea. Specifically, we (1) address the impact of Confucianism on societal interactions today and (2) analyze how educational policies have influenced educational opportunities for men and women of all social classes.

The Fall of the Korean Empire

The decline of the Joseon Dynasty during the last 20 years of the nineteenth century was hastened by economic pressures brought on by famine and years of excess spending by ruling officials. This economic situation forced the Korean government to enact a set of reform measures that opened the country to even more external contacts. Responding to a growing rebellion by peasant farmers who were protesting heavy taxes and abuses of human rights and a desire to stem the influx of foreign influences in Korea, the Joseon government asked the ruling Qing Dynasty government in China for assistance. Neighboring Japan felt the Qing presence in Korea was a threat to their nation, leading to the First Sino-Japanese War (1894–1895), which was fought mostly on Korean soil. Fearing the Japanese after their defeat of the Qing army, Korean ruling monarch Queen Min sought help from Russia. In an attempt to prevent Russia from expanding their authority in the region, the Japanese government removed Queen Min from power by having her assassinated in 1895. After her death, Korean society experienced upheaval as remaining government officials engaged in political, economic, and social interactions with foreign influences, primarily Japan, China, and Russia to stabilize the regime and remain in control of

Korea. A series of events forced Korea's remaining government officials to enter treaty agreements with other nations for trade agreements and to broker access for other nations to use Korea's waterways and to provide valuable access to Korea's northern border with China. After winning the Russo-Japanese War (1904–1905), Korea became a forced protectorate of Japan (Takaki 1989; Cho'oe 2006). In 1910, the Japan–Korea Annexation Treaty officially marked the beginning of Japanese rule in Korea, making Korea an official colony of Japan until the end of World War II.

Japanese Colonization and Education in Korea

Japan's colonization of Korea had a profound impact on education in Korea, the effects of which can still be seen in today's educational system. During this 35-year period of colonial rule, the Japanese government is credited by some scholars with the modernization of Korea through development of infrastructure and industry and the implementation of a systematic education program (Reinfeld 1997; Seth 2010). However, during the Japanese occupation, Korean citizens were forced to endure harsh assimilation tactics as subjects of the Japanese empire. Cultural suppression was a critical aim of the Japanese government at this time. For example, symbols of Korean culture (e.g., historic artworks, literature, public buildings, and monuments) were altered, destroyed, or replaced with Japanese cultural artifacts. Laws abolished ownership of private property, citizens were forced to worship Japanese religious figures, and any resistance to assimilation was met with severe punishment, imprisonment, and even death (Buzo 2007; Schmid 2002). Regarded by some historians as the birth of modern Korean nationalism (Seth 2010), a nationwide student-led demonstration protesting Japanese rule occurred on March 1, 1919. Men and women of all classes and ages took part in the demonstration. National leaders of the anti-Japanese resistance movement sought to "save the nation through education" (MEST 2012) with a primary focus on indoctrinating the youth to work in solidarity with the national independence movement.[5] Following the protests of 1919, Japanese rulers attempted to mitigate continued unrest with new reform efforts aimed at controlling civil discontent. In spring of 1920, the Japanese government facilitated a symbolic merger of the two nations through a royal marriage between the Korean crown prince and Japanese princess (Seth 2010). The marriage, along with the amnesty of several thousand Korean political prisoners, was meant to signify the beginning of a more tolerant attitude toward Korean culture. This led to an explosive growth in youth, religious, social, and political activities among the oppressed Korean population. At the same time, the Japanese attempted to maintain a tighter control through the expansion of administrative offices and greater police occupation throughout all of the provinces of Korea (Seth 2010).

---

[5] March 1st is now observed as Independence Day, to commemorate the first display of Korean resistance during the Japanese occupation.

Of particular interest to our discussion is the impact of these events on the established social norms, which had permeated Korean society since the introduction of Confucian philosophy over 1,500 years earlier. For example, as young women participated in the resistance movement, they found new avenues to express their views regarding inequality. As a result, the Korean women's liberation movement was closely entwined with the fight for nationalism. Even some men who participated in the resistance insisted that education of Korean women was critical for strengthening the nation. However, they advocated for educational advancements that would improve women's ability to educate their own children, not necessarily for equality of education for women (Kim and Choi 1998). Gender barriers were not the only aspect of Confucian ideology that was challenged during the resistance as men and women from different social strata collaborated to resist the Japanese occupation. This solidarity of social classes that emerged from the resistance movement of the early 1920s has been credited in large part to the rise of nationalism in Korea. As the nationalist movement grew, competing visions about how best to move the country forward would have profound consequences for Korea when Japanese rule ended at the close of World War II.

Education in Postcolonial Korea

After liberation from the Japanese in 1945, Korea was artificially divided into two occupied administrative zones along the 38th parallel. The political division of the country was the result of an agreement between Allied forces at the end of World War II resulting in the US military occupying the southern half of the peninsula and the Soviet military occupying the northern half. Deeply held philosophical and ideological differences, first emerging at the latter half of the nineteenth century and strengthened during the nationalist movements during colonial occupation, resulted in a stalemate for all reunification efforts. In 1948, two separate nations, the Republic of Korea (South Korea) and the Democratic People's Republic of Korea (North Korea) were established. However, tensions between the governments of the newly formed nations regarding sovereignty of the peninsula erupted into a 3-year civil war from 1950 to 1953.[6]

The Korean people endured 3 years of war, leaving millions of soldiers and civilians dead, the land devastated, and the country still divided. The war destroyed the infrastructure in both countries, including a large percentage of the transportation systems and irrigation systems, electric power networks, and social services networks, including hospitals, schools, and universities, that had been built by Korean laborers during colonial occupation (Reinfeld 1997). By some estimates, more than 75 % of infrastructure was damaged or demolished in the South, so the majority of the government's allocation of economic and human resources in the 1950s was

---

[6] Officially, the war has never ended. The armistice agreement to end the war was not signed by all parties. Instead, a ceasefire has been in effect since 1953.

spent on reconstruction efforts for all sectors of society. This included everything from building schools, establishing teacher-training programs, replacing the Japanese colonial curriculum with a national Korean curriculum, and developing policies for how to effectively extend education to all Korean citizens regardless of class or gender. High achievement in the areas of mathematics and science (by both males and females) becomes even more remarkable when we consider that in postcolonial Korea in the 1950s, "the very vocabulary to talk about modern science and mathematics hardly existed in the Korean language and had to be invented before textbooks could even be written" (Sorenson 1994, p. 11).

Since 1953, Korea has undergone one of the fastest industrial revolutions in history, mainly by developing human resources through formal education. In doing so, Korea has quickly developed from a poor agrarian-based economy to one of the world's largest industrial economies (Morris 1996). During this same time period, Korea's modern education system was built from the ground up—the results of which are widely praised in the international community and the focus of much research and attention by local and international educational researchers (Akiba 2007; Ilon 2011; Sorenson 1994). Both during colonial occupation and during the reconstruction period of the 1950s, education was widely viewed as a means of both personal advancement and as necessary for the advancement of the nation. We argue that this context is essential for understanding the policies that have enabled the Korean government to rapidly expand educational opportunities for most citizens within just two generations and for questioning the ways in which these historical factors continue to play a role in achievement and educational inequality in Korea today. In the sections that follow, we offer a brief review of some major educational policies we feel have contributed to this success, and we discuss policies and achievements within the sociohistorical context we have established.

Rebuilding a Nation Through Educational Reform

The removal of the Japanese colonial government and the devastation of the country by the war created an equality of poverty among citizens that fractured the centuries-old hierarchical class structure fostered by neo-Confucian ideologies (Morris 1996). Starting in the late 1950s, both men and women from different class levels had greater opportunities to advance through education than ever before in the Korean history. Many education and economic historians point to the policy initiatives of the 1950s and 1960s as key factors in establishing education as a central pursuit for all citizens, which has had a lasting impact on the educational landscape of Korea today. Postwar Korea embraced some of the more traditional tenets of Confucian philosophy regarding the importance of education for cultivating morality, virtue, and wisdom. However, in comparison with the Korea of the Joseon Dynasty, some of the more oppressive aspects of neo-Confucianism, including class-based social strata and traditional attitudes toward women, were relaxed as people came together to rebuild the country.

Key to sustaining and developing the new republic was the government's provision of educational access for all citizens as a means to both foster a sense of patriotism and encourage citizens to support the goals of the government in the interest of national advancement (Morris 1996). It is important to note that the Japanese colonizers educated the very nationalists in charge of the post-liberation nation building. In addition, the neoliberal, hegemonic force superimposed on the Korean government by the US military, which has occupied Korea since 1953, heavily influenced decisions regarding the creation of the new government. Thus, we acknowledge that the issues of political and social reform in Korea are incredibly complicated and are beyond the scope of this work. Rather it is our intention to point out some of the broad reform policies that we believe have contributed to Korean students' academic success and to draw attention to challenges that these policies have wrought.

Universal Elementary Education

From the beginning of the Republic of Korea, the government invested heavily in education in an effort to capitalize on the most plentiful resource available, the Korean people. In 1954, the government launched its first "six-year plan for free compulsory elementary education," aimed at increasing school attendance and combating low literacy and numeracy rates (Kim 2002). By 1959, enrollment rates increased to 95.4 % (Kim 2002, p. 31), with student demand outpacing resources so much in urban areas that some class sizes exceeded more than 90 students and over 40 % of the schools were forced to utilize a "two-shift" school operation system, holding classes in both day and evening to accommodate students (KEDI 2007, p. 82).

During this time period, less funding and resources were allocated toward secondary and tertiary education and more toward building schools and reducing class sizes by hiring teachers to support the "two-shift" policy. By 1965, there was an increasing demand for secondary education and growing competition for access and enrollment in secondary and tertiary education because there were so few institutions available. At this time, secondary education was selective, with only 35 % of students continuing to secondary school and 6 % enrolling in tertiary schools (Morris 1996). As a result, only the brightest and wealthiest students advanced into higher education. In 1967, the second plan for universal education was instituted, providing more classrooms, free textbooks, and resources for all elementary schools. In the late 1960s, resources were scarce for secondary and tertiary spending, so although elementary education was free, private resources were needed to pay tuition and fees for students pursuing education past the primary levels (Kim 2002). Introduced as a necessity in the 1960s, the expectation that families supplement secondary and tertiary education with private funds has become institutional policy. In the 1960s, economic disparity among families was less than it is today, which has important implications for growing inequities in educational attainment based on differences in social class.

Equalization of Secondary Education Policies from the 1970s Through the 1990s

By the 1970s, the economy was thriving, the living standard was increasing, and families who were previously members of the underclass were, for the first time, positioned to financially support their children to be educated. During this time period, educational expenditures increased, and secondary and tertiary schools began to expand to meet the needs of students exiting the elementary schools. At the same time, Korea began to see a marked decline in fertility rates, meaning that families had fewer children, making each child more precious with regard to securing access to top secondary schools. In Korean society, there is an expectation that children will take care of their parents when they are older, thus it is important that parents position their children to be as successful as possible so they can shoulder this responsibility in adulthood. As a result, competition was incredibly fierce among elementary school students, as the demand for admittance to secondary schools had outpaced the development of secondary education facilities. Many private schools began to be built to increase seating capacity for students who were not served by the public schools. Parents began to invest resources in private tutoring to help their child pass the entrance exams of top high schools. The problem became so great that in 1974, the government instituted the *High School Equalization Policy,* aimed at reducing the burden families had to pay for extracurricular lessons supporting children's admittance to top schools. This policy abolished the entrance exams for secondary schools and instituted a lottery system for enrolling students in neighborhood middle schools (Kim 2002). This policy has been criticized by some as having put into effect a system that limited school choice and served to strengthen government control over schools by reducing the number of private institutions (Lee 2004).

Today, Korea has a single-track, 6–3–3–4 system, which maintains a single line of school levels in order to ensure that every citizen can receive primary, secondary, and tertiary education according to one's ability (MEST 2012). In addition to 6 years of elementary school, 3 years of middle school, 3 years of high school, and 4 years of university education, Korea's education system is also expanding to include both preschool and lifelong education. By 1986, enrollment in secondary education had risen to an incredible 95% of the total student population (Morris 1996). In addition, Korea graduates 94% of secondary school students with high school diplomas (OECD 2011). Both men and women have benefited from these policies and have made incredible strides in education in the last 20 years. Additionally, in 2008, nearly 70% of all Korean youth were enrolled in either a 2- or 4-year university or vocational program (Morgan 2010), and it is estimated that by 2025, 80% of all 25–44-year-olds will have participated in higher education, making Korea's population the most educated students in the world (McNeill 2011).

By some measures, Korea continues to lag behind other OECD nations with regard to gender equality. However, enrollment levels at the primary and secondary levels are nearly 100% for both boys and girls, and enrollment in undergraduate universities by females is about 44 % and rises each year (Lee et al. 2010). In most OECD countries, women outnumber men in undergraduate enrollment, but that is not the case in Korea, as a substantially higher number of men have attained tertiary levels of education in Korea (OECD 2005). Young women in Korea are much more likely to be educated at secondary and tertiary levels than were their mothers and grandmothers, more so than any other OECD nation (OECD 2006). Current trends suggest that females will eventually equal males in tertiary enrollments, even though Korea has a slight higher male-to-female birth ratio (OECD 2005). Interestingly, as educational equality increases for women in Korea, fertility rates have continued to decline. In fact, Korea has the lowest birth rate of all OECD nations. Many researchers attribute this decline to improved educational opportunities as young women are delaying marriage and having children to pursue education and careers. In addition, while educational opportunities are expanding for women, of all OECD nations, the widest gap in wages based on gender is experienced by Korean women (OECD 2006), meaning that the "glass ceiling" is a still very much a reality in Korea and that women are less likely to get into higher-paying jobs or managerial positions. Since the introduction of legal protections for women in the labor market in 1987, women have increasingly entered the workforce, but men are still employed at much higher rates than women. As Confucian ideology has been such an integral part of social ethics in Korea, it continues to influence interactions in the workplace. For example, even today, women are expected to be subordinate to men in the workplace and some societal traditions, such as restrictions on women's participation in social activities, continue to negatively affect advancement in employment as opportunities for social networking are limited. As the influence of Confucian virtues wanes, women are increasingly being integrated in the workplace, which is expected to have an impact on educational needs as women prepare for more advanced employment opportunities in a wider range of careers.

Although gender inequality is still a persistent issue in many aspects of Korean society, with regard to access to educational opportunities, there are few discernable differences in either school attendance or academic achievement in terms of gender or socioeconomic status. When we examine gender and class in relation to performance in the PISA international assessment, Korean students show no significant difference in the average science scores for males and females (537 and 539 respectively) (Cho et al. 2011; OECD 2010a). In addition, Korea shows a below-average impact of socioeconomic background on performance in science (OECD 2010b) and lower-than-OECD-average (OECD 2010c) on between-school variation in student performance and in-school variation meaning that there is no statistically meaningful performance gap between students in urban schools and those in rural schools, even after accounting for differences in socioeconomic background (KICE 2012). Analysis of Korean students' performances in science on past PISA and TIMSS exams demonstrates that the percentage of top performers (above Level 6) is similar to the OECD average (1.1 %) and that the percentage of students scoring below Level 2 in science (6.3 %) is remarkably lower than the OECD average (18.0 %) in science (OECD 2010a). Thus, compared with other top-performing countries in the area of science, Korea has relatively smaller proportions of students at the two extremes—suggesting smaller gaps in achievement across gender and class than are seen in many other OECD nations. Gender disparities favoring male participation in mathematics and science (especially in mathematics-based disciplines such as physics) continue to persist, and studies continue to reveal that women encounter high levels of gender discrimination in the workplace and in social life (Lee et al. 2010). Interestingly, gender differences in attitudes toward science were more striking, as assessments show that Korean females are less likely than males to be motivated to pursue science at the tertiary levels (OECD 2006). Although the disparities in achievement in relation to class and gender are not as large as those in other nations, such as the USA, Korean science educators are interested in moving all students toward higher levels of achievement, so researchers are concentrating their efforts at these "gaps."

Responding to Inequity Through Policy and Programs in the Twenty-first Century

Viewed from a historical perspective, government policies in postwar Korea have clearly served to foster a more egalitarian approach to education, which has had an important impact on gender and class equality. However, as the above-mentioned figures suggest, international assessments have revealed some small gaps in student achievement, which have become an important focus for research and policy development in Korea in the last 10 years. Coupled with some persisting issues in gender inequality at the tertiary level, the areas of gender and class remain a challenge for educators and policy makers in the twenty-first century. In this section, we both introduce new challenges facing Korean society and highlight some of the policies implemented in the last decade aimed at minimizing achievement disparities.

Teacher Education and Government Policies

High student achievement on international assessments in mathematics and science has been associated with excellence in teacher education and professional development (Kang and Hong 2008). Korea continues to lead many OECD nations, including the USA, in the area of teacher quality as measured by attainment of advanced degrees in content areas and extremely low incidence of teaching out of field of certification. In addition, the entrance requirements to enter teacher education programs are extremely high, and the recruitment policies (including hiring, tenure, and compensation) are more successful at retaining teachers in the workforce. Job security also plays an important role in maintaining high levels of competition for teaching jobs. Teachers are automatically tenured until retirement at the age of 62, effectively assuring teachers' lifetime employment with full benefits as civil servants (Kang and Hong 2008). These characteristics are often attributed to the remaining influence of Confucian philosophy, which espouses reverence for teachers as leaders of the people. These factors contribute to the relatively high sociocultural status that teaching continues to hold in Korea relative to many other countries.

Coupled with high standards for preparing teachers, the Korean government has also implemented policies seeking to diminish disparities in students' ability to access highly qualified teachers. The government-mandated "teacher rotation" policy makes sure schools do not accumulate disparities in access to teachers who have been educated in the top teacher education programs or who have the most experience. This policy is responsible for randomly re-assigning teachers to a new school every 4 years, meaning that students attending schools in neighborhoods or regions that are economically depressed are as likely to have a highly qualified, experienced teacher as are students in neighborhoods or regions with greater economic wealth (Kang and Wong 2008). Even with these policies in place, a recent national assessment found that ninth-grade students in the urban areas tended to achieve higher levels in science than did those students in suburban and rural areas (Kim et al. 2011a, b). The students who live in rural areas tend to be from families with lower socioeconomic status. Because this gap is not easily explained by differences in student access to teacher and materials resources, researchers are focusing on the impact of what is referred to as private or "shadow" instruction in Korea. In the following section, we discuss this topic in more detail.

Shadow Instruction: A Legacy of Confucian Ideals and Postcolonial Reform Efforts?

National assessments have demonstrated that there are increasing gaps in achievement between males and females, students in different social class levels, and between students in different regions of the country. This gap has been partly attributed to inequities in student access to private tutoring. Indeed, a large percentage of a student's instructional time each day is actually received outside of public-funded schools. In 2005, as many as 75 % of students in grades 7–9 reported that they engaged in some form of private tutoring, via after-school academic programs or "cram" schools (Lee 2007). Recent studies have shown that families at higher income levels spend considerably more than low-income households and that students with parents who hold college degrees are more likely to receive private tutoring than are those whose parents received only high school education (Kang and Hong 2008). Another study (Lee 2007) found that only 51 % of low-achieving students (those scoring in the bottom 20 percentile) received private tutoring compared with 89 % of high-achieving students (those scoring in the top ten percentile). This same study found that the expenditure on private instruction by the high-achieving students was 3.1 times that of the low-achieving students (Lee 2007). Economic disparity was not the only difference affecting private tutoring, as Kang and Hong (2008) note that students in rural areas were less likely to access private instruction compared with their urban counterparts.

Thus, just as in the Joseon Dynasty, families in Korea are expected to invest a great deal of their personal finances into the education of their children as they compete for positions in higher education. We believe this "zeal or fever" (Kang and Hong 2008) for education in present-day Korea is a testament to the continued influence of Confucian beliefs about gaining success and upward mobility through hard work, as well as the government's economic emphasis on developing human resources in postcolonial Korea (Lee 2006). Each November, high school seniors all sit for a universal college entrance exam known as the College Scholastic Ability Test (CSAT), which is reminiscent of the civil service examination practice from a century ago. A strong sense of meritocracy continues to permeate Korean society with the belief that if a student, male or female, rich or poor, performs well on the CSAT, he or she will have the opportunity to succeed at the top national universities. Unlike a century ago, today all students have the right to sit for the exam; however, growing economic disparities between families means that class has emerged yet again as a major factor limiting opportunities for academic equality.

Another factor that has potentially contributed to the growth in private tutoring are the government policies instituted during postcolonial educational reform efforts. At the inception of Korea's economic development, the government was unable to finance school education beyond the primary level, so there is a long history of fiscal responsibility placed on families for educating their own children (Morris, 1996). Although this model was necessary to help offset funds needed for other recovery efforts at the time, it has remained unchanged. Today, about 6% of the gross domestic product (GDP) is spent on education (Ilon 2011), meaning that Korea spends more on education as a proportion of its economy than do all other nations in the OECD except for Iceland (McNeill 2011). However, about 2.5% of the GDP is contributed in private funds by households to provide for tutoring, afterschool programs, test-preparation services, and tertiary education (Ilon 2011), which means that families spend more on private education in Korea than anywhere else in the world (McNeill 2011). This issue continues to plague the Korean education system, and finding ways to minimize negative effects on student achievement is a top priority for the nation. In this final section, we describe two types of programs that are effecting positive change.

After-School Programs and Gifted Education

In Korea, "after-school" programs refer to school activities conducted outside the regular school curriculum. The educational purpose of the after-school programs includes supplementation of the regular school curriculum, cost reduction of private education, and contribution to local education efforts offered by neighborhood schools. As noted earlier, Korean students rely on private education to enhance their school performance. Since 2006, almost all the schools (99.9%) in the nation have implemented after-school programs, and their effects have dramatically changed the activities of students after regular school hours. In 2010, more than 63% of students participated in after-school programming. Based on statistical data and regional surveys, participants of the after-school programs have shown improvement in school performance as well as reduction in private education cost. Thus, after-school programs may be helping to reduce achievement gaps related to economic disparity.

Gifted education programs are also being introduced to help provide schools and families with an additional means for enhancing science and mathematics learning. Today, more than 25 universities provide gifted education programs in the areas of science and mathematics. These programs are a bit unique in that they are geared toward fostering an "in-depth approach" to learning, in contrast with the more commonly used "acceleration approach." In-depth learning approaches, with a focus on project-based curriculums, are designed to support students to exercise their creativity, engage in problem solving and decision making, and develop science presentations and essay writings. Consequently, the contents of the gifted education deviate from the conventional scope of the science curriculum that is delivered K–12 classrooms. These programs are designed to both enhance knowledge and address concerns regarding Korean students' interest in and attitudes toward science.

Cyber Home Learning System

The Cyber Home Learning System (CHLS) is an online education system that was first introduced in 2004. The CHLS is a government-sponsored program meant to enhance public education, narrow educational achievement gaps among students from different regions and socioeconomic classes, and reduce family expenditures on private tutoring by serving as a supplement for student after-school education (Kim 2005; KERIS 2008). Unlike in some other OECD nations, in Korea there is a relatively high rate of home-Internet use (98.5 %) (KISA 2011), so this system is accessible to school-age children in nearly all homes. CHLS supports students by providing learning programs based on the national curriculum for no charge. Students can either study independently or select cyber teachers or tutors who can support and manage students' learning through the CHLS program.

Since 2007, the government has targeted efforts to expand CHLS use by students in rural areas and from low-income families as a way to directly assist students who are socially and economically disadvantaged. Research examining the effectiveness of CHLS use on achievement by students from low-income families and from rural areas has been positive, suggesting CHLS is fostering confidence in learning, increasing academic achievement, and reducing the need for parents to pay for private education (Lee 2009). In 2008, over 3 million students were subscribed to the service, and more than 300,000 accessed the program each day (KERIS, 2008). More than 74 % of users reported a high level of satisfaction (Shin and Shin 2012), and about 82 % of respondents reported positive changes in their academic performance, interest in subjects, and level of confidence after using CHLS (KERIS 2008). CHLS continues to evolve, adding supplementary and advanced programs to existing basic programs and continually revising the program offerings based on changes to the national curriculum. Additional improvements are still needed with regard to technology updates and the quality of instructional design and delivery. CHLS also faces challenges related to differentiating materials to be able to satisfy the needs of students with wide range of abilities. However, these are areas for on-going development.

Conclusions and Implications

There is an ever-changing need for innovative policies, research initiatives, and changes in science teacher practices and teacher education to address inequities in achievement at local and international levels. Clearly, researchers must continue to address many issues related to gender discrimination and gender inequality, especially in the areas of science and mathematics education. In this chapter, we chose to focus on some sociohistorical events that have had lasting effects on education and academic achievement. In the last seven decades, education has become a central structure in Korean society. As detailed in our chapter, research-ers are currently dealing with some developing problems, including inequities in achievement due to economic disparity, as well as continuing to struggle with gender inequality.

In our examination of achievement and education in Korea, the influence of Confucianism became a central focus for our discussion about practices and beliefs in Korean schools. Confucian philosophy has long served as a structure for Korean society, which we believe has had lasting effects on education in many positive ways, including the promotion of high academic standards and a zeal for education, which has served to position Korean students as leaders on several international assessments. At the same time, the Confucian ideals that promote ideologies of meritocracy, reliance on rote memorization, and use of teaching practices that stifle creative thinking may erode some of the gains made over the last two generations, meaning that Korean students might lose their competitive edge. Other countries with less rigid systems can more easily reform their school systems, re-tool teacher education programs, and create new curricula that help to position their students to be more flexible and to be better able to adapt to the shifting needs of an in-creasingly globalized economy. Thus, Korean researchers and policy makers must continue to question and examine the historical impact of Confucian philosophy on education in Korean society if they are to determine which governmental policies and practices will most effectively position Korea to continue to grow and develop as a leader in the global arena in this century.

As evidenced by the policy examples shared in this chapter, work is already being done to address some of these issues, including addressing aspects of social interactions that are inherently discriminatory toward women and people in lower social class levels. More research and teacher education needs to focus on how to promote gender equity in science education and boost achievement of students from different socioeconomic backgrounds. In this chapter, we discussed policies that are expanding opportunities for high-achieving students to enhance their science education experiences with enrichment in mentally gifted programs. In addition, we noted an innovative project leveraging Korea's near-universal access to technology to help support students in rural areas and those whose parents cannot afford private instruction to ensure student success on high-stakes exams. However, additional resources and efforts need to be spent to enhance learning for low-achieving students and students with special learning needs. In addition, teacher education programs need to be revised to address some of the challenges facing the next generation of science teachers. One such example is the increasing need for multicultural education and support for Korean language learners in K–12 classrooms (Cho 2010). However, the issue of immigration is complicated for Korean society, as Korea's population is not very diverse in terms of race, and historically they have experienced little immigration. This lack of diversity means it is difficult for non-Koreans to be integrated into the fabric of society, including the school environment.

However, this is an issue that must be dealt with because Korea has a negative birthrate and an aging population, meaning that the country will have to face some difficult decisions regarding the need to increase immigration if the country is to remain economically productive over time. Currently, there are nearly 1 million registered long-term foreign residents in Korea and an estimated 500,000 illegal foreign residents (KBS World 2012). In a country of 49,000,000 people, the total number of immigrants makes up less than 3 % of the population, but this population has increased rapidly in only 5 years. The government estimates that by 2020, there will be about 5 million immigrants living in Korea (KBW World 2012), making up more than 10 % of the population. So, although the number of immigrant students residing in Korea is small compared to other countries, Korea is experiencing a rapid influx of multicultural families whose children will need to be educated in Korean schools (Cho and Yoon 2011). Currently, no assessment information is available to examine potential differences in science achievement among students from multicultural families. However, many see the welfare and achievement of these children and the successful integration of their families into society as being of critical importance for Korea's future.

Korean society has a well-organized, highly developed system in place for educating their citizenry, and they have a long and powerful history on which they can reflect to apply lessons learned from the past to help continually reshape their future. In addition, Korean researchers can capitalize on the opportunities they have to learn from other countries that have been addressing achievement-related problems resulting from differences in gender, class, and race. The challenge for today's educational reformers is to not only maintain these gains but also continue to expand equitable opportunities for educational advancement. Finding ways to prevent the achievement gap from widening and to implement innovative polices that expand opportunities for all students to pursue science, technology, engineering, and mathematics (STEM) careers will be important areas for research in Korea over the next two decades.

References

Akiba, M., LeTendre, G. K., & Scribner, J. P. (2007). Teacher quality, opportunity gap, and national achievement in 46 countries. *Educational Researchers, 36*, 369–387.

Buzo, A. (2007). *The making of modern Korea*. New York: Taylor & Francis.

Choi, K., Lee, H., Shin, N., Kim, S.-W., & Krajick, J. (2011). Re-conceptualization of scientific literacy in South Korea for the 21st century. *Journal of Research in Science Teaching, 48*(6), 670–697.

Cho, Y.-D. (2010). The actual conditions of the multicultural education in elementary and secondary schools. *Theory and Research in Citizenship Education, 42*(1), 151–184.

Cho'oe, Y. (2006). *From the Land of Hibiscus: Koreans in Hawai'i, 1903–1950*. Hawaii: University of Hawaii Press.

Cho, Y.-D., & Yoon, H. (2011). Korea's initiatives in multicultural education: Suggesting "reflective socialization". In *Global Approaches to the Multicultural Education* (International Joint Research Project of the International Alliance of the Leading Education Institutions [IALEI Research] (pp. 1–31). Seoul: Center for Multicultural Education, Seoul National University.

Cho, J., Kim, S., Lee, S., Kim, M., Ok, H., & Rim, H. (2011). *Comparative analysis of the characteristics of reading achievement based on the results of PISA 2009. KICE Research Report RRE 2011-4-3*. Seoul: Korea Institute for Curriculum and Evaluation http://kice.re.kr/en/board.do?boardConfigNo=138&menuNo=409&action=view&boardNo=29658 Accessed 5 Aug 2012.

Connor, M. E. (2009). *Koreas*. Santa Barbara: ABC–CLIO.

Eckert, C. J., Lee, K.-B., Lew, Y.I., Robinson, M., & Wagner, E.W. (1990). *Korea old and new: A history*. Korea: Ilchokak Publishers.

Elman, B. A. (2000). *A cultural history of civil examinations in late Imperial China*. Berkeley: University of California Press.

Ilon, L. (2011). Can education equality trickle-down to economic growth? The case of Korea. *Asia Pacific Educational Review, 12*, 653–663.

Kang, N.-H., & Hong, M. (2008). Achieving excellence in teacher workforce and equity in learning opportunities in South Korea. *Educational Researcher, 37*(4), 200–207.

KBS World. (2012, March 19). *Registered foreigners near one million*. KBS World news. http://world.kbs.co.kr/english/news/news_zoom_detail.htm?No=6658. Accessed 23 May 2012.

KERIS. (2008). *Overview of the CHLS (Cyber Home Learning System) in Korea*. KERIS Issue Report RM 2008-8.

Kim, G.-J. (2002). Education policies and reform in South Korea. In *Secondary Education in Africa: Strategies for renewal* (pp. 29–39). Washington, DC: World Bank.

Kim, E. H., & Choi, C. (1998). Introduction. In E. H. Kim & C. Choi (Eds.), *Dangerous women: Gender and Korean nationalism* (pp. 1–9). New York: Routledge.

Kim, K., Kim, W., Choi, I., Sang, K., Kim, H., Shin, J., & Kim, S. (2011a). Characteristics of national assessment of educational achievement result and school progress. *KICE Research Report RRE 2011-2-4*. Seoul: KICE. http://kice.re.kr/en/board.do?sortName=boardEtc01&boardConfigNo=138&page=2&menuNo=409&sortOrder=DESC&action=view&boardNo=28506 Accessed 5 Aug 2012.

Kim, H., Lee, I., Jeong, J.S., Shin, M.K., & Kim, M.K. (2011b). The national assessment of educational achievement in 2010: analysis of the science achievement test results. *KICE Research Report RRE 2011-3-5*. Seoul: KICE. http://kice.re.kr/en/board.do?boardConfigNo=139&menuNo=410&action=view&boardNo=29542 Accessed 5 Aug 2012.

Kim, Y. (2005). *Discussion in a network seminar between educational research and development: Establishing Cyber Home Learning System*. Seoul: Korea Education & Research Information Service Report. http://www.keris.or.kr/data/dt_research.jsp?bbsid=board01&gbn=view&gp=2&ps=10&sp=subject&sw=%BB%E7%C0%CC%B9%F6%B0%A1%C1%A4%C7%D0%BD%C0&ob=sor1&ix=2830&orderkey=2 Accessed on 7 June 2012.

Korean Educational Development Institute (KEDI). (2007). *Understanding Korean education: Volume 3: School education in Korea*. Seoul: KYUNGHEE Information Printing.

Korea Institute for Curriculum and Evaluation (KICE). (2012). *Korea–United States bilateral seminar: Study examining turning around low-performing schools*. KICE Research Report ORM 2012-39. Seoul: Korea Institute for Curriculum and Evaluation.

Korean Internet Security Agency (KISA). (2011). *Internet usage statistics*. http://isis.nic.or.kr/. Accessed 4 July 2012.

Lankov, A. (2012, April 12). Confucianism in Korea. *Korea Times*. http://www.koreatimes.co.kr/ www/news/opinon/2012/07/165_108831.html. Accessed 7 July 2012.

Lee, J. (2007). Two worlds of private tutoring: The prevalence of after-school mathematics tutoring in Korea and the United States. *Teachers College Records, 109*, 1207–1234.

Lee, J. (2009). *A study on satisfaction and effectiveness of the Cyber Home Learning System for low-income students*. Unpublished master's thesis, Graduate School of Education, Konkuk University, Republic of Korea.

Lee, J.-H. (2004, Spring). The school equalization policy of Korea: Past failures and proposed measure for reform. *Korea Journal, 221–234*.

Lee, J. K. (2006). Educational fever and South Korean higher education. *Revista Electronica de Investigation Educativa, 8*(1), 1–14. http://redie.uabc.mx/vol8no1/contents-lee2.html. Accessed 27 June 2012.

Lee, K.-B. (1988). *A new history of Korea*. (trans: E. W. Wagner). Cambridge: Harvard University Press - Harvard-Yenching Institute Publications.

Lee, K. H., Lee, E. J., Paik, S. H., & Lee, H. (2010). Discovering the potential of gifted females in mathematics. In H. J. Forgasz et al. (Eds.), *International perspectives on gender and mathematics education* (pp. 287–314). Charlotte: Information Age.

McNeill, D. (2011, November 27). After decades of building colleges, South Korea faces a lack of students. *The Chronicle of Higher Education*, Global News Section. http://chronicle.com/ article/article-content/129896/. Accessed 1 July 2012.

Ministry of Education, Science and Technology (MEST). (2004). *History of education: National history compilation*. Seoul: MEST.

Ministry of Education, Science and Technology (MEST). (2009). *Introduction: Education system*. Seoul: MEST.

Ministry of Education, Science and Technology (MEST). (2012). *National report on the development and state of the art of adult learning and education*. Seoul, Republic of Korea: MEST. http://www.unesco.org/fileadmin/MULTIMEDIA/INSTITUTES/UIL/confintea/pdf/National_ Reports/Asia%20-%20Pacific/Rep._of_Korea.pdf. Accessed on 23 June 2012.

Morgan, J. (2010, December 9). Appetite for Education. *Times Higher Education*. http://www. timeshighereducation.co.uk/story.asp?storycode=414509. Accessed 1 July 2012.

Morris, P. (1996). Asia's Four Little Tigers: A comparison of the role of education in their development. *Comparative Education, 32*(1), 95–109.

National Center for Education Statistics (NCES). (2009). *Highlights from TIMSS 2007: Mathematics and science achievement of U.S. fourth and eighth-grade students in an international context*. NCES 2009-001 Revised document prepared for the U.S. Department of Education. Washington, DC: Inst. of Education Sciences.

OECD. (n.d.). *About the Organisation for Economic Co-operation and Development (OECD)*. http://www.oecd.org/about/. Accessed 27 July 2012.

OECD. (2005). *Society at a glance: OECD social indicators*. Paris: OECD. http://www.oecd-ilibrary.org/social-issues-migration-health/society-at-a-glance-2005_soc_glance-2005-en Accessed 27 July 2012.

OECD. (2006). *Education at a glance: OECD Indicators*. Paris: OECD. http://www.oecd-ilibrary. org/education/education-at-a-glance-2006_eag-2006-en Accessed 27 July 2012.

OECD. (2010a). *PISA 2009 results: What students know and can do: Student performance in reading, mathematics and science* (Vol. I). Paris: OECD. http://www.oecd-ilibrary.org/education/ pisa-2009-results-what-students-know-and-can-do_9789264091450-en Accessed 27 July 2012.

OECD. (2010b). *PISA 2009 results: Overcoming social background: Equity in learning opportunities and outcomes* (Vol. II). Paris: OECD. http://www.oecd-ilibrary.org/content/ book/9789264091504-en Accessed 27 July 2012.

OECD. (2010c). *PISA 2009 results: What makes a school successful? Resources, policies and practices* (Vol. IV). Paris: OECD. http://www.oecd-ilibrary.org/education/pisa-2009-results-what-makes-a-school-successful_9789264091559-en Accessed 27 July 2012.

OECD. (2011). *Strong performers and successful reformers in education: Lessons from PISA for the United States*. Paris: OECD. http://dx.doi.org/10.1787/9789264096660-en. Accessed 27 July 2012

Reinfeld, W. (1997). Tying infrastructure to economic development: The Republic of Korea and Taiwan (China). *Infrastructure strategies in East Asia: The Untold Story*. In A. Mody (Ed.), *Infrastructure strategies in East Asia: The untold story* (pp. 3–26). Washington, DC: World Bank.

Seth, M. J. (2006). *A concise history of Korea: From the Neolithic period through the nineteenth century*. New York: Rowman & Littlefield.

Seth, M. J. (2010). *A concise history of modern Korea: From late nineteenth century to the present*. Plymouth: Rowman & Littlefield.

Schmid, A. (2002). *Korean between empires, 1895–1919*. New York: Columbia University Press.

Shin, S., & Shin, S. (2012). The analysis of the factors affecting the impact of Cyber Home Learning on elementary grade. *Journal of Korean Association of Information Education, 16*(1), 91–98.

Sorenson, C. W. (1994). Success and education in South Korea. *Comparative Education Review, 38*(1), 10–35.

Takaki, R. (1989). *Strangers from a different shore: A history of Asian Americans*. Boston: Little, Brown & Co.

Chapter 12
Closing the Achievement Gap in Singapore

Jason TAN

Unifying a Fragmented Set of Schools, 1959–1979

When the ruling People's Action Party (PAP), which has governed for an uninterrupted period of 44 years since 1959, first came into power at the head of a self-governing state, one of the key items on its policy agenda was education. This was because education was seen as a way to provide the manpower needed urgently for Singapore's industrialization plans. In addition, education was viewed as playing a crucial role in developing social cohesion in a multilingual, multiethnic, and multireligious society. This latter role gained greater prominence when Singapore became a full-fledged independent nation in 1965. The departing British colonial authorities had left behind an incoherent education system that was divided into four separate media of instruction: English, Chinese, Malay, and Tamil. Each of these school systems had its individual textbooks, curricula, examinations, and teacher qualifications and salaries. Those enrolled in Malay and Tamil medium primary schools lacked access to secondary and postsecondary schooling in these languages.

The PAP embarked on a series of measures during the 1960s and the 1970s in order to unify a fragmented set of compartmentalized education systems. These included standardizing textbooks, curricula, examinations, and teacher qualifications and salaries. In addition, the party built primary and secondary schools at a rapid rate in order to increase student enrollments. By 1966, primary education had become universal. This was a major milestone toward redressing a situation where schooling opportunities were relatively scarce.

Right from the beginning of its reign, the PAP declared that Singapore would operate on the founding principles of multiracialism and meritocracy. These ideals were supposed to ensure that all the major "races" (the official nomenclature was Chinese, Malay, Indian, and Others) would be treated fairly and equally, and that social mobility and advancement would be through one's individual merit as measured by examination performance. Students were supposed to compete fairly

J. TAN (✉)
Policy and Leadership Studies, National Institute of Education, Singapore, Singapore
e-mail: engthye.tan@nie.edu.sg

J. V. Clark (ed.), *Closing the Achievement Gap from an International Perspective*,
DOI 10.1007/978-94-007-4357-1_12, © Springer Science+Business Media B.V. 2014

on equal grounds, and the best performers would justly deserve rewards in the form of opportunities for subsequent educational advancement and better-paying jobs. The hidden message was that this system of meritocracy would invariably result in unequal educational and job market outcomes, but that these inequalities were just and fair.

Increasing Diversity (and Elitism) in Education from 1979 onward

After 2 decades of policies aimed largely at standardizing the school system and providing students with a one-size-fits-all curriculum, a major watershed occurred in early 1979 with the publication of a report that had been commissioned by the then Prime Minister, Lee Kuan Yew. The report tried to address major flaws such as high student dropout rates at both primary and secondary levels, which were compared unfavorably with those in France, Taiwan, Britain, and Japan (Ministry of Education 1979). The recommended solutions included instituting streaming policies at the end of the third year of primary schooling and at the end of the sixth year of primary schooling. Streaming was supposed to better address the diversity in students' learning capacities, by moving away from the rigidity of a one-size-fits-all curriculum. Students across different streams would be exposed to curricula of differing levels of difficulty (with provision being made for lateral movement across streams) in the hope that more of them would be able to remain longer in school and attain better literacy outcomes. As a result of the report's recommendations, streaming was institutionalized in primary schools at the end of 1979 and in secondary schools at the end of 1980. A subsequent Education Ministry report in 1991 recommended refinements to streaming while leaving the underlying premise of streaming untouched.

The 1979 report marked a new phase in Singapore's education development, namely, diversifying the education landscape after a two-decade experiment with providing a common set of experiences for all school students. Though the streaming of students was supposed to lead to improvements in the learning outcomes of all students, it also meant not only the institutionalization of unequal learning outcomes but also the de facto segregation of students, both within and across schools. For instance, a small number of more well-known secondary schools were allowed to enroll only students in the more prestigious streams. Further policy moves were made during the 1980s, this time to provide top-performing students with superior learning opportunities. These included the Gifted Education Programme (which allowed for greatly reduced class sizes), Art Elective Programme, Music Elective Programme, and Language Elective Programme. In addition, a select number of prestigious secondary schools were allowed to become independent schools, enjoying increased operating autonomy in matters such as fees, class size, enrollments, and teacher deployment, while continuing to receive substantial government grants.

Beginning in the 1990s, in response to a public outcry that such schools were elitist in nature, the government turned a number of secondary schools into autonomous schools, which would offer high-quality education while charging lower fees than the independent schools. By this time, the government was proclaiming the success of its streaming policies in reducing school dropout rates and ensuring the universality of secondary schooling. However, the introduction of various initiatives targeted expressly at students at the top end of the achievement spectrum further institutionalized the growing inequality of student learning outcomes. The Ministry of Education's official rhetoric about desired outcomes of education (Ministry of Education 1998) and twenty-first-century competencies (Ministry of Education 2010) begged the question about whether, in fact, all students were expected to attain equal outcomes and equal levels of competency (see, e.g., Ho 2012a).

In retrospect, these initiatives aimed at identifying and selecting the top layer of students were understandable in the light of Lee's entrenched elitist philosophy, which involved having a tiny educated elite of "top leaders" at the apex of what he termed a "pyramidal structure" governing a middle layer of "good executives" and the "well-disciplined and highly civic-conscious broad mass" (Lee 1966, p. 13). The education system had to be segmented accordingly so as to nurture the talents of the "top leaders," develop "high-quality executives" to help the leaders implement their plans, and train the "broad mass" to "respect their community and…not spit all over the place" (ibid., p. 13). An additional layer in Lee's thinking was his eugenicist beliefs and his abiding concern that the well-educated Singaporeans were failing to reproduce themselves in adequate numbers compared to their less-educated counterparts. This concern was so pressing that Lee attempted a short-lived policy that provided the children of female university graduates priority in admission to primary schools beginning in 1984. This policy was revoked after 1 year due to extreme public unhappiness (Saw 2012). Nevertheless, Lee persisted in his deeply held views and made periodic statements about the genetic basis of intelligence, creativity, and leadership qualities (see, e.g., "Entrepreneurs are born, not made" 1996; "How Singapore grooms its leaders" 2005; Lee 2011; Parliamentary Debates, 66, 1996, Cols. 331–345; Parliamentary Debates, 70, 1999, Cols. 1651–1653). Such remarks raise the question of the role of schools in addressing the achievement gap, and whether the gap is in fact bridgeable.

Yet another move in the direction of greater diversity and segmentation within the school system came in 2002 with the publication of an Education Ministry report that recommended that the top-performing secondary school students be allowed to bypass the national secondary school examinations and enjoy, instead, 6 years of secondary education before sitting for their university entrance examinations. The report also recommended offering these students a greater variety of terminal examination qualifications. The underlying idea behind bypassing the national examinations is to reduce the amount of time spent on coaching students for these examinations, thereby providing more time for these students to develop higher-order thinking skills, as well as nonacademic outcomes such as leadership skills. In the wake of the publication of this report, a few independent schools and autonomous schools began offering what are termed in local parlance "integrated programs." These "integrated programs," which involved 6 years of secondary schooling before sitting for a major examination, were also offered in the newly established independent schools specializing in such fields as sports, mathematics and science, and the arts.

Interschool Competition: Fueling Inequality

The impact of all of these policy initiatives favoring the top-performing schools and students has been heightened, since the 1990s with the publication of the annual league tables of the academic and nonacademic outcomes of secondary schools. In addition, the introduction of a quality-assurance mechanism known as the School Excellence Model, along with an associated system of annual awards to schools based on their achievement in academic and nonacademic outcomes, has led to a strategic decision making on the part of some school leaders in terms of such matters as admitting students who are likely to prove to be "assets" to the school, and concomitantly reducing the intake of students who are likely to prove to be "liabilities." Furthermore, some schools have taken steps to reduce student enrollments in subjects that are deemed more difficult to do well in (Tan 2008). Anecdotal evidence would also suggest that some school leaders have reduced the number of cocurricular activities in order to better focus on activities that win awards and have also reduced opportunities for students to participate in activities purely for recreational as opposed to competitive purposes. These trends, which would appear to limit the opportunities of some students for development in both academic and nonacademic domains, have been given new life with the advent of the Direct School Admission (DSA) scheme in 2004. This scheme allows secondary schools, especially the independent schools and autonomous schools, discretion in admitting a certain percentage of their annual student intakes before the students receive their Primary School Leaving Examination (PSLE) results. The DSA scheme has intensified interschool competition for students with proven academic and nonacademic track records, and consequently limited opportunities for non-DSA students to enroll in cocurricular activities. In addition, parents and students have to engage in strategizing well ahead of the annual DSA exercises in order to chalk up a personal portfolio of success.

Another serious consequence of all the intense interschool competition, which is part of an overall marketization of education as a commodity, is that of a growing prestige hierarchy of schools and social stratification. There is already evidence that students from wealthier home backgrounds are overrepresented (as are students from the majority ethnic Chinese community) in independent schools (see, e.g., Tan 1993). In addition, the expansion of educational enrollments across the board has done little to reduce the intense competition for places in the more prestigious primary and secondary schools (as displayed, for instance, in the inflated property values in the vicinity of popular schools), which has in turn fueled the growth of the private tutoring industry (Tan 2009). The growth of this industry has consequences for closing the achievement gap, as there is evidence that poorer households find private tutoring less affordable than wealthier households (Blackbox Research 2012). The growth of the tutoring industry is one manifestation of the phenomenon that some researchers have termed a "parentocracy," where the role of parental, financial, social, and cultural capital becomes increasingly crucial in terms of active strategizing in order to ensure children's success in school.

Trying to Reduce the Achievement Gap: Egalitarian Impulses

Reference has already been made earlier in this chapter to some of the adverse consequences of streaming students at the primary and secondary levels. Other consequences include de facto (although unintended) ethnic segregation within and across schools (see, e.g., Kang 2004). This is due to the fact that disproportionately large percentages of ethnic Malay and Indian students are streamed into the slower-paced streams at both the primary and secondary levels. These disparities result in ethnic Malay and Indian students (and working-class students) being underrepresented in most of the most prestigious schools and being correspondingly overrepresented in some of the least prestigious schools. There is also evidence that streaming has contributed to prejudice on the part of students in faster-paced streams, and on the part of teachers as well, toward students in slower-paced streams (see, e.g., Ho 2012b; Kang 2004). The public concern over the elitist trend in educational policymaking has been given added impetus since the mid-1990s because of the growing income disparities and the prospect of diminished upward social mobility for the less affluent sectors of the population (Ho 2010). The rapid influx of new immigrants over the past 2 decades, as a result of deliberate government policy, has heightened worry among parents, teachers, and local students about the added competitive element that talented foreign students are perceived to represent (see, e.g., Quek 2005). All of this simmering discontent boiled over during the 2011 general election campaign and contributed in part to a drop in the number of votes for the PAP (Chong 2012).

The PAP government's response to the growing public disquiet over streaming and other elitist trends in educational policymaking has been mixed. On the one hand, it has claimed that all schools are good schools (see, e.g., Parliamentary Debates, 63, August 25, 1994, Col. 398) (a claim belied by the intense competition to gain admission into more prestigious schools), while on the other hand, it has stated that the independent schools are to be developed into "outstanding institutions, to give the most promising and able students an education matching their promise" (Parliamentary Debates, 59, January 6, 1992, Col. 18).

At a more concrete level, the PAP government has instituted measures that it claims will have a "leveling up" effect in helping students from less advantaged home backgrounds attain school success. One of these is the provision of financial assistance schemes, which have been in place since the 1960s, and which cater to the payment of school and examination fees, as well as the purchase of uniforms and textbooks. Second, there are financial subsidies for kindergarten students and after-school student centers (Ministry of Finance 2012). Third, each school receives annual grants for activities such as arts appreciation and overseas learning experiences.

A major funding initiative known as Edusave was launched in 1993 by the then Prime Minister Goh Chok Tong. Goh claimed that the scheme would help equalize opportunities for all Singaporeans, with education as the main means of socioeconomic mobility, regardless of their family background. Furthermore, he claimed that the scheme would "temper our meritocratic free market system with compassion and more equal opportunities" to ensure that "all children, rich or poor, are brought to the same starting line, properly equipped to run" (Goh 1990, p. 25). The government declared that the Edusave Endowment Fund would provide each child between the ages of 6 and 16 with an Edusave account, into which the government would make annual contributions. The money in these accounts was to be used for educational purposes. In addition, all nonindependent secondary schools would receive annual per capita grants. Each school would establish an Edusave Grants Management Committee to decide on the allocation of its annual grants. These grants could be used for the purchasing of resources and equipment, the conducting of enrichment programs, and the hiring of administrative support services. Next, three new scholarships were announced. The first, the Edusave Entrance Scholarships for Independent Schools, would be awarded to the top 25 % of the students who qualified for admission to independent schools each year. The second, known as the Edusave (Independent Schools) Yearly Awards, would be awarded to the top 5 % of each year cohort in these schools. The third one, the Edusave Scholarships for Secondary Schools, would be awarded to the top 10 % of the students in nonindependent secondary schools. In subsequent years, the scheme was extended to students in special needs schools, primary schools, privately run Islamic schools, and institutes of technical education. Another extension of the original Edusave idea involved awarding Edusave Merit Bursaries to students from lower-middle- and low-income families who had performed well in school. The workings of the Edusave Scheme reveal government attempts to balance its meritocratic precepts with a healthy dose of social compassion.

Besides the Edusave scheme, various other government schemes have been put in place, especially in the light of the ongoing public disquiet over the prospect of a permanent underclass forming (Tharman 2012). These include the Infocomm Development Authority of Singapore's financial subsidies for disadvantaged and disabled students to own a computer and to obtain broadband access (Infocomm Development Authority of Singapore 2012). In addition, the Ministry of Community Development, Youth and Sports administers a Child Development Account scheme for every child under the age of 12. This account can be used to pay for childcare, kindergarten fees, medical services, spectacles, and computers (Ministry of Community Development, Youth and Sports 2012).

Beyond the provision of financial assistance, the Education Ministry instituted the Learning Support Programme (LSP) in the 1990s in all primary schools. This program aims to help students in the first 2 years of primary schooling, who have been identified by their teachers as lacking basic numeracy and literacy skills. The students are taught separately in pull-out sessions in an attempt to bring these skills up to par. Apart from the LSP, no other official school-based schemes are in place specifically to address STEM achievement gaps at the primary or secondary levels of schooling.

Other Education Ministry policies have attempted to blunt the raw divisive edge of some elitist policies. For instance, over the past decade, there have been moves to blur some of the boundaries across different academic streams at the primary and secondary levels, to encourage greater interaction between primary students enrolled in the Gifted Education Programme and their other schoolmates, and to provide a greater semblance of upward mobility from lower- to higher-prestige academic streams. Recent official reviews of primary and secondary education have recommended the provision of additional resources, such as after-school study facilities in order to help students from disadvantaged home backgrounds. After years of a relatively hands-off attitude toward special needs schools, two Enabling Masterplans have been published in the past 5 years, calling for greater government involvement in the funding and running of special needs schools, improved professional opportunities for students in these schools, and more integration of students with special needs within mainstream and special schools (Poon 2012). Yet another sector of education that has received renewed official attention is the preschool sector, especially after the publication of an Economist Intelligence Unit report, in which Singapore scored relatively weakly in terms of preschool quality (Lien Foundation 2012). The report's findings have renewed public calls for the government to play a more interventionist role in the provision, funding, and regulation of preschools.

The Ethnic Malay Minority: Catching Up

Earlier on in the chapter, reference was made to the problems faced by ethnic Malay minority students. Forming 13.4 % of the Singapore population, the gap between them and their ethnic Chinese counterparts (forming 74.1 % of the population) has been the focus of considerable attention by both the PAP as well as the Malay community leaders for over 50 years. The constitution in the newly self-governing Singapore recognized the Malays as the indigenous people and explicitly proclaimed the government's responsibility "to protect, safeguard, support, foster and promote their political, educational, religious, economic, social and cultural interests and the Malay language" (Singapore Government 1958, p. 1). Limited affirmative action policies were introduced in the early 1960s, including the provision of free secondary and tertiary education, special bursaries and scholarships, free textbooks, and transport allowances. However, the government refused to accede to requests by the opposition Members of Parliament (MPs) for special Malay quotas in employment and trading licenses. Instead, Lee Kuan Yew claimed that such quotas would not benefit the majority of Malays. The Malay MPs kept urging the Malays to adopt correct mental attitudes in order to succeed and compete with the non-Malays (Parliamentary Debates 25, March 16, 1967, Col. 1337).

The issue of Malay educational underachievement reassumed center stage on the political scene in 1981, when the 1980 population census results revealed a growing underrepresentation of the Malays in the professional/technical and administrative/managerial sectors of the workforce. In addition, the Malays formed only 1.5% of the total number of adults with university degrees. In August 1981, Lee urged the Malay leaders and educationists to give top priority to upgrading the educational level and training of the large number of Malays without a secondary education. As a result, the Council on Education for Muslim Children (or Mendaki, in the Malay language) was established in October that year. In his opening address at the Mendaki congress in May 1982, Lee claimed that "it is in the interests of all [Singaporeans] to have Malay Singaporeans better educated and better qualified" (Lee 1982, p. 6). He also promised government assistance in the form of making premises available for Mendaki, and by urging non-Malay teachers to help Mendaki. Lee also claimed that a government-run scheme would not match community-run efforts because the latter would be able to "reach them through their hearts, not just their minds" (Lee 1982, p. 9).

Over the past 30 years, Mendaki's efforts have revolved mainly around three main areas: conducting tutoring classes from primary to pre-university levels with a focus on examination preparation; providing scholarships, bursaries, and study loans for undergraduate and postgraduate students; and promoting Islamic social values that will support educational success. Ethnic Malays were allowed by the government to make voluntary monthly contributions from the Central Provident Fund accounts. In addition, a Mendaki-Ministry of Education Joint Committee was set up in 1989 as another visible gesture of support.

Less than a decade after the formation of Mendaki, a rival organization, the Association of Muslim Professionals (AMP), was set up in 1991 in order to address the lack of popular support for Mendaki due to its overly close political ties with the PAP Malay MPs. The AMP has focused on conducting educational programs, preschool education, family education, and promoting greater Malay economic participation. Shortly after the formation of the AMP, the government collaborated with ethnic Indian community leaders to establish the Singapore Indian Development Association in 1991 to tackle the problem of Indian students' educational underachievement. In the following year, the Chinese Development Assistance Council was established with government assistance in the wake of Goh Chok Tong's comments that the PAP's unsatisfactory performance in the 1991 general elections reflected discontent by the poorer ethnic Chinese, who felt neglected by the government's focus on helping the Malays. At the same time, the pre-existing Eurasian Association received government financial support for its endowment fund to finance education and welfare programs. The Mendaki-Ministry of Education Joint Committee was expanded to embrace these newer ethnic-based self-help groups.

The idea of using ethnic-based self-help groups has been controversial as critics have decried their incompatibility with multiracial ideals and have claimed that many of the issues facing educational underachievers might in fact be socioeconomic in nature rather than specifically ethnic. In response, the PAP has stuck to its assertion that a national body would not be sensitive enough to the needs of each ethnic community. Community-based efforts are more effective because they draw on and mobilize deep-seated ethnic, linguistic, and cultural loyalties (see, e.g., Parliamentary Debates, 86, 2009, Cols. 1174–1176).

Thirty years after the formation of Mendaki, despite reductions in Malay students' dropout rates from primary and secondary schools and improvements in their performance in national examinations, quantitative gaps persist between the Malays and the ethnic Chinese majority. The limited official data on STEM achievement have revealed the steady gaps in ethnic Malay mathematics and science achievement at the national-level PSLE. (No STEM achievement data are available that highlight the effect of socioeconomic status.) For instance, the Malay and Chinese pass rates in mathematics in 2002 were 56.5 and 90.2%, respectively. Almost a decade later, the respective pass rates in 2011 were 60.1 and 89.4%. Likewise, the Malay and Chinese pass rates in PSLE science were 77.5 and 95.0% in 2002, and 73.8 and 94.3% in 2011, respectively. Mathematics pass rates in the national-level General Certificate of Education Ordinary Level examination show similar gaps, with Malay students' pass rates in 2002 and 2011 being 66.9 and 71.2%, respectively, vis-à-vis ethnic Chinese students' pass rates of 92.2 and 92.8%, respectively (Ministry of Education 2012).

In addition, the Malays continue to be grossly underrepresented at the universities even amid tremendous government efforts to expand higher education enrollments over the past two decades. Anecdotal evidence suggests that the Malay students are overrepresented in the slower-paced streams or achievement bands in primary and secondary schools and correspondingly underrepresented in the more prestigious streams or achievement bands. To date, neither the Mendaki nor the AMP has been able to show conclusively what impact, if any, they have had on improving the Malay educational achievement in general, or STEM achievement in particular. Nor have there been any research studies that establish the precise nexus of factors—cultural, educational, or structural—that account for the persistence of the interethnic educational achievement gap. The task of improving the Malay educational achievement has not been made any easier by Lee's entrenched view that "the Malays are not as hardworking and capable as the other races" (Plate 2010, p. 53) and his belief in the genetic basis of the Malays' educational shortcomings (Lee 2011, p. 188; Plate 2010, p. 53). Lee has claimed that despite official efforts to help the Malays,

> They will never close the gap with the Indians and the Chinese, because as they improve, the others will also improve. So the gap remains. They are improving but they are not closing that gap. That's a fact of life. (Lee 2011, p. 206)

Conclusion

This chapter has discussed the existence of educational achievement gaps along mainly social class and ethnic lines despite Singapore's much-flaunted international success in STEM assessment measures. It has discussed the PAP government's claims that the education system plays a key role in maintaining a meritocratic society. It has also shown how certain education policies such as streaming and the diversification and segmentation of the education landscape have been motivated largely by the press to identify and select the future elite that will assume leadership roles in society. Official rhetoric has claimed that these efforts are logically superior to a one-size-fits-all system as they better cater to different learning needs. Since the 1990s, the marketization of education and the increasing competition among schools for awards in both academic and nonacademic domains has intensified the rush to recruit students who are "assets" instead of "liabilities," a trend that tips the balance in favor of students who have the requisite home support to do well in school. The proliferation of the private tutoring industry has further implications for the extent to which the achievement gap between the financially better-off and the financially disadvantaged can be bridged.

In response to the growing public disquiet over the elitist nature of some of these policies and widening income disparities, the PAP has claimed that "we cannot narrow the [income] gap by preventing those who can fly from flying....Nor can we teach everyone to fly, because most simply do not have the aptitude or ability" (Goh 1996, p. 3). At the same time, it has instituted various egalitarian policy measures that provide financial or pedagogical assistance in a concerted bid to reduce achievement gaps. These include the Edusave scheme, the LSP, and the Enabling Masterplans for special needs students. More recently, the PAP has come under pressure to improve the quality of preschool provision. To date, no specific school-based programs have been instituted to address the STEM achievement gaps, whether they be ethnic- or social class-based. Official STEM achievement data are scant and provide no hints of socioeconomic gaps, but of the ethnic Malay minority falling behind the ethnic Chinese majority at both the primary and secondary levels of schooling.

This chapter also highlighted the particular case of the Malays, who form the largest ethnic minority. It discussed the historical evolution of government thinking on addressing the Malays' educational problems and highlighted the formation of Mendaki in 1982. The official endorsement of an ethnic-based self-help approach paved the way for the formation of other such ethnic-based self-help groups such as the AMP. Despite criticism over the efficacy of these groups, the PAP has insisted that ethnic-based efforts are far superior to an ethnically neutral approach as they harness deeply seated ethnic loyalties. Evidence of the efficacy of the Mendaki and the AMP is equivocal to date, as no data have been presented about the contribution of their private tutoring schemes to improving ethnic Malay STEM achievement.

The chapter has highlighted the intensely political nature of education policy-making. It has shown the recurrent tensions between the PAP's elitist and egalitarian impulses, between Lee's deeply held eugenicist beliefs and the need to respond to voter discontent. Not only has Lee gone on record as saying that the Malays will never close the achievement gap, but he has also claimed that "we are trying to reach a position where there is a level playing field for everybody which is going to take decades, if not centuries, and we may never get there" (Parliamentary Debates, 86, 2009, Col. 1173). This latter statement would appear somewhat at odds with the PAP's espoused meritocratic ideals. It is, however, a frank admission that the achievement gaps, including those in STEM subjects, in Singapore's education system are not amenable to quick-fix solutions but are, rather, permanent features of the landscape.

References

Adekile, A. (2012). Achievement gap. In J. A. Banks (Ed.), *Encyclopedia of diversity in education* (pp. 16–21). Los Angeles: Sage.

Blackbox Research Pte Ltd. (2012). Private tuition in Singapore: A whitepaper release. http://www.blackbox.com.sg. Accessed 25 July 2012.

Chong, T. (2012). A return to normal politics: Singapore general elections 2011. In D. Singh & P. Thambipillai (Eds.), *Southeast Asian Affairs 2012* (pp. 283–298). Singapore: Institute of Southeast Asian Studies.

Entrepreneurs are born, not made. (1996, July 11). *The Straits Times*, 25.

Goh, C. T. (1990). Edusave–a dividend to our citizens, an investment in our future. *Speeches, 17*(5), 8–13.

Goh, C. T. (1996). Narrowing the income gap. *Speeches, 20*(3), 1–4.

Ho, K. W. (2010). Social mobility in Singapore. In T. Chong (Ed.), *Management of success: Singapore revisited* (pp. 217–241). Singapore: Institute of Southeast Asian Studies.

Ho, L. (2012). Sorting citizens: Differentiated citizenship education in Singapore. *Journal of Curriculum Studies, 44*, 403–428.

Ho, P. S. K. (2012b). "I have won a world championship and now I can retire": Exploring normal technical students' ways of unpacking academic expectations in Singapore. *International Journal of Educational Development, 32*, 111–119.

How Singapore grooms its leaders. (2005, November 18). *The Straits Times*, 22.

Infocomm Development Authority of Singapore. (2012). *Enriching a child's future.* Singapore: Author.

Kang, T. (2004). Schools and post-secondary aspirations among female Chinese, Malay and Indian normal stream students. In A. E. Lai (Ed.), *Beyond rituals and riots: Ethnic pluralism and social cohesion in Singapore* (pp. 146–171). Singapore: Eastern Universities Press.

Lee, K. Y. (1966). *New bearings in our education system.* Singapore: Ministry of Culture.

Lee, K. Y. (1982). Mendaki's task is to raise education of all Malays. *Speeches, 5*, 5–25.

Lee, K. Y. (2011). *Hard truths to keep Singapore going.* Singapore: Straits Times Press.

Lien Foundation. (2012). *Singapore's preschool education placed 29ᵗʰ amongst 45 countries on the Starting Well Index.* http://www.lienfoundation.org. Accessed 20 July 2012.

Ministry of Community Development, Youth and Sports. (2012). *Enhancements to the child development account (CDA).* http://www.mcys.gov.sg. Accessed 28 June 2012.

Ministry of Education. (1979). *Report on the Ministry of Education 1978.* Singapore: Author.

Ministry of Education. (1998). *Desired outcomes of education.* Singapore: Author.

Ministry of Education. (2010). *Nurturing our young for the future: Competencies for the 21st century.* Singapore: Author.

Ministry of Education. (2012). Performance by ethnic group in national examinations 2002–2011. http://www.moe.gov.sg. Accessed 28 Feb 2013.

Ministry of Finance. (2012). *To the resident.* Singapore: Author.

Parliamentary Debates, 25, 1967; 59, 1992; 63, 1994; 66, 1996; 70, 1999; 86, 2009.

Plate, T. (2010). *Conversations with Lee Kuan Yew: Citizen Singapore: How to build a nation.* Singapore: Marshall Cavendish Editions.

Poon, K. K. (2012). The education of children with special needs: History, trends and future directions. In J. Tan (Ed.), *Education in Singapore: Taking stock, looking forward* (pp. 101–111). Singapore: Pearson.

Quek, T. (2005, February 13). China whiz kids: S'pore feels the heat. *The Straits Times*, 3–4.

Saw, S. H. (2012). *The population of Singapore* (3rd ed.). Singapore: Institute of Southeast Asian Studies.

Singapore Government. (1958). *Singapore (Constitution) Order in Council, 1958*. Singapore: Government Printer.

Tan, J. (1993). Independent schools in Singapore: Implications for social and educational inequalities. *International Journal of Educational Development, 13*, 239–251.

Tan, J. (2008). The marketisation of education in Singapore. In J. Tan, & P. T. Ng (Eds.), *Thinking schools, learning nation: Contemporary issues and challenges* (pp. 19–38). Singapore: Pearson.

Tan, J. (2009). Private tutoring in Singapore: Bursting out of the shadows. *Journal of Youth Studies, 12*, 93–103.

Tharman, S. (2012). Transcript of remarks by Mr Tharman Shanmugaratnam Deputy Prime Minister, Minister for Finance & Minister for Manpower at the PPIS 60th anniversary charity gala dinner. http://www.mof.gov.sg. Accessed 28 June 2012.

Part VI
Africa

Chapter 13
Equity Deferred: South African Schooling Two Decades into Democracy

Nick Taylor and Johan Muller

Introduction

Apartheid as a formal political system in South Africa lasted for just over four decades—from 1948, when the white National Party came to power, reaching an official end with the first democratically held elections of 1994. Apartheid was a formal policy of racial segregation and political and fiscal inequality that, brief as it was, rested on centuries of colonialism that had established the pattern of inequality in a less formal but nevertheless deeply ingrained manner. Consequently, the apartheid legacy casts a long and pernicious shadow over present attempts to construct a just and equal society in South Africa.

Nowhere were the inequities of South Africa's policies of racial discrimination more apparent than in the sphere of schooling. In the case of African children, provision occurred in geographical silos—in departments run by either the nominally "independent" homeland areas of Bophuthatswana, Transkei, Ciskei and Venda or the six self-governing territories. The rapidly growing cohorts of African children in the urban township were schooled by the Department of Education and Training. All other children were schooled by the national departments under the Houses of Assembly (whites), Representatives (coloured) or Delegates (Indian). Whereas access to schooling in the junior grades for African children was catered to (albeit inadequately), access to secondary schooling was restricted. At the height of the apartheid era, public spending on white children was around 5 times the amount spent on Africans, with per capita allocations for coloured and Indian pupils falling somewhere in between (Buckland and Fielding 1994; SAIRR 2011). Resistance to

N. Taylor (✉)
National Education Evaluation and Development Unit,
Ministry of Basic Education, Pretoria, South Africa
e-mail: ntaylor@jet.org.za

J. Muller
School of Education, University of Cape Town, Cape Town, South Africa
e-mail: johan.muller@uct.ac.za

J. V. Clark (ed.), *Closing the Achievement Gap from an International Perspective*,
DOI 10.1007/978-94-007-4357-1_13, © Springer Science+Business Media B.V. 2014

apartheid education became an obvious rallying point for resistance to apartheid in general.

Under the dual pressures of political opposition and demands from the public and private sectors for a literate workforce, the Nationalist government began belatedly but rapidly to expand access to secondary schooling for Africans in the early 1970s, especially in the urbanizing areas, paradoxically swelling the ranks of the articulate disenfranchised whose anger boiled over in the 1976 school riots with their epicenter in Soweto. Though on a political level, this undoubtedly paved the way for apartheid's demise, this rapid expansion of secondary schooling ran in advance of the capacity to run it efficiently, and a pattern of pre-university teacher qualifications became entrenched, which remains the norm to this day. Despite formal remuneration parity for all teachers by 1992 (Edupol 1993), less than 5% of teachers had university qualifications in their teaching subjects by 2006 (Reddy and Kanjee 2006, p. 110). The schooling system serving 80% of the population was thus built on weak foundations and continues to deliver poor-quality learning outcomes, despite nearly two decades of redistributive spending since the advent of democratic government.

In 1994, an improvement strategy was initiated, redistributing the budget toward the poorest provinces and toward historically disadvantaged schools. This included providing funding for daily meals in the poorest 50% of schools. Since 2006, the poorest two quintiles of schools have been classified as "no-fee schools," and this was later extended to include the third quintile (DBE 2011). Non-personnel spending is redistributive: public spending on the poorest fifth of schools is roughly six times higher than spending on the richest fifth of schools (van der Berg et al. 2011). As a result, recurrent per capita public spending today is higher for African than for white children, although schools in more affluent communities remain better resourced due to the practice of charging school fees (NPC 2011). Nevertheless, the country's per pupil expenditure, at USD 1383, compares favourably with that of the sub-Saharan African (USD 167) and Latin American (USD 614) averages (DBE 2011). South Africa's school system is relatively well financed when compared to those of the large majority of developing countries. Indeed, South Africa is usually classified as a middle-income country. Nonetheless, as the chapter illustrates below, its average comparative test scores continue to fall below the average of many of its lower-income regional neighbours[1].

---

[1] Since the focus in this chapter is on comparisons with other African countries where, TIMSS data aside, science outcome data is not available, we have not discussed science outcome data further below. For a reason that might yet require explanation, testing data in Africa have focused on literacy and mathematics, not on science.

Achievement Gaps

Until relatively recently, the only national attainment test written by South African learners has been the school-leaving "matriculation" examination at the end of grade 12. Since attainment on this examination is hardly a basis for international comparison, until the advent of international tests, educational authorities were unaware of the relative performance of South African learners and consequently had no tool to measure and compare the performance of the schooling system as a whole with that of comparable countries. The Third International Mathematics and Science Study (TIMSS), which ran international tests for grade-8 learners in 1995, 1999 and 2003, included South Africa from the outset (Mullis et al. 2000, 2004). South Africa did poorly in the first round in 1995, but because there were only two other African countries in the sample—Morocco and Tunisia—local comparisons were not available, and anyway, reforms had started in earnest only in 1998 with the introduction of an outcomes-based curriculum, so the education authorities were not unduly worried.

Poor performance of the country's school system first registered in a graphic way in 1999 with the results of the United Nations Educational, Scientific and Cultural Organization (UNESCO)-coordinated Monitoring Learning Achievement (MLA) study, which tested all grade-4 learners in numeracy and literacy in 18 African countries (Taylor et al. 2003). South Africa recorded mean country scores of 48% for literacy and 30% for numeracy (Chinapah et al. 2000; Chinapah 2003). It was a shock to find South Africa performing below all 17 other African countries, below even Botswana and impoverished Malawi, which scored 43% for numeracy (Reddy and Kanjee 2006).

In addition to the MLA, South Africa has participated in seven cross-country comparative studies: the TIMSS (grade-8 mathematics and science, 1995, 1999, 2003 and 2011), Progress in International Reading Literacy (PIRLS) (grade-4 and grade-5 reading, 2006) and Southern and Eastern Africa Consortium for Monitoring Education Quality (SACMEQ) (grade-6 reading and mathematics, 2005 and 2007). The message coming from all these was unambiguous: the country performs poorly compared to many of its more impoverished neighbours, and very poorly in relation to developing countries in other parts of the world (Taylor et al. 2008). For example, in the round of SACMEQ testing conducted in 2000, of the 14 southern and eastern African countries participating, South Africa was placed ninth in both reading and mathematics. South Africa scores lower than a number of countries whose per capita gross national income (GNI) figures are around one-tenth of South Africa's figures. Matters are not getting better: results from the SACMEQ III exercise conducted in 2007 again place the country in the bottom half of the 15 African-country samples (SACMEQ 2011; Spaull 2011).

There is a thus an achievement gap between South Africa and seven SACMEQ countries, but the within-country gap is much larger. Poorer children receive schooling inferior to that of their more affluent peers. Disaggregating the 2007 SACMEQ results by poverty quartile, Spaull (2011) shows that for the wealthiest

Table 13.1 Mean literacy scores (3-year average) and mean SES by former education department: 2007–2009. (Source: Taylor forthcoming)

Former department	Mean literacy over 3 years	Mean SES[a]	Composition of sample	
			Observations	Percent
African (DET & homelands)	25.19	1.70	6776	80.8
Coloured (HOR)	39.12	2.97	880	10.5
Indian (HOD)[b]	43.86	2.81	108	1.3
White (HOA)	58.78	3.35	619	7.4
Total	29.16	1.95	8383	100.0

[a] A 5-point asset-based index of poverty was calculated, using data derived from pupil questionnaires
[b] Only four historically Indian schools were surveyed in the NSES, making this group too small to warrant meaningful analysis

25% of students, South Africa ranks 4th out of 15 for reading.[2] However, when ranked by the performance of the poorest 25% of students, South Africa ranks 14th out of 15 for reading. For mathematics, the figures are 6th out of 15 for wealthy students and 12th out of 15 for poor students. Thus, the average poor South African student performs worse at reading than the average poor Malawian or Mozambican student, in spite of the fact that the average poor South African student is less poor than the average poor Malawian or Mozambican student (van der Berg et al. 2011). Although the top end of the system is deracializing, with white pupils making up only 40% of the population in former whites-only schools in 2010 (DBE 2011), the poorly performing bottom end continues to serve only impoverished African students.

The nature of the South African achievement gap is starkly illustrated by data from the National School Effectiveness Study (NSES), a longitudinal study that tracked a random national cohort of learners for 3 years, commencing in grade 3 in 2007 (Taylor et al. forthcoming). Table 13.1 shows that mean literacy scores for children in former African schools are less than half of those for white and black children in historically white schools.

Figure 13.1 shows how literacy scores on the same test for the NSES cohort of children changed in successive years. The three solid lines are for former African schools, and the three broken lines for former white schools. For both groups, the distribution of achievement improved with each year, but the distribution for grade-5 students in historically black schools was still considerably weaker than that of grade-3 students in historically white schools. It is clear that by the fifth grade the educational backlog experienced by children in poor and poorly performing schools is equivalent to well over 2 years worth of learning, when compared with their peers in better-performing schools.

---

[2] This itself is a noteworthy finding, showing that even the wealthiest quartile of SA schools is outperformed not only by Seychelles and Mauritius, which is to be expected given the higher SES of these island nations, but also by Tanzania, which, along with Malawi, has the poorest poverty rating in the sample (Hungi et al. 2010).

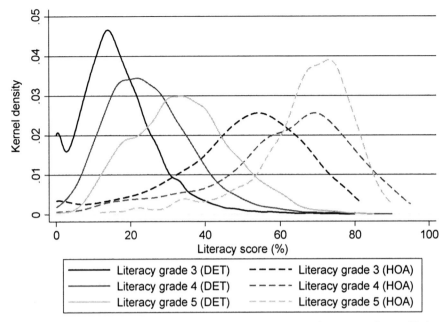

Fig. 13.1 Kernel density curves of grades 3, 4 and 5 literacy by ex-department. (Source: Taylor forthcoming)

The picture for numeracy is similar. Figure 13.2 differs from Fig. 13.1 in that the distributions for historically black schools are more widely spread and the distributions for historically white schools are more concentrated at the top end, evidently with little room for improvement with scores in 2007 already concentrated at the high end of the spectrum.

Thus, despite concerted effort since at least 1998:

- South Africa is lagging behind the rest of Africa.
- There are continuing large disparities in the outcomes produced by different kinds of schools linked to past racial affiliation. In other words, the few exceptions notwithstanding—and there are indeed striking if isolated exceptions—African learners stand a dramatically better chance of scholastic success, all things being equal, in a mixed (ex-white) school than in an African school.

In the next section, the chapter will examine some of the studies that have sought the roots for these continuing disparities, and we discuss some of the interventions based on their findings.

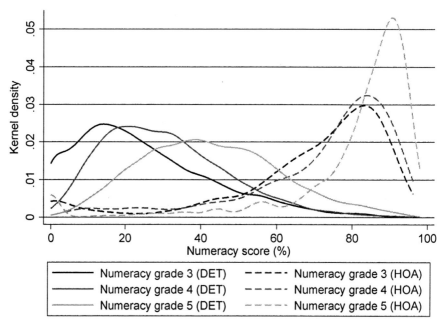

Fig. 13.2 Kernel density curves of grades 3, 4 and 5 numeracy by ex-department. (Source: Taylor forthcoming)

School Improvement Initiatives

Prior to 1994, school improvement was largely ignored by the government, with nongovernment bodies or organisations (NGOs), often with foreign donor funding, setting themselves in opposition to the apartheid state and striving to counter the ruling ideology by means of teacher in-service programs. Pupil-centred classrooms were seen as a route to democracy and liberation from apartheid schooling, which was identified with a repressive form of traditional schooling. These programs have a long history in South Africa, and many continue to exist alongside a host of interventions that have developed in the last two decades. In a survey conducted in 1995, 99 teacher in-service projects were recorded. One-third of the projects were found to have been the subject of evaluations of one or another kind, but only one used objective measures of learning outcomes to assess impact (Taylor 1995). Until the fall of apartheid, these programs were generally small in scale, and more often than not consisted of training for teachers in progressive teaching methods.

Although a number of studies had begun to describe the problems existing in schools serving poor communities (e.g., MacDonald 1990; Chick 1996; Muller 1989; Walker 1989), it was only when the report of the President's Education Initiative (PEI) was published in 1999 that these conditions gained public attention (Taylor and Vinjevold 1999). Although the report was a collation of findings from more than 30 small-scale qualitative studies, which therefore did not lend themselves to reliable generalisation, the findings have been confirmed by virtually all subsequent school- and classroom-focussed research (Hoadley 2010). The PEI described schools in which loose time-keeping practices, poor subject knowledge on the part of teachers, and infrequent reading and writing in class were ubiquitous. These studies were suggesting that the poor performance of the majority of schools required more than pedagogical reorientation of teachers. Nevertheless, many initiatives continued to focus on pedagogical issues, as the local corporate sector and international donor community began to take a serious interest in school improvement and launched several major programs (Taylor 2007).

The Imbewu project (1998–2001) was the first of these major donor-funded initiatives. Working in 523 rural schools in the Eastern Cape province, training for teachers and principals concentrated on the principles and methods of child-centred teaching and outcomes-based education, as defined by the new curriculum introduced in 1998. Perold (1999) found an enthusiastic response to Imbewu on the part of parents, principals and teachers. However, in a 3-year longitudinal study, Schollar (2001) concluded that, although changes in school management and classroom-teaching practices could be discerned, pupil tests revealed no learning gains in reading, writing or mathematics.

The *District Development and Support Project* (DDSP) (2000–2002) was the first initiative in South Africa based on a systemic school improvement design, which attempted to align curriculum, teaching and assessment through the coordination of activity at the levels of the classroom, school, and district offices (Taylor 2007). Working in 453 primary schools in the four poorest provinces, interventions sought to improve the operational efficiency of district offices and schools and to improve classroom teaching in language and mathematics through teacher training courses in subject content. Changes in test scores were recorded, and an analysis by Schollar (2006) concluded that these gains were associated with two measures: heightened expectations of improved test results and the introduction of support measures in the form of detailed specifications of the curriculum, pupil workbooks, item banks of exercises, and monitoring of classroom level activities.

The Quality Learning Project (QLP) (2000–2004) was an example of a systemic program at the high-school level. Working in 524 high schools selected by the nine provincial departments of education, the QLP delivered training and support programs aimed at achieving better management of districts and schools and improved classroom teaching. The Dinaledi project, working in 102 poor high schools across the country, was also structured as a systemic initiative and is driven by the national Department of Education for the first time. Training was provided, and materials were supplied to teachers and principals (Human 2003). Although both were designed in a broad outline as systemic initiatives, Dinaledi and QLP were very different in the details of their initial school profiles and are, therefore, not strictly comparable. However, both projects showed impressive average gains on the national school-leaving Senior Certificate (SC) examination compared with the national mean. However, at the same time, a high proportion of schools in each program benefited not at all from the respective interventions (Kanjee and Prinsloo 2005; Taylor 2007). In other words, there were measurable effects of the interventions, but the improvement was not systemic; that is, it was not spread throughout the schools in the project.

With one exception (see below), this latter feature has proved to be an intractable problem besetting school improvement initiatives, none of which has been able to effect system-wide change in the sample of schools targeted for intervention. The source of this problem is at least twofold. First, the dosage of program interventions has been too light. Given the poor knowledge foundations of most South African teachers, interventions targeting the training of teachers will have to be far more intensive than the short workshops that have passed for developmental training if there is any hope of making a difference to teachers' conceptual understanding of their teaching subject. Second, although designed as systemic interventions in which program activities are linked with those of district officials, the large majority of South African district offices are staffed with officials whose knowledge resources are no stronger than those of teachers. Thus, districts are unable to effectively play their designated role in school improvement projects, or indeed in normal day-to-day support activities.

So far, then, alarmed at the continuing achievement gaps in South African schooling, the government has redirected funding to poor schools and equalised teacher salaries, and the private and international donor agencies have targeted teacher-upgrading initiatives by NGOs. None of these has made a measurable impact on the systemic features of the inequities. Something more was clearly called for.

Accountability Measures

Following the general election of 2009 and the splitting of the Department of Education into separate departments for schools (Department of Basic Education, or DBE) and further and higher education (Department of Higher Education and Training, or DHET), the DBE addressed the problem of poor school outcomes with renewed vigour. A number of the measures put in place were designed to strengthen the accountability of teachers and school leaders toward parents and pupils. Foremost among these is the Annual National Assessment (ANA) exercise, which consists of a set of tests in language and mathematics administered annually at grade levels 1 through 6. ANA has three principal goals: to signal to teachers what should be covered in the curriculum and how best to assess it; to measure the performance of the school system and how this changes over time; and to empower parents with important information about the performance of their own children and that of their school (DBE 2010, p. 10).

The first full administration of ANA occurred in 2011 and consisted of two components. The "universal" component was administered and scored by teachers themselves and applied to all learners in all primary schools. The "verification" component was applied at grade 6 in 1800 schools, where more rigorous moderation procedures were followed during administration, and test scripts were re-marked centrally after being marked by teachers. Below, we assess the extent to which each of these two components of the program is likely to achieve its goals.

Regarding the first goal of ANA, the DBE (2010) reports that during the 2-year pilot phase, evidence emerged that the ANA did assist teachers to employ better assessment practices in their classroom, by exposing them to well-constructed tests and marking memoranda, and also by encouraging district offices and provincial departments to review their own initiatives aimed at supporting teachers. From a theoretical perspective, the "universal ANA" would be expected to provide a valuable resource for teachers, being that it is designed as an assessment instrument *for learning*, to use Black and Wiliam's (1998) classic definition. The assessment theory thus supports the DBE's contention that the tests are likely to assist teachers to understand how good tests are constructed and to judge the standards required at the respective grade levels. The responses of their pupils to the test items also provide teachers with invaluable information on student learning and the effectiveness of their own teaching strategies.

However, whether any such improved knowledge and understanding on the part of teachers leads to improved teaching and learning is open to question. The experience of the Early Grade Reading Assessment (EGRA) initiative in Liberia is instructive in this regard (Piper and Korda 2010; Bruns et al. 2011). EGRA is an instrument for assessing early reading, and the Liberian study attempted to establish its effects on reading performance in grades 2 and 3, under conditions of high poverty and low school outcomes. Implementation of the program in 2008 consisted of two treatments. In the "light" version, the tests were administered and the results disseminated to schools and parents. This group also received teacher training in completing report cards, with instructions to complete and distribute them four times a year. The full intervention consisted of the "light" treatment plus intensive teacher training in reading instruction. In the "light" intervention, scores improved on only the lowest reading skill (letter fluency). The full treatment, on the other hand, effected highly significant improvements in all seven reading measures assessed by EGRA.

EGRA holds two important lessons for the ANA initiative and for school improvement generally. First, tests set externally and administered, scored and disseminated by teachers can be useful, but only when *combined with intensive teacher training*. Given the poor state of teachers' subject knowledge (see below), we might expect the same conditions to apply in South Africa. Second, improvement in both the full and light treatments was associated with improved mathematics scores, despite the fact that neither treatment included any reference to mathematics. Here too, classes whose teachers received the "full" treatment did better than those who experienced the "light" intervention. This second lesson gives support to the contention that *language proficiency is central* to making progress in all other subjects, including mathematics.

Regarding the second goal of ANA, in her Foreword to the report on the 2011 round of administration, the Minister of Basic Education states that the program is intended to monitor the improvement of the quality and levels of educational outcomes in the schooling system, toward the target of 60% achievement by 2014 (DBE 2010, p. 1). The "verification" component of ANA lends itself well to this systemic purpose, provided the tests are well constructed. Further, provided administration, scoring and analysis are rigorously standardised, the results will be reliably comparable across schools and over time. Both provisos would need to be scrupulously fulfilled if the test scores are to enjoy any legitimacy in the eyes of a public that has become sceptical of the government's ability to improve school outcomes.

However, with respect to the third goal of the program—providing parents with information to hold schools and teachers accountable—the DBE is on less firm ground. Because the "universal ANA" tests are administered and scored by teachers with little done to standardise the process, this component cannot produce data that is reliable enough to be used comparatively. Under these conditions, the margins of error are too great to sustain credible comparisons. This feature precludes the use of "universal ANA" as a tool for "information for accountability" purposes, to use Bruns et al.'s (2011) term. Nevertheless, the DBE does envisage ANA being directed to such goals: "ANA can provide parents on the School Governing Body, as well as parents in general, with a better picture of the grades and subjects where special attention is needed. This can assist both efforts in the school and efforts in the home aimed at ensuring that learning occurs as it should" (DBE 2010 p. 11). This is at least arguable, and government would do well to heed the warning of Bruns et al. (2011), who caution against using metrics that are not widely regarded as providing valid, reliable and fair measures of school quality.

Furthermore, noting that most accountability schemes are of a recent provenance and that evidence for their medium-term effects and scalability is not yet available, Bruns and her colleagues speculate that gaming practices are likely to increase as teachers become acquainted with the rewards and sanctions associated with such programs and learn to exploit their loopholes. There is growing evidence that, where test results have high-stakes consequences for teachers or schools, scores are prone to perverse incentives, even cynical manipulation. For example, accounts of schools and districts in the USA cheating in the tests used to measure progress on the No Child Left Behind program are on the increase (Jacob and Levitt 2003; Ravitch 2010; Jonsson 2011; Ravitch 2011).

Test scores do offer the most objective information for holding schools accountable, but several conditions must be met for such programs to play a meaningful part in school improvement initiatives. In order to gain legitimacy, tests must be seen by the participating parties and the general public to be a valid metric of school quality; be administered, analysed and reported with technical efficiency; provide reliable evidence of school performance; be planned in consultation with teacher unions; and be recognised as fair in their application, with due regard paid to the poverty levels of the feeder community and the human and technical resources of the school. The National Senior Certificate (NSC) examination at the end of grade 12 has been used for many decades as a distributor of postschool opportunities into further and higher education and the labour market. The NSC can be said to possess a number of the conditions listed above, with the glaring exception of the last. But even here, the annual release of the results is often accompanied by criticism and controversy.

Table 13.2 Grade-6 teacher knowledge on literacy skills, SACMEQ III teacher test, 2007 (percent correct). (Source: Taylor and Taylor (forthcoming))

Processes of comprehension				
Retrieve info explicitly stated in text	Inferential reasoning	Interpretation	Evaluation	Total
75.06	55.21	36.61	39.73	62.99

It would be folly to assume that summative assessment exercises such as the NSC are not prone to manipulation. System-wide gaming of the process is known to have occurred in the years 1999–2003 (Umalusi 2004; Taylor 2009) in the wake of intense pressure from the then Minister of Education for the score profiles to show improvement. It is suspected that practices, such as the exclusion of high-risk candidates or advising candidates to register for easy subject options at the expense of mathematics and science, continue to be used by principals and teachers to improve school pass rates (Taylor 2011). Pressure to manipulate results is bound to be exacerbated should the stakes be raised, such as the scheme proposed by the province of KwaZulu/Natal to make school grants dependent on NSC scores (*Business Day*, April 4, 2011). In the public mind, school performance on the NSC is judged on pass rates, a statistic that is particularly easy to manipulate. The metric used in Brazil to rate the quality of school performance combines scores on the national *Provo Brasil* tests, with measures of student flow (grade progression, repetition, and graduation rates) (Bruns 2010; Bruns et al. 2011), thus discouraging the "culling" practice, which is apparently prevalent in South Africa.

Teacher Knowledge

Over the years, the supposition has been growing that a core feature of the achievement gap in South Africa is the low level of teacher competence, especially with regard to content subject knowledge, but because of teacher resistance, it has proved difficult to test this competence directly. The SACMEQ III[3] data provide the first opportunity to systematically assess the nature of the subject knowledge of 6-grade teachers. In the reading test (Table 13.2), South African teachers performed best on items requiring the retrieval of information stated explicitly in the text. Performance declined as soon as higher cognitive processes were invoked to answer a question. Good scores were recorded on items requiring straightforward inferences, but questions involving interpretation and evaluation were generally poorly done.

Similarly, the subject knowledge of the majority of South African grade-6 mathematics teachers is inadequate for effective teaching (Table 13.3). Although many

---

[3] The third round of the Southern and Eastern African Consortium for Monitoring Education Quality was conducted in grade-6 language and mathematics in 2007, where South Africa was one of 14 participating countries.

Table 13.3 Grade-6 teacher knowledge on mathematics skills, SACMEQ III teacher test, 2007 (percent correct). (Source: Taylor and Taylor (forthcoming))

Mathematical strand					
Arithmetic operations	Fractions, ratio and proportion	Algebraic logic	Rate of change	Space and shape	Total
67.15	49.68	46.51	42.30	56.44	52.39

of the items in the teacher test draw on knowledge not present in the primary school curriculum, it seems reasonable to suppose that it is through a good understanding of elementary algebraic reasoning and working with simple equations and graphs that teachers gain the background knowledge necessary to provide even young children with sound conceptual understanding in key topics such as proportional reasoning. Teacher performance on these items is poor and is not much better on a number of critically important topics specifically listed in the grade-6 curriculum.

If it is justified to generalize from the results of the SACMEQ III test (see also Carnoy et al. 2012), then it seems plausible to infer that, without intensive training in the foundations of the subject of the sort usually provided at university, learning gains effected through the accountability measures currently being implemented by the DBE (if any appear at all) will reach a low ceiling in most schools, since (based on evidence of the SACMEQ III test) teachers do not have the knowledge to teach key cognitive skills such as inferential and evaluative reasoning or the foundational mathematical concepts of number operations, fractions and ratio.

Teacher Training

The knowledge gaps described in Tables 13.2 and 13.3 continue to exist, despite the flood of teacher training programs in operation over the last two decades. For example, between 1990 and 2008, the proportion of South African teachers certified as "qualified" increased from 53 to 94.4% (DBE 2010), largely through the widespread provision of courses leading to an Accelerated Certificate in Education (ACE), a certificate at a pre-university degree level. The Department of Education acknowledged that the billions spent on in-service training from state funds over the last decade were not well spent:

> The fact that the cognitive performance of children remains low, even though qualifications have increased, casts some doubt on the importance of paper qualifications as a determinant of ultimate impact—at least in the way that the improvement of such qualifications has been implemented up to now.
> DOE 2009, p. 65

Regarding the initial training of teachers, the Higher Education Qualifications Council of the Council on Higher Education recently undertook a review of teacher qualifications. The review concluded that the quality of a substantial proportion of teacher education programs is questionable, with few meeting minimum standards in the areas of program organisation, design, coordination and work-based learning. Also, the quality of staff, especially in postgraduate programs, was suboptimum in areas like staff development, research output and orientation of part-time staff. This was echoed by an evaluation of the ACE programs, which concluded that the majority of programs providing Accelerated Certificates in Education over the last decade have been of mediocre quality at best (CHE 2010).

It is clear, with respect to both initial teacher education and in-service training, that there is a gap between qualifications and subject expertise. In particular, it would seem that the majority of qualified teachers have serious shortcomings in their subject knowledge expertise. Although sound subject knowledge might not be a sufficient criterion for effective teaching, it is surely a prerequisite and, therefore, a minimum requirement for basic teaching competence.

South Africa is not the only country to exhibit the gap between qualifications and subject expertise described above. An interesting approach to addressing this problem can be found in Brazil, where the central government has instituted an exam, the *Exame Nacional de Ingresso na Carreira Docente*, for all new teacher candidates (Bruns 2010). The exam covers both content and pedagogy but is not required for existing teachers. Although taking the exam is binding on all new graduates who wish to enter the profession, the degree of decentralisation in Brazil ensures that states can choose to use the results in various ways. The example set by the national ministry has led two states, São Paulo and Minas Gerais, to put in place more rigorous, content-based tests that graduates must pass to gain entry into the profession. In São Paulo, tests of content mastery are also used to regulate the conversion of temporary teachers into permanent contracts. Furthermore, the Brazilian case is instructive with regard to schemes designed to reward teachers already in service for higher levels of knowledge. For example, in 2009, São Paulo adopted the *Prova de Promoção* to create a new, high-paid career track for top teachers; teachers gain entry to the new salary scale by passing a difficult test of content mastery (Bruns 2010).

These programs for improving content knowledge of both new and existing teachers are currently on trial in Brazil, and it is important that these be rigorously evaluated before transferable policy lessons can be drawn for other developing countries. In South Africa, with its strongly unionised and militant teacher body, it would be wise to await the results of the Latin American experiments, or at least to pilot such programs under local conditions, before going to scale with radical new approaches. However, the problem of poor teacher subject knowledge is severe, and systems change cannot be expected without improvement of this critical teaching resource.

Table 13.4 Literacy scores grade 3, Western Cape province. (Source: Constructed from Mourshed et al. (2010, p. 41))

	Poverty quintile				
	Q1	Q2	Q3	Q4	Q5
Scores 2004–2008 (percentage attaining 50%)	27–42	23–42	30–43	44–45	75–80
Gain (percentage point)	+25	+19	+13	-1	-5

Signs of Progress

The McKinsey report on 20 school systems that have registered sustained and wide-spread student outcome gains lists South Africa's Western Cape province[4] as having made a "promising start" (Mourshed et al. 2010). The evidence cited by McKinsey refers to steady gains in literacy scores at grades 3 and 6 since the province brought in system-wide testing in 2002. Furthermore, there was a dramatic reduction in variation across the poverty spectrum, as illustrated by the grade-3 scores in Table 13.4.

McKinsey leaves unexplained why grade-3 and grade-6 mathematics scores on the Western Cape tests failed to increase over the same period. Scores in literacy and mathematics in the SACMEQ tests between 2000 and 2007[5] similarly failed to increase. The most likely explanation for the lone literacy gains is that, to cater to the very wide range in performance across the population, a relatively large number of low-level items (word recognition) was included in the tests in order to ensure that scores for even the weakest students were registered. In 2010, acknowledging that the large majority of schools are now able to teach these elementary reading skills, the province changed the literacy tests to contain a larger proportion of intermediate and higher-level skills effectively bumping up the standard. The gains exhibited by the poorest schools in the province show that the most disadvantaged pupils now attain intermediate levels of reading proficiency, a very significant advance on the situation prior to 2002. McKinsey ascribes the reading progress exhibited by Western Cape schools to a package of interventions, including the dissemination of test scores to parents, engagement with low-scoring schools by district officials, supply of books to schools, mandating a daily reading period in schools, cash prizes for top performing schools in each quintile in each of the eight districts, and teacher training.

Though each of these elements of the package of reforms instituted by the province almost certainly played a part in the progress made in the last decade, evidence from an evaluation of the teacher development programs offered by the Cape Teaching and Leadership Institute (CTLI) in 2010 indicates that training could be a key element. These are block release courses of at least a week in duration. Substitutes

---

[4] The most highly developed of the country's nine provinces responsible for the administration of schools.

[5] The SACMEQ scores show a slight decline for the WC in both literacy and mathematics (SACMEQ 2011).

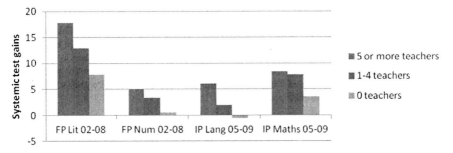

Fig. 13.3 Comparison of gains on Western Cape provincial tests, according to the number of teachers per school trained at CTLI between 2002 and 2009. Key: *FP* grade 3; *IP* grade 6. (Source: Dechaisemartin 2011)

are hired to replace teachers on course. Teachers spend the entire week in residence at the CTLI, where intensive training focuses on subject content. The evaluation concluded that gains on the annual provincial tests in both literacy and mathematics, at both grade-3 and grade-6 levels, are associated with increased numbers of teachers attending the program (Fig. 13.3).

The trends are clear: the greater the number of teachers from any one school attending CTLI training over the years, the greater the gain scores exhibited by the school on the provincial grade-3 and grade-6 literacy and numeracy tests. The effects are statistically significant, are substantial for foundation phase (FP) literacy and intermediate phase (IP) mathematics, and are smaller but still significant for FP numeracy and IP language. This suggests that *immersion courses of subject content* are a more effective model of in-service training than are either the ubiquitous workshop format adopted by NGO interventions or the training leading to ACE qualifications.

Conclusion: Elusive Equity or Equity Deferred?

In a well-balanced and sober assessment of the educational balance sheet in the early 2000s, Fiske and Ladd (2005) described how the new democratic government in South Africa had pursued the "equity imperative," only to find it elusive. This elusiveness is ascribed principally to three factors:

- The apartheid legacy, which had established geographical patterns of residence and patterns of affluence and poverty based on them, has proved incorrigibly enduring.
- The limited availability of financial and human resources.
- Political factors derived from the political settlement, which had awarded a certain decentralised independence to the provinces, rendering nationally driven reforms and standards difficult to enforce.

Fiske and Ladd concluded:

> An adequacy measure of equity need not require that whites and blacks exhibit similar outcomes. What it does require, though, is that outcomes for black students be raised to a minimum threshold that will equip them to function as workers and citizens in the new democratic era.... South Africa has not yet made either the social or educational investments that this standard would require. Overall, then, racial equity in education remains elusive. (Fiske and Ladd 2005 p. 233–234)

By this assessment, what is required is greater investment for elusive to become actual equity. This chapter has suggested that greater investments have indeed been made to little avail with respect to the achievement gap. The studies reported on suggest that the intervention that will best repay the investment will be investment in teacher training, both degree-level initial qualifications for entry into the teaching profession and intensive subject-content-based in-service training courses of durations of at least a week at a time.

Of course, this will not be the whole solution. We have not considered here the science achievement gap, nor the poor quality of leadership and management of the principals, district- and provincial-level staff. Indeed, a part of the solution must be to train at least the principals in sound curricular and financial management. However, the knowledge and competence of the teachers puts an *absolute cap* on the attainment levels of students, so that *a threshold level of subject competence is the first necessary condition for systemic improvement* in the South African system. Until this is done, it will be a case not of elusive equity but of equity deferred.

References

Black, P., & Wiliam, D. (1998). *Inside the black box: Raising standards through classroom assessment*. London: King's College.

Bruns, B. (2010). *Achieving world class education in Brazil: The next agenda*. Washington, DC: World Bank.

Bruns, B., Filmer, D., & Patrinos, A. (2011). *Making schools work: New evidence on accountability reforms*. Washington, DC: World Bank.

Buckland, P., & Fielden, J. (1994). *Public expenditure on education in South Africa, 1987/8 to 1991/2; An analysis of the data*. Johannesburg: Centre for Education Policy Development/ Washington DC: World Bank.

Carnoy M., Chisholm L., & Chilisa B. (Eds.). (2012). *The low achievement trap: Comparing schooling in Botswana and South Africa*. Cape Town: HSRC Press.

CHE. (2010). Report on the national review of academic and professional programmes in education. *HE Monitor No. 11*. Pretoria: Council on Higher Education.

Chick, J. K. (1996). Safe-talk: Collusion in apartheid education. In H. Coleman (Ed.), *Society and the language classroom* (pp. 21–39). Cambridge: Cambridge University Press.

Chinapah, V. (2003). *Monitoring Learning Achievement (MLA) project in Africa*. Report presented to the Association for the Development of Education in Africa (ADEA) Biennial Meeting 2003, Mauritius, December 3–6. Paris: International Institute for Educational Planning. http://www. adeanet.org/adeaPortal/adea/biennial2003/papers/2Ac_MLA_ENG_final.pdf. Accessed 23 March 2012.

Chinapah, V., H'ddigui, W., Kanjee, A., Falayajo, W., Fomba, C., Hamissou, O., Ratalimanana, A., & Byamugisha, A. (2000). *Towards quality education for all*. Cape Town: HSRC Press.

DBE. (2010). *Report of the Annual National Assessments of 2011*. Pretoria: Department of Basic Education.

DBE. (2011). *Macro indicator trends in schooling: Summary report 2011*. Pretoria: Department of Basic Education.

Department of Education. (2009). *Trends in macro indicators report South Africa 2009*. Pretoria: Department of Education.

Dechaisemartin, T. (2010). *Evaluation of the cape teaching and leadership institute*. Johannesburg: JET Education Services. www.jet.org.za.

Edupol. (1993). *Teacher salaries in South Africa: A policy perspective*. Johannesburg: Urban Foundation.

Fiske, E., & Ladd, H. (2005). *Elusive equity: Educational reform in post-apartheid South Africa*. Cape Town: HSRC Press.

Hoadley, U. (2010). *What do we know about teaching and learning in primary schools in South Africa? A review of the classroom-based research literature*. A report for the Grade 3 Improvement Project. Stellenbosch: Department of Economics, University of Stellenbosch.

Human, P. (2003). *The 102 schools project, also referred to as the Dinaledi Project: A brief survey of project activities in the period May 2002–January 2003*. Mimeo.

Hungi, N., Makuwa, D., Ross, K., Saito, M., Dolata, S., van Capelle, F., et al (2010). *SACMEQ III Project results: Pupil achievement levels in reading and mathematics*. Paris: Southern and Eastern Africa Consortium for Monitoring Educational Quality.

Jacob, B., & Levitt, S. (2003). Rotten apples: an investigation of the prevalence and predictors of teacher cheating. *Quarterly Journal of Economics, 118*(3), 843–77.

Jonsson, P. (2011). America's biggest teacher and principal cheating scandal unfolds in Atlanta. *Christian Science Monitor*, July 5. http://news.yahoo.com/americas-biggest-teacher-principal-cheating-scandal-unfolds-atlanta-213734183.html. Accessed 10 July 2011.

Kanjee, A., & Prinsloo, C. H. (2005). *Improving learning in South African schools: The quality learning project (QLP) summative evaluation (2000–2004)*. Pretoria: Human Sciences Research Council.

MacDonald, C. (1990). *Crossing the threshold into standard three. Main report of the Threshold Project*. Pretoria: Human Sciences Research Council.

Mourshed, M., Chijioke, C., & Barber, M. (2010). *How the world's most improved school systems keep getting better*. London: McKinsey & Co.

Muller, J. (1989). "Out of their minds": An analysis of discourse in two South African science classrooms. In D. A. Roger, & P. Bull (Eds.), *Conversation: An interdisciplinary approach* (pp. 313–337). Philadelphia: Multilingual Matters.

Mullis, I. et al. (2000). *TIMSS 1999 international Mathematics report–findings from the IEA's report of the Third International Mathematics and Science Study at the 8th grade*. Chestnut Hill: Boston College.

Mullis, I. et al. (2004). *TIMSS 2003 International Mathematics report–findings from the IEA's report of the Third International Mathematics and Science Study at the 4th and 8th grade*. Chestnut Hill: Boston College.

NPC. (2011). *National development plan–vision for 2030*. Presidency. Pretoria: National Planning Commission.

Perold, H. (1999). Reflections on the Imbewu Project. Mimeo. http://www.jet.org.za

Piper, B., & Korda, M. (2010). *EGRA Plus: Liberia. Program evaluation report*. Research Triangle Park: Research Triangle International.

Ravitch, D. (2010). *The death and life of the great American school system: How testing and choice are undermining education*. New York: Basic Books.

Ravitch, D. (2011, September). School "reform": A failing grade. *New York Review of Books, 58*, 4.

Reddy, V., & Kanjee, A. (2006). *Mathematics and science achievement at South African schools in TIMSS 2003*. Pretoria: HSRC Press.

SACMEQ. (2011). *Education in South Africa*. Paris: IIEP. http://www.sacmEq.org/education-south-africa.htm. Accessed 28 April 2011.

SAIRR. (2011). *2010/2011 South Africa Survey*. Johannesburg: South African Institute of Race Relations.

Schollar, E. (2001). *Final Report of the Evaluation of the School-level Impact of Imbewu*. Mimeo. http://www.jet.org.za.

Schollar, E. (2006). Analysis of the impact on pupil performance of the District Development Support Program (DDSP). http://www.jet.org.za.

Spaull, N. (2011). *A preliminary analysis of SACMEQ III South Africa*. Stellenbosch Economic Working Papers: 11/11. Stellenbosch: University of Stellenbosch, Department of Economics.

Taylor, N. (1995). *Inset, NGOs and evaluation: A review*. Paper presented at the Kenton-at-Settlers conference, October 27–30, 1995, Rhodes University, Grahamstown.

Taylor, N. (2007). Equity, Efficiency and the Development of South African Schools. In T. Townsend (Ed.), *International handbook of school effectiveness and school improvement*. Dordrecht: Springer.

Taylor, N. (2009, April). Standards-based accountability in South Africa. *School Effectiveness and School Improvement, 1–16*.

Taylor, N. (2011). Balance criteria should replace perverse pass-rate incentives. *Transformation Audit 2011: 58–61*. Cape Town: Institute for Justice and Reconciliation.

Taylor, S. (forthcoming). Modeling educational achievement in the NSES. In N. Taylor et al. (Eds.), *What makes schools effective? Report of South Africa's national schools effectiveness study*. Cape Town: Pearson.

Taylor, N., & Taylor, S. (forthcoming). Teacher knowledge 2 professional habitus. In N. Taylor et al. (Eds.), *What makes schools effective? Report of South Africa's national schools effectiveness study*. Cape Town: Pearson.

Taylor, N., & Vinjevold, P. (1999). (Eds.). *Getting Learning Right: Report of the President's Education Initiative Research Project*. Johannesburg: Joint Education Trust.

Taylor, N., Muller, J., & Vinjevold, P. (2003). *Getting schools working*. Cape Town: Pearson.

Taylor, N., Fleisch, B., & Shindler, J. (2008). *Changes in education since 1994*. Review commissioned for the Office of the President. www.jet.org.za.

Taylor, N., van der Berg, S., &, Mabogoane, T. (Eds.). (forthcoming). *What makes schools effective? Report of South Africa's national schools effectiveness study*. Cape Town: Pearson.

Umalusi. (2004). *Have the standards of the senior certificate examination declined? Summary report on the evaluation of the senior certificate examination*. Pretoria: Umalusi.

van der Berg, S., Burger, C., Burger, R., de Vos, M., du Rand, G., Gustafsson, M., Shepherd, D., Spaull, N., Taylor, S., van Broekhuizen, H., & von Fintel, D. (2011). *Low quality education as a poverty trap*. Stellenbosch: University of Stellenbosch, Department of Economics.

Walker, M. (1989). *Action research as a project*. Cape Town: Primary Education Project.

Part VII
Australia

Chapter 14
Securing STEM Pathways for Australian High School Students from Low-SES Localities: Science and Mathematics Academy at Flinders (SMAF)

Debra Panizzon, Martin Westwell and Katrina Elliott

Introduction

A skilled workforce in science, technology, engineering, and mathematics (STEM) is considered a high priority for guaranteeing Australia's future economic prosperity in a competitive global economy. Importantly, it is not just about creating a workforce for current boom areas (e.g., mining and defense), but also about ensuring that these individuals have the intellectual capacity, knowledge, and employability skills for an ever-changing workplace (Office of the Chief Scientist 2012). Aligned with this component, we need a scientifically literate population to ensure that social and environmental aspects are not ignored or overlooked in the quest for greater economic prosperity (Tytler et al. 2008). Education is central and pivotal to this agenda. Hence, it is not surprising that reports published in Australia over the last decade have emphasized the need to nurture and enhance the pipeline of students undertaking mathematics and science in high schools and universities if the country has to meet the plethora of careers emerging in a range of STEM-related areas (Australian Academy of Science 2011; DEST 2006a, 2006b; FASTS 2002; Office of the Chief Scientist 2012).

The quandary is that although much political attention has focused on the STEM arena, the reality is that we have experienced ever-decreasing numbers of students selecting science and mathematics in the senior years of schooling, resulting in fewer graduates with degrees in engineering and other scientific fields requiring physics and specialist mathematics (Ainley et al. 2008; Goodrum et al. 2011; Pearce

D. Panizzon (✉)
Faculty of Education, Monash University, Clayton, Victoria, Australia
e-mail: debra.panizzon@monash.edu

M. Westwell
Flinders Centre for Science Education in the 21st Century,
Flinders University, Bedford Park, Adelaide, Australia

K. Elliott
Department of Education and Child Development, Adelaide, SA, Australia

J. V. Clark (ed.), *Closing the Achievement Gap from an International Perspective,*
DOI 10.1007/978-94-007-4357-1_14, © Springer Science+Business Media B.V. 2014

et al. 2010). Critically, Australia is not alone; many other OECD countries (OECD, Organisation for Economic Cooperation and Development) are experiencing similar trends (OECD Global Science Forum 2006). For example, the high value placed on STEM in the USA is exemplified in the following quote extracted from a weekly address by President Obama in 2011:

> Over the next ten years, nearly half of all jobs will require education beyond high school, many requiring proficiency in math and science. And yet today we've fallen behind in math, science, and graduation rates. … If we want to win the global competition for new jobs and industries we've got to win the global competition to educate our people. (Obama 2011)

In thinking about the workforce implications of STEM more deeply, a critical issue emerges in Australia in relation to the number of teacher graduates with the necessary discipline knowledge qualified to teach senior high school physics, specialist mathematics, and (to a lesser extent) chemistry (Harris et al. 2005). For example, in a recent study involving 600 high school science teachers (grades 7–12) conducted in South Australia, Panizzon, Westwell, and Elliott (2010) identified that at the time of survey completion,

- 42 % of senior high school physics teachers of grades 11–12 were unqualified (due to a lack of appropriate tertiary qualifications), with 61 % of these teachers being under 40 years of age, compared with 37 % in the over-40 age range.
- 28 % of senior chemistry teachers were unqualified, with approximately 36 % of these teachers being under 40, compared with 24 % of teachers being over 40.
- 25 % of senior teachers of biology were unqualified to teach at this level. Interestingly, 24 % of teachers under 40 years of age were unqualified, which compared favorably with 25 % of teachers in the over-40 age bracket.

Clearly, there is a major issue highlighted by these findings about the age of the most qualified teachers of physics in South Australia, when compared with chemistry and biology. Importantly, these differences become more acute when considered in relation to geographical location (Panizzon 2009) and socioeconomic status (SES; Lyons et al. 2006), with the numbers of unqualified teachers increasing in rural schools and/or schools located in low-socioeconomic locations.

In order to address this situation, what is required is a clear understanding of the factors impacting students' choices regarding these subject areas. Fortunately, there is an extensive research basis available that identifies not only the factors affecting subject choice but also the complexity of the issue (Goodrum et al. 2011; Osborne and Collins 2001; Sjøberg and Schreiner 2005; Tytler et al. 2008). However, when this literature is considered collectively, an interesting phenomenon is highlighted, namely that developing countries experiencing economic expansion and growth over the last decade (e.g., China and India) have an oversupply of citizens with qualifications in STEM-related fields, thereby counteracting the OECD trend. One of the key explanations for this finding is recognition by the populace in these developing countries that STEM careers provide the means of improving SES to escape the poverty cycle (Lowell and Salzman 2007).

Table 14.1 Summary of comparative results for PISA 2006 and 2009

	PISA	OECD mean	Australian mean	Countries with significantly higher scores
Mathematics	2006	498	520	8
	2009	496	514	12
Science	2006	500	527	3
	2009	501	527	6

The latter point raises pertinent questions about the impact of SES on student aspirations and achievement in science and mathematics. Whereas Australia is considered an equitable country with immense opportunities, there is mounting evidence that inequities and educational gaps exist in relation to its indigenous students (Thomson et al. 2010), geographical location (Panizzon 2009), and low SES (McConney and Perry 2010). Importantly, these potential gaps become even more divisive when compounded by factors concerning the attraction of qualified science and mathematics teachers, which is already problematic in Australia.

In this chapter, the inequity with regard to student achievement in mathematics and science in relation to SES is explored. Initially, a context is provided through an analysis of Australian data from the Programme for International Student Assessment (PISA) for 2003 and 2006. Aligned with this is a discussion of school-related factors, such as the difficulty of attracting and retaining qualified staff often associated with schools in lower-SES localities. With this background established, the Science and Mathematics Academy at Flinders (SMAF) is presented as a program specifically designed and implemented to meet a need within the community to address the lack of teacher expertise in senior sciences and mathematics in a number of schools with lower-SES clientele. In closing, a number of interesting educational dilemmas are considered when successful educationally-purpose-fit programs to address specific needs are upscaled and applied to other contexts to meet very different stakeholder agenda.

Student Achievement and Socioeconomic Status

International comparisons of student achievement in scientific and mathematical literacy indicate that Australian students are outperforming their peers in many OECD countries (see Table 14.1) (Thomson and De Bertoli 2008; Thomson et al. 2010).

In reflecting on these results, Australian students achieved more consistently in science with a decrease in performance evident for mathematics. Yet, a more critical observation is that while our student performance remained above the OECD mean, other countries "raised the bar" during the intervening years, improving their overall student performance markedly. As a consequence, the number of countries achieving significantly higher PISA results than Australia has increased from 2006 to 2009.

Use of these large-scale PISA data sets has highlighted a number of patterns regarding Australia's performance, which have caused particular concern in educational and political circles, with the most significant being in relation to student SES (Fig. 14.1).

As observed in these data for PISA 2006 and 2009, it is clear that as student SES background increases from the lowest to highest quartiles, their achievement in mathematical and scientific literacy also improves. However, the actual extent of this difference is more obvious when considered in relation to years of schooling. For example, the 96 points for scientific literacy between students in the lowest and highest SES quartiles for 2009 represents "two-and-a-half years of schooling or more than one proficiency level" (Thompson et al. 2010, p. 235). Mathematical literacy demonstrates the same level of inequity. So, although Australian students overall achieved significantly above the OECD mean score in each domain for each of the PISA testing rounds since 2000, the inequity in terms of SES has remained. Hence, our quality may be high, but our equity is low (McGaw 2006).

The discussion so far has focused on student SES background, which is a composite index determined by the information provided by students as part of the PISA testing cycle. Specifically it relates to the highest parental educational attainment (i.e., years of education) and occupational status, along with a range of economic and cultural resources provided within the home. In a recent article, McConney and Perry (2010) completed a secondary analysis of the PISA 2006 data for Australia by comparing student SES background with school SES. For their analysis, school SES was determined by calculating an aggregated mean score of the student SES data provided for Australian students by PISA. Additionally, as part of their analysis, they established cutoff points to produce five bands or quintiles for school SES and student SES (although PISA identified four). Hence, within each quintile of school SES (e.g., lowest school SES), there are five quintiles representing students with the lowest SES to students with the highest SES in these schools. Although questions may be raised about the mechanism for deriving the school SES, the authors considered this procedure a "stable proxy measure for school SES given the absence of the latter variable in the Australian data" (McConney and Perry 2010, p. 433).

A subset of the data provided in their article was used to illustrate results for mathematical and scientific literacy based upon the first (i.e., lowest), third, and fifth (i.e., highest) student SES quintiles against the mean school group SES (Figs. 14.2 and 14.3). Reflecting upon Fig. 14.2 and mathematics, these results corroborate the findings from PISA in that as the student SES increases, so do their achievement levels. The difference with the data produced by McConney and Perry (2010) is that the impact of the school becomes noticeable. For example, for mathematics the difference between the average low-SES student (i.e., first quintile) in a school with a low SES (i.e., first quintile) and the average high-SES student (i.e., fifth quintile) in the same school is 60 points, representing a standard deviation of 0.70.

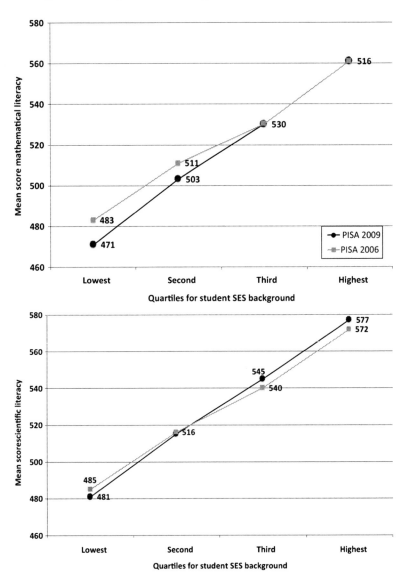

Fig. 14.1 Student achievement in relation to student SES backgrounds for PISA 2006 and 2009. (Thomson and De Bortoli 2008; Thomson et al. 2010)

In terms of scientific literacy (Fig. 14.3), this same pattern of findings emerges with the difference in achievement between the student with a low SES (i.e., first quintile) in a school with a low SES (i.e., first quintile) and a student with a high

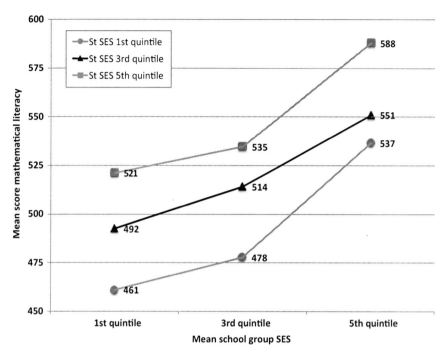

Fig. 14.2 Australian mean scores for mathematical literacy PISA 2006 by student SES and mean school group SES. (data extracted from McConney and Perry 2010)

SES (fifth quintile) in the same school being 68 points, which represents a standard deviation of approximately 0.69 (McConney and Perry 2010). As such, these findings corroborate what is already evident in the literature about the impact of student SES background on educational achievement in a range of subject areas (Gorard et al. 2001; Kieffer 2010; Lara-Cinisomo et al. 2004; Williams 2010).

However, what is particularly pertinent here is the impact of school SES on student achievement, which is clearly demonstrated in the presented data. For example, a student with a low SES (in first quintile) achieved a mean score of 458 for mathematics in a school with a low SES (first quintile) compared with a mean score of 533 in a school with a high SES (fifth quintile). This represents a difference of 76 points with a standard deviation of 0.89 (McConney and Perry 2010). Importantly, the same results emerge for scientific literacy as demonstrated in Fig. 14.3. Consequently, the performance of Australian students in relation to scientific and mathematical literacy is significantly impacted by the SES of the school they attend (Thomson et al. 2010).

In order to understand the factors aligned with school SES, it is important to gain some appreciation of the educational context in Australia. Perhaps the most unusual characteristic is Australia's long history of private schooling, with the numbers of students enrolled in private schools steadily increasing over time (Ryan and Watson 2004). These private schools include two separate groups: (1) Catholic schools that

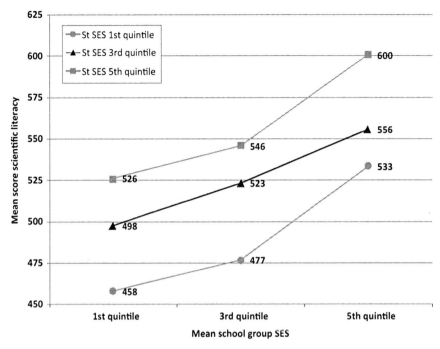

Fig. 14.3 Australian mean scores for scientific literacy PISA 2006 by student SES and mean school group SES. (data extracted from McConney and Perry 2010)

comprise their own system and (2) independent schools that include a range of religious denominations, and which essentially are administered as distinct entities. In 2004, one-third of government funding was allocated to private schools with an enrollment of one-third of all students in Australia (Ryan and Watson 2004). Critically, these private schools also require the payment of additional fees by parents. The result is that, in general, students with higher-SES backgrounds are more likely to attend independent schools, those in the middle-to-high-SES ranges enroll at Catholic schools, and a higher proportion of students from lower-SES backgrounds attend public schools (Campbell 2005). However, this is only a general trend, and there are exceptions (e.g., selective public schools in New South Wales usually attract students from higher-SES backgrounds).

So why is it that the school SES factor has such an impact on student achievement? While this is not an easy question, research in Australia demonstrates that a range of associated factors impact these low-SES schools. For example, many struggle to attract and retain highly qualified and experienced teachers (Lyons et al. 2006). Not surprisingly, with access to greater financial and associated resources (e.g., computers), both independent and Catholic schools are often able to entice and retain a pool of more capable teachers. This position becomes especially difficult in relation to science and mathematics, with the decreasing numbers of qualified senior physics, chemistry, and advanced-level mathematics teachers entering the teaching workforce (Harris et al. 2005; Panizzon et al. 2010). So, while competition for qualified teachers is already fierce, it becomes unobtainable for schools with a predominance of low-SES students.

Meeting the Needs of the Local Community

In 2010, the chance to enhance the opportunities for students attending schools in lower-SES localities arose when a group of concerned principals met at Flinders University to discuss issues facing schools in the southern part of Adelaide in South Australia. As a result of this meeting, a collaborative program titled the Science and Maths Academy at Flinders (SMAF) emerged, which involved a range of staff from Flinders University and local government schools. In this section, the background of SMAF is outlined, followed by a description of the program, including the contributions provided by the various partners. Following this, an overview of outcomes for the project is discussed along with the ways in which the program is continuing to evolve to meet the ongoing demands of participating students.

Background

Seven South Australian government high schools located in the southern region of Adelaide approached the South Knowledge Partnership Transfer (SKPT) to assess the interest of Flinders University in developing a collaborative project to meet a community need. The focus was to provide the teaching of grade-12 physics, chemistry, and specialist mathematics in the university campus using experienced high school teachers from the schools comprising the partnership. In a number of cases, this need arose because of a lack of teachers with the required discipline knowledge and expertise to conduct grade-12 classes in these subject areas. However, in other schools, it was due to the school decision not to fund classes with small numbers of students (i.e., under five). Importantly, grade 12 is the final credentialing year of high school, with students required to complete an externally set and externally marked examination so that scores can be collated to provide the Australian Tertiary Admission Ranking (ATAR) score, which is used for university entrance. The SKPT immediately established a steering committee to explore a possible program with a team from the Flinders Centre for Science Education in the twent-first Century to act as coordinators for what was titled the Science and Mathematics Academy at Flinders (SMAF).

The key objectives of SMAF were to

- increase the participation rate of students undertaking grade-12 physics, chemistry, and advanced mathematics subjects;
- improve and develop innovative teaching and learning in these subjects;

- increase university admissions especially in degree areas requiring these subjects; and
- increase aspirations of students in STEM-related career pathways.

Hence, in terms of *closing the gap*, SMAF ensured that schools with higher proportions of students from low-SES backgrounds could retain access to these subjects as part of their official subject offerings. This was deemed critical given that once a subject becomes unavailable within a school community, the likelihood of being able to attract students keen to pursue these subjects into the future diminishes. The result is often a downward spiral, making it increasingly difficult to reintroduce these subjects as time progresses, given that the "collective knowledge of the school" is weakened—making it impossible to attract qualified and highly competent subject teachers (Darling-Hammond 2004, p. 1944). Ultimately, this leads to inequitable participation in the senior years of schooling, along with decreased future participation in STEM-related fields (Harris et al. 2005; Kelly and Sheppard 2009).

SMAF Structure and Program

Participation in the inaugural year of the program in 2011 included six schools with one of the schools withdrawing from the program. The SMAF teachers were selected from these schools with interested teachers requiring to submit an application that provided evidence about the following criteria:

- experience in teaching the subject,
- ability to establish positive relationships with a diversity of students and peers (teachers and academic staff),
- proficient organizational and management skills,
- degree of flexibility and responsiveness in meeting the needs of a wide range of students,
- level of communication skills,
- experience with electronic curriculum development and delivery,
- willingness to work as a team, and
- commitment to the success of all students.

Based upon the student enrollments, three chemistry teachers (for 35 students collectively), two physics teachers (for 38 students collectively), and one specialist mathematics teacher (for 21 students) were selected. Additionally, each school was required to provide a tutor to work with SMAF students at appropriate times throughout the week; however, there was a high degree of variation here. Finally, a number of university academic staff (both academics and PhD students) agreed to teach particular components of the courses based upon their own scientific research. Overseeing SMAF was a highly experienced chemistry teacher who was seconded to the Flinders Centre for Science Education in the twenty-first Century from the Department of Education and Children's Services (DECS).

Table 14.2 SMAF teaching program	Subject	Flinders university (on campus)		Within school
	Physics	Wednesday	9 am–12 noon	Access to in-school support provided by academic tutor
	Chemistry	Wednesday	1 pm–4 pm	
	Specialist mathematics	Friday	9 am–12 noon	

A major contribution to SMAF was the provision of financial support by the SKPT team at Flinders University. These monies were used to fund the SMAF co-ordinator for 0.8 of her salary along with the employment of a part-time laboratory technician. Furthermore, some schools provided in-kind support, such as laboratory equipment or, in rare instances, laptops for individual students. Flinders University supplied free parking to teachers and students over the course of the program.

The overall schedule for SMAF (Table 14.2) required students and teaching staff to be on campus for at least 3 h per week, followed by in-school support by an academic tutor. However, students undertaking combinations of these subjects (e.g., physics and chemistry) had to remain on campus for the entire day.

Before teaching began, each team of teachers met on at least five occasions to consider the curriculum (i.e., content and skills requirements) and the pedagogies that might successfully cater to the diversity of students involved in SMAF. This point was especially critical to the success of the program given that students came from six different schools with quite divergent prior experiences. At another level, there was the need to ensure that lines of communication between the SMAF teachers and tutors in the schools were clear, particularly in regard to the expectations and requirements of students between the formal teaching sessions each week. This was initially achieved with SMAF sending an email to tutors after each lesson explaining what the students had learned and what students needed to complete or revise during their home school lesson. However, a tutor *pro forma* was developed during the year to enhance communication between the SMAF teachers and home school tutors for each subject.

To enhance ongoing communication not just between teachers but also between students, the decision was made to incorporate an online platform. As such, all SMAF participants had access to Flinders Learning Online (FLO), through which a range of teaching materials and information were provided, along with the opportunity to engage in electronic forums (blog and wiki) at times that suited and supported own learning of students.

The SMAF coordinator was driving these meetings, discussions, and opportunities for collaboration—scheduling lessons, allocating and booking teaching spaces, and dealing with arising issues. In this manner, a high degree of consistency and continuity was possible along with an ongoing monitoring of the program so that concerns and difficulties could be dealt with before they became unmanageable. Clearly this was a pivotal role to the success of SMAF.

Evidence of Success Around SMAF

To ensure a degree of objectivity, an external evaluator to Flinders University was employed to gauge the success and challenges facing SMAF. Data were collected from (1) a student survey consisting of Likert-scale and open questions about their teaching and learning experiences of SMAF, along with some items regarding student demographics; (2) interviews with five teachers, six tutors, two principals, the SMAF coordinator, and a student focus group; (3) participating schools about subject enrollment trends over time; and (4) the student retention in the program. These data are used to highlight evidence about the success of SMAF along with some of the future challenges for improving the program.

Participation and Involvement of Lower-SES Students

One of the key goals of SMAF was to enhance the participation rates of students from lower-SES backgrounds in physics, chemistry, and specialist mathematics. Student SES is determined using a number of key aspects, such as the highest level of schooling experienced by parents, highest parental qualification, and residential postcode. Although it was not possible to derive the actual SES categorization for each of the individual students participating in SMAF, the data obtained regarding the educational background of parents and residential postcode provided informative background knowledge (Fig. 14.4).

In terms of parents, 8–15 % had completed only up to grade 10 of high school, whereas 17–18 % held a university degree, with a dramatic decline in the proportion of parents with postgraduate qualifications. The modes occurred for mothers at grade 12 (i.e., our final credentialing year of high school) and for fathers with a university degree. Given this background, it is interesting to observe the trend in relation to educational expectations of students. The majority clearly aimed to undertake a university degree with 16–18 % considering a masters or PhD to be their ultimate goal. Hence, these SMAF students appear to have high educational aspirations. However, it is important to also note the high proportion of students who were unaware of the educational attainment of their parents (i.e., between 26 and 31 %) or "not sure" of their own aspirations (i.e., 33 %). The positive aspect for these indecisive students is that involvement in SMAF may help to solidify their future career directions and focus their attention on their next educational steps.

Another key factor for determining student SES in Australia is residential postcode. The Australian Bureau of Statistics (ABS) ranks postcode areas from the lowest to the highest based upon a complex index determined by measures, such as parental occupation, education, and salary. Within this ranking system, the lowest 10 % of postcode areas are allocated a decile number of 1, whereas the highest 10 % of areas receive a decile of 10. Hence, all postcode areas for South Australia are divided into ten groups, with decile 1 representing the most disadvantaged. By

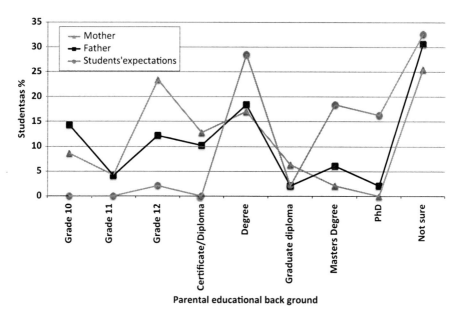

Fig. 14.4 Educational expectations of SMAF students and parental educational background

accessing these data, it was possible to align students and their postcodes with these deciles to provide further insight about SES (Fig. 14.5).

In looking at these data, it is clear that there was considerable diversity in terms of disadvantage for the SMAF cohort. For example, approximately 33 % of the students lived in areas designated in the bottom 30 % of disadvantage for South Australia (i.e., deciles 1–3), whereas another 40 % resided in areas in the top 30 % (i.e., deciles 8–10) for the state. Consequently, it does appear that SMAF provided the opportunity for students from lower-SES backgrounds to continue in physics, chemistry, and specialist mathematics. Importantly, given that a similar cohort of students is currently involved in the program for 2012, this goal is seemingly being attained.

Focusing upon student participation in SMAF specifically, Fig. 14.6 indicates that this has improved in the second year in all three subjects. Importantly, this increase has occurred in the original six schools and by the addition of another two schools in the program in 2012.

A more detailed comparison of these trends in relation to individual schools is presented in Fig. 14.7, with schools labeled A–H to ensure anonymity. Clearly, with the exception of school A, only small groups of students were involved in SMAF from each school. However, it was for this very reason that SMAF was established. Schools labeled A–F represent the initial schools, whereas G–H are the two new schools participating in 2012. In looking at these data, school A is particularly interesting. Here is a school that has the qualified staff to conduct these classes *in situ*, yet has been a key driver in the development of SMAF. In terms of contribution,

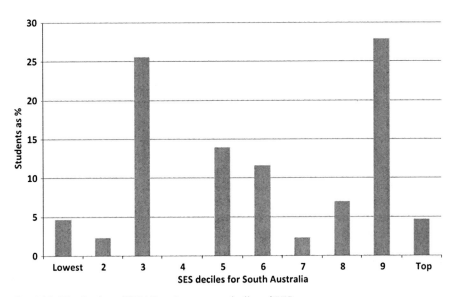

Fig. 14.5 Distribution of SMAF students across deciles of SES

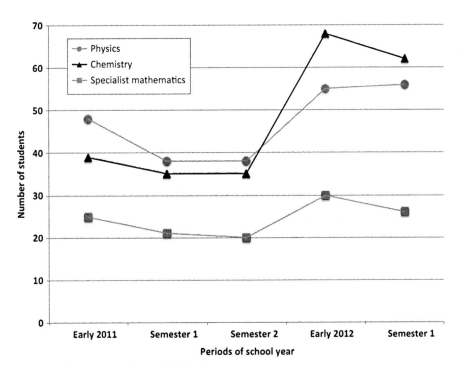

Fig. 14.6 Enrolment trends for SMAF students

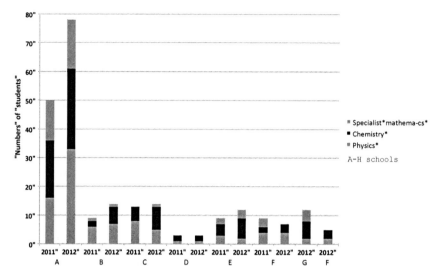

Fig. 14.7 SMAF student participation across individual schools

three teachers from the school were SMAF teachers for both years. Equally though, there may be educational challenges for the other participating schools regarding the ongoing involvement of this school or a similar school over the longer term of the program. This point will be discussed in the last section of this chapter.

Enhanced Teaching and Learning Experiences

Results from the student surveys and interviews demonstrated a range of positive responses about the ways in which learning was supported through the SMAF program. A sample of the items used in the survey and responses for each of the three subjects is presented in Fig. 14.8. It is clear from these results that the majority of students selected the "agree options" of the Likert scales. For example, in relation to *I enjoy learning [subject] at SMAF*, 65 % of physics, 75 % of chemistry, and 67 % of specialist mathematics students agreed with this statement. Similarly, a strong positive agreement was attained for statements regarding the student–teacher relationships (see statements 6, 16, and 26) and student–student relationships (see statements 8, 18, and 28). Importantly, inclusion of negative statements, such as *Chemistry learning is boring at SMAF,* ensured a degree of validity and reliability of the survey results, as the consistency of students' selections could be checked. Importantly, a comparison across subjects highlighted that students appeared more positive in relation to their learning in chemistry and specialist mathematics than in physics. This type of information provided valuable feedback for the coordinator and subsequently the teachers of the SMAF program.

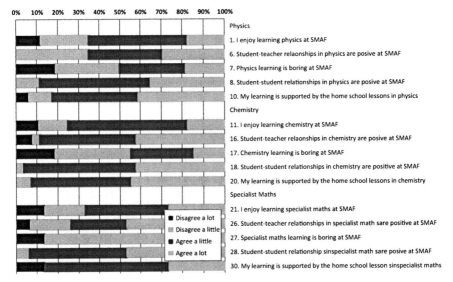

Fig. 14.8 Students' perceptions of learning through SMAF

In terms of students' general experiences of SMAF (items 70–74 from the survey), teaching on the university campus facilitated access to additional resources and expertise along with providing students with greater insights about university life. Given the high proportion of students selecting the "agree options" for these statements, SMAF has been successful in allowing students to gain a firsthand perspective of what being a university student was. As articulated by one of the SMAF teachers:

> The students feel comfortable at university because they've been here, they see the university kids around them—it is the transition. Which ever university they go to it will be quite easy—they know how to go about parking, how to use the online program for courses, the library and how to access resources. They also have some idea about how universities are structured and the building layout.

This is an invaluable experience for all students and especially for those from low-SES backgrounds who may be the first in their families to attend a university. Involvement in a structured program on university space shared by university students and academics supports the transitioning of students from small-school contexts to a much larger and impersonal environment (Hillman 2005).

The other aspect highlighted by these data is in relation to enhancing awareness and aspirations of students' about STEM-related careers and further study. As reflected by the consistently positive agreement by students for items 75–82, SMAF appears to have also met this goal (Fig. 14.9).

Though much of this evidence is quantitative, comments made during the interviews with students, teachers, and the coordinator identified other benefits of the SMAF program. For example, one of the most critical components of being on

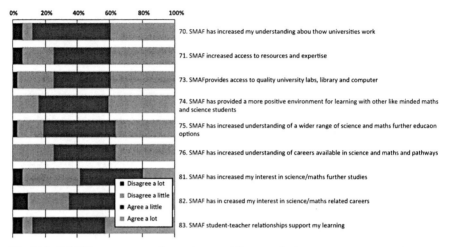

Fig. 14.9 SMAF impact on overall experience and future aspirations

campus was that it ensured that teachers and students could access topical research from academics and PhD students to make the curriculum especially relevant and meaningful while allowing teachers to update their own scientific expertise and understanding. Two specific examples mentioned by teachers about the type of academic support provided by university staff and how it helped enhance student learning included:

- Professional development for all chemistry teachers from SMAF schools about the latest use of absorption spectrometry, mass spectrometry, and gas chromatography. Once teachers were proficient with the instruments, they were invited to an SMAF class to work with students on a number of chemical investigations that incorporated a range of these analytical techniques.
- The environmental chemistry curriculum requirements were taught using a student-centered and peer-assessment approach. Essentially, the students were allocated time in class to develop their understanding of the key ideas from the curriculum statement. In their own time, students across schools collaborated to share their learning via a wiki with the goal being to organize a formal group presentation to the rest of the class. However, to ensure that up-to-date data and information were incorporated, a number of environmental scientists attended two Wednesday-afternoon SMAF teaching sessions to answer student questions while explaining many of the broader impacts and ramifications. The importance here is that chemistry became meaningful to these students while they gained a greater awareness of the way in which scientists conduct, interpret, and discuss their research.

In addition to the actual connections with university personnel, students and the SMAF teachers identified other improvements about teaching and learning as a consequence of being on campus. These included:

- Access to physics and chemistry laboratories provided potentially greater availability to a range of resources. As explained by the physics students, instead of being in a class with one linear air track and having to wait to use the track or watch a teacher demonstrate, students were able to investigate the physics principles in groups because there were eight linear air tracks in the university laboratory. Additionally, each air track was connected to a computer so that students could observe the data being recorded in real time while making decisions about the variables to be investigated. For physics students, this provided a more realistic and meaningful learning experience.
- Access to the library facilities provided the opportunity to read science journals online, which helped in the development of research questions for scientific investigations. They were also able to use the computer facilities with a range of software not available in their schools.
- Collaboration with academics in a variety of ways allowed students to explore the range of educational options available in physics, chemistry, and specialist mathematics. This included asking about possible careers in these fields, along with the pathways for achieving these goals. For many, there was a real connection between the science and mathematics they were learning and their application to the real world.

Educational Implications for the Future

The indications are that SMAF has addressed a critical need in the local community—enabling students from lower-SES localities to continue studying physics, chemistry, and specialist mathematics albeit on a university campus and not within their own schools. Importantly, access to these subjects in their final year of high school ensures that STEM pathways remain an option for these students as they consider post-compulsory education. Without access to SMAF, many of the students would have either selected other subjects available at their schools in grade 12, thereby moving away from STEM pathways, or moved to another school to continue with the science and mathematics subjects.

However, in addition to this critical outcome, SMAF has provided the opportunity for students to meet a larger group of like-minded individuals interested in and keen to pursue physics, chemistry, and specialist mathematics. This aspect is especially important for students from low-SES backgrounds who often lack suitable role models in their immediate family environments to support their academic achievement while easing their transition from school to university (Ainsworth 2002; Fullarton 2002). Given the observations of the teachers and the coordinator, SMAF facilitated a high degree of collaboration, sharing, and interaction among students regardless of school or SES. In specific instances, teachers from schools with students from lower-SES backgrounds commented to the SMAF evaluator about the positive changes noted in the motivation of these students as a consequence of their participation in the program.

Similarly, the engagement of teachers and school tutors in SMAF provided an opportunity to develop a professional learning community. The SMAF evaluator documented evidence that the sharing of resources, ideas, pedagogies, and academic discussions facilitated momentum for change in participating schools in the middle school. An important component to increase the broader involvement of teachers in this learning community has included the reselection of SMAF teachers at the beginning of the school year, along with positions vacated due to the natural attrition of SMAF teachers to other schools.

Critically, in thinking about SMAF more objectively in relation to the future, there are a number of educational issues to be considered in clarifying the longer-term goals and aspirations of the program. This is especially critical given the range of stakeholders involved and the potential for their individual agenda to cloud the intended purpose of SMAF. Initially, SMAF was devised in a manner such that the number of students involved was manageable, forming a sub-school (in a sense) on campus with the group creating their own microculture and structural unit. Importantly, all the needs of both SMAF students and teachers were provided to create a seamless transition between schools and the university environment. For example, the SMAF coordinator organizes the overall structure (timetable, teaching spaces, and access to technical staff for practicals), liaises between teachers and schools and between students and teachers, collects absentee lists of students for distribution to schools, and ensures that each SMAF teaching session runs smoothly. At another level, the coordinator negotiates university procedures, working with relevant personnel to address issues regarding the availability of appropriate teaching spaces (when there is already fierce competition with university classes) as well as student and teacher access to the library (e.g., need for a library card) and online electronic platform (e.g., with a university number required to enter the system) along with the various university protocols required for chemical and laboratory safety. A knowledge and understanding of secondary students, the way in which schools operate, and the expectations of the curriculum are vital to anticipating possible problems and to overcoming them in a seamless manner. Hence, the concern becomes: *What happens without the necessary money to fund this pivotal position?* It is clear from the current situation that schools are not in a place to contribute to this salary, and, although SKPT is funding it presently, there is no commitment to continue this indefinitely.

It might be expected that if the university (as a stakeholder) deemed that SMAF made a valuable contribution to the future of the university (e.g., greater numbers of students enrolling in STEM-related degrees), then a financial commitment to fund part of the salary of the coordinator would ensue. Although this is positive, there might be a push attached to this to further increase the numbers of students enrolled in SMAF as a means of building the cohort of university recruits. The critical issue here is that SMAF works because it has maintained relatively small numbers of students with a low teacher-to-student ratio. However, already there has been an increase in this ratio from approximately 1:19 (teachers to students in physics) in the initial year to 1:27 in 2012 (see Fig. 14.6). In the longer term, the impact of increased numbers of students requires additional teaching staff along with greater access to teaching spaces, which are already highly competitive within the university. Quite simply, up-scaling SMAF with greater student enrollment is highly likely to result in a program very different from the one initially intended. As such, the purpose for which SMAF was originally developed and implemented may be in a sense *highjacked* for another purpose, namely to increase university enrollments.

Although these are potentially significant educational concerns, another issue concerning SMAF requires deeper reflection. Intended to provide an opportunity for students who would otherwise not be able to continue with physics, chemistry, or specialist mathematics within their own schools, SMAF meets this agenda. Subsequently, schools might be expected to participate in SMAF for a few years while student engagement in these subject continues or grows, allowing the school to employ an appropriate subject-specialist teacher to oversee the teaching of these subjects in the school. At this time, they could withdraw from the SMAF program, making way for another school in need. Hence, it was not intended that schools abdicate their responsibility to provide grade-12 students with access to these subjects by allocating the required funds elsewhere, relying on SMAF to fill this gap. It is clear from the literature that, though a short-term solution, the ramifications for these schools in the longer term are substantive, as articulated by Linda Darling-Hammond (2004, p. 1943):

> A lack of mentors; high turnover of the untrained teachers, which creates continual hiring needs and instability; an erosion of professional development for other teachers in the building; and the instructional burden that results for other teachers to make up for the shortcomings of their colleagues.

As articulated in this quote, it is not just about the value contributed by specialist teachers in relation to student learning, but also equally about the impacts on other science and mathematics teachers within the school. For example, while junior high school teachers in Australia are expected to have science degrees (i.e., at least 3 years of university science), this may be predominantly in the biology area, not physics. So, these teachers may be confident in teaching biology-related topics to the junior students (e.g., genetics) while requiring guidance and mentoring when teaching physics-related topics (e.g., linear motion). Having access to discipline specialists in these fields within the school creates a productive learning environment and "collective knowledge" for both teachers and students. Therefore, while the short-term benefits for the schools participating in the SMAF program are favorable, it is crucial for the longer term that they are encouraged to develop the necessary expertise for these subjects "within their own schools". Of course, this might require the implementation of particular conditions with regard to involvement with SMAF, such as the expectation that schools might participate in the program only for a maximum of 4 years. Clearly, this places the Flinders Centre for Science Education in the twenty-first Century, as managers of the program, in a potentially precarious situation albeit an important educational stance to make for the "greater good."

Related to this point is the need for greater clarification of the criteria used for school inclusion in the SMAF program, given that it is likely to attract increasing interest with time. For example, one of the schools currently involved in SMAF is extremely large in that it provides the highest proportion of student participants. The same school also contributes at least three of the six SMAF teachers. Interestingly however, this school has the capacity to conduct senior classes in physics, chemistry, and specialist mathematics, yet has been a staunch supporter of and advocate for SMAF. In fact, without the involvement and in-kind support provided by this school in the initial stages, the program might have evolved quite differently. Essentially, the contribution of this school has been extremely positive even though there is a degree of domination within the program. Clearly, this creates a precarious balancing act between the valuable contribution provided by the school and the degree to which the school may be influencing the goals, values, and intentions of the program, given that it has the specialist capacity in its own right. For schools outside of SMAF with little understanding of the way in which the program evolved, there may be confusion about the criteria used for school selection. As such, careful consideration must be given to developing and articulating these criteria so that the process for selection and ongoing involvement in SMAF is clear to all schools and stakeholders.

Conclusions

Enhancing student involvement and participation in STEM-related subjects and careers is crucial to the economic prosperity of any country. Importantly, there must be a degree of equity here so that all students have the opportunity to pursue these pathways into the future. Unfortunately, for Australia and many other Western countries, equity is not assured for students with indigenous backgrounds, attending rural schools, or those from low-SES backgrounds (McGaw 2006). Although a variety of programs have been implemented in many countries *to close these equity gaps*, success has been mostly sporadic and random. However, it must start somewhere.

The Science and Mathematics Academy at Flinders (SMAF), discussed in this chapter, was devised and implemented from the ground up to enhance equity for students from low-SES locations by facilitating access to grade-12 physics, chemistry, and specialist mathematics. Based upon educational models of best practice, SMAF provides a high degree of ownership for the principals, teachers, students, and parents involved. Though successful in meeting its initial agenda, as articulated here, SMAF will face a pivotal challenge in the not too distant future: *How to maintain the original intent and integrity of the program as the pressure to up-scale and increase school participation increases.* In this instance, this challenge becomes particularly difficult given the range of stakeholders involved and the various pressures and expectations driving their agenda. However, there is also a strong commitment by those concerned to help close or at least reduce the gap between low SES and STEM involvement. Hence, this is merely the beginning.

Acknowledgements Appreciation is extended to the Southern Knowledge Transfer Partnerships at Flinders University for providing the finance necessary to fund the administration and resources required for the Science and Mathematics Academy at Flinders. We thank Dr. Suzanne Owen for undertaking the evaluation of SMAF, providing data for use in this chapter. Finally, we acknowledge the ongoing and sustained support of the principals, teachers, and students involved in SMAF.

References

Ainley, J., Kos, J., & Nicholas, M. (2008). *Participation in science, mathematics and technology in Australian education.* ACER research monograph, 63. Camberwell, Victoria: ACER. http://www.acer.edu.au/documents/Mono63_MathsSciTechSept08.pdf. Accessed 19th March 2012.

Ainsworth, J. W. (2002). Why does it take a village? The mediation of neighborhood effects on educational achievement. *Social Forces, 81*(1), 117–152.

Australian Academy of Science. (2011). Australian science in a changing world: Innovation requires global engagement. Position paper. www.science.org.au/reports. Accessed 3rd March 2012.

Campbell, C. (2005). Changing school loyalties and the middle class: A reflection on the developing fate of state comprehensive high schooling. *Australian Educational Researcher, 32*(1), 3–24.

Darling-Hammond, L. (2004). Inequality and the right to learn: Access to qualified teachers in California's public schools. *Teachers College Record, 106*(10), 1936–1966.

DEST. (2006a). *Audit of science, engineering and technology skills.* Canberra, ACT: Department of Education Science and Training.

DEST. (2006b). *Audit of science, engineering and technology skills.* Canberra, ACT: Department of Education Science and Training.

FASTS. (2002). Australian Science: Investing in the future. http://www.the-funneled-web.com/Response/FASTS-policy-2002.pdf. Accessed 2nd March 2012.

Friedlaender, D., & Frenkel, S. (2002). *School equity study documentation.* Los Angeles: UCLA Institute for Democracy, Education, and Access.

Fullarton, S. (2002). *Student engagement with school: Individual and school influences.* Longitudinal Surveys of Australian Youth (LSAY) research report, number 27. Melbourne: Australian Council for Education Research. http://research.acer.edu.au/lsay_research/31. Accessed 18th March 2012

Goodrum, D., Druhan, A., & Abbs, J. (2011). *The status and quality of year 11 and 12 science in Austrlralian schools.* Report prepared for the Office of the Chief Scientist. http://www.science.org.au/reports/documents/Year-1112-Report-Final.pdf. 3rd Feb 2012.

Gorard, S., Fitz, J., & Taylor, C. (2001). School choice impacts: What do we know? *Educational Researcher, 30*(7), 18–23.

Harris, K., Jensz, F., & Baldwin, G. (2005). *Who's teaching science? Meeting the demand for qualified science teachers in Australian secondary schools.* Report prepared for Australian Council of Deans of Science by the Centre for the Study of Higher Education. Melbourne, VIC: University of Melbourne.

Hillman, K. (2005). The first year experience: The transition from secondary school to university and TAFE in Australia. Longitudinal Surveys of Australia Youth Research Report, Number 40. http://research.acer.edu.au/lsay_research/44. Accessed 3rd May 2012.

Kelly, A. M., & Sheppard, K. (2009). Secondary school physics availability in an urban setting: Issues related to academic achievement and course offerings. *American Journal Physics, 77*(10), 902–906.

Kieffer, M. J. (2010). Socioeconomic status, english proficiency, and late-emerging reading difficulties. *Educational Researcher, 39*(6), 484–486.

Lara-Cinisomo, S., Pebley, A. R., Vaiana, M. E., Maggio, E., Berends, M., & Lucas, S. (2004). *A matter of class: Educational achievement reflects family background more than ethnicity or immigration.* Santa Monica, CA: RAND Corporation.

Lowell, L. B., & Salzman, H. (2007). Into the eye of the storm: Assessing the evidence on science and engineering education, quality, and workforce demand. http://www.urban.org/publications/411562.html. Accessed 19th March 2012.

Lyons, T., Cooksey, R., Panizzon, D., Parnell, A., & Pegg, J. (2006). *Science, ICT and mathematics education in rural and regional Australia: Report from the SiMERR National Survey.* Canberra: Department of Education, Science and Training.

McConney, A., & Perry, L. B. (2010). Science and mathematics achievement in Australia: The role of school socioeconomic composition in educational equity and effectiveness. *International Journal of Science and Mathematics Education, 8,* 429–452.

McGaw, B. (2006). Achieving equality and equity in education. http://www.unisa.edu.au/hawke-centre/events/2006events/ BarryMcGaw_presentation_Aug06pdf. Accessed 15th Dec 2011.

Obama, B. (2011). Weekly address: To win the future, America must win the global competition in education. February 19th, 2011. http://www.whitehouse.gov/the-press-office/2011/02/19/weekly-address-win-future-america-must-win-global-competition-education. Accessed 20th March 2012.

OECD Global Science Forum. (2006). *Evolution of student interest in science and technology studies: Policy report.* http://www.oecd.org/dataoecd/16/30/36645825.pdf. Accessed 3rd Oct 2011.

Office of the Chief Scientist (2012). Health of Australian science. http://www.chiefscientist.gov.au/wp-content/uploads/Report-for-web.pdf. Accessed 14th April 2012.

Osborne, J., & Collins, S. (2001). Pupils' views of the role and value of the science curriculum: A focus group study. *International Journal of Science Education, 23*(5), 441–467.

Panizzon, D. (2009). Science education in rural areas: Exploring the issues, challenges and future directions. In S. Richie (Ed.), *The world of science education: Handbook of research in Australasia* (pp. 137–162). Rotterdam: Sense Publishers.

Panizzon, D., Westwell, M., & Elliott, K. (2010). Exploring the profile of teachers of secondary science: What are the emerging issues for future workforce planning? *Teaching Science, 56*(4), 18–41.

Pearce, A., Flavell, K., & Dao-Cheng, N. (2010). Scoping our future: Addressing Australia's engineering skills shortage. Australian National Engineering Taskforce. http://www.anet.org.au/wp-content/uploads/2010/12/Scoping-our-futureWEB.pdf. Accessed 20th Dec 2011.

Ryan, C., & Watson, L. (2004). The drift to private schools in Australia: Understanding its features. Discussion Paper No. 479. Canberra, ACT: Australian National University.

Sjøberg, S., & Schreiner, C. (2005). How do learners in different countries relate to science and technology? Results and perspectives from the project Rose. *Asia Pacific Forum on Science Learning and Teaching, 6*(2), 1–17. http://www.ied.edu.hk/apfslt. Accessed 4th March 2012.

Thomson, S., & De Bortoli, L. (2008). Exploring scientific literacy: How Australia measures up. Resource document. http://www.acer.edu.au/news/2007_PISA.html. Accessed 20th Dec 2007.

Thomson, S., De Bortoli, L., Nicholas, M., Hillman, K., & Buckley, S. (2010). *Challenges for Australian education: Results from PISA 2009.* http://research.acer.edu.au/ozpisa/9. 8th June 2011.

Tytler, R., Osborne, J., Williams, G., Tytler, K., & Clark, J. C. (2008). Opening up pathways: Engagement in STEM across the primary–secondary school transition. http://www.deewr.gov.au/Skills/Resources/Documents/OpenPathinSciTechMathEnginPrimSecSchTrans.pdf. Accessed 7th Nov 2011.

Williams, D. J. (2010). School composition and contextual effects on student outcomes. *Teachers College Record, 112*(4), 1008–1037.

Chapter 15
The Road to Excellence: Promoting Access and Equity to Close the Achievement Gap Internationally

Julia V. Clark

*At the desk where I sit, I have learned one great truth.
The answer for all our national problems—the answer for all
the problems—the answer for all problems of the world comes
to a single word. The word is education.*

Lyndon Johnson

*If you can solve the education problem, you don't have to do
anything else. If you don't solve it, nothing else is going to
matter all that much.*

Alan Greenspan

The premise of this book is that all children deserve quality education. Each of the preceding chapters has endeavored to uncover the problems and issues associated with the achievement gap and the role education plays in closing it. Throughout the book, each chapter author has highlighted their country's problem and proposed what might be done and, in some instances, given images of how to do it according to research conducted, often focusing on a particular initiative, policy reform, or intervention. There is no single way to close the achievement gap.

The achievement gap refers to the disparity on a number of educational measures between the performance of groups of students, especially groups defined by gender, race/ethnicity, and socioeconomic status. The achievement gap can be observed on a variety of measures, including standardized test scores, grade point average, dropout rates, and college-enrollment and -completion rates. Various gaps exist between groups all over the globe. Closing the achievement gap has become a focal point of education reform efforts, and many nations have made it their mission to close the gap. Efforts to combat the gap have been numerous but fragmented, and have ranged from affirmative action and multicultural education to finance equal-

J. V. Clark (✉)
5600 Wisconsin Avenue, Suite 1205,
Chevy Chase, MD 20815, USA
e-mail: jvclark@starpower.net

J. V. Clark (ed.), *Closing the Achievement Gap from an International Perspective,*
DOI 10.1007/978-94-007-4357-1_15, © Springer Science+Business Media B.V. 2014

ization, improving teacher quality and school testing and accountability programs to create equal educational opportunities.

The achievement gap, and the challenges in closing it, is one of the most perplexing problems confronted by educational systems today. However, the achievement gap is not new. For decades, ethnic minority and poor students have disproportionately faced conditions that are a hindrance to achieving levels reached by the majority of students in areas of education in their cognitive development. In many countries, gender inequality is also an issue. There are differences in what happens in schools that are associated with differences in student achievement, including high standards with rigorous curriculum, and qualified and experienced teachers. Differences in such key components of schooling go along with differences in achievement among different student populations, at least when average achievement is compared. Research conducted around the world shows inequity in access to qualified teachers, facilities, resources, challenging mathematics and science curricula, and opportunities, and too few students enrolled in advanced coursework all contribute to the achievement gap in mathematics and science. School characteristics such as family income and mathematics and science course taking are all correlates of academic achievement. In addition, policies regarding teacher qualifications and curriculum vary from country to country, resulting in differences in access to high-quality teachers and higher-level mathematics and science courses.

From rural Australia to inner-city Washington, DC, education is the path to progress, both for individual citizens and for the nations they inhibit. Ensuring equality of educational opportunity, and the improved life chances that education can bring, is a matter of social justice. It is also an aid to political stability and, increasingly, an economic necessity. Education provides the basis for infrastructure development, adequate sustenance, health care, healthy and sustainable environments, civic and social order and growth, productive civil order and growth, and productive civil and international relations. Yet, across the globe, both rich and poor nations struggle with stubborn educational and social-mobility gaps that divide members of disadvantaged groups from their more privileged peers. The problem is not uniform. The size of the gaps, the severity of the deprivation, and the identity of the disadvantaged vary from culture to culture. Achievement levels that seem troublingly low in Canada look exceedingly high in Mexico.

Ensuring students' access to qualified teachers is an important goal of educational policy and reform in many countries. There is a lack of highly qualified teachers, especially in mathematics and science and other science, technology, engineering, and mathematics (STEM) fields; low social status and salary of teachers and their poor working conditions (as in Turkey); a lack of systemic induction programs; and inequitable distribution of qualified teachers between high-poverty and low-poverty schools. Many countries show major gaps in students' access to qualified teachers between wealthy and high-poverty students, and White and ethnic minority students. High-poverty students and ethnic minority students are twice as likely as wealthy and White students to be assigned novice teachers. They are also more likely to be taught by uncertified teachers, as in Africa and the United States.

The achievement gap is being addressed in various ways in many countries. High-achieving countries—Korea, Singapore, and England—have centralized systems of teacher education and certification with tighter regulatory control by the central government. Many countries around the world, like Australia, have centralized teacher hiring and distribution policies.

In Korea, inequities in achievement are due to economic disparity and gender inequities. In regard to STEM education, in comparison to students in other countries, Korean students routinely outperform students on mathematics and science standardized examinations. Even Korea's lowest-performing students score higher on mathematics and science standardized exams than the lowest-performing students in other countries. In addition, fewer than 6% of Korea's students fail to complete high school and more than 70% of students go on to enroll in two- or four-year university or vocational programs upon completing high school. With regard to access to elementary and secondary education opportunities, there are few discernible differences in either school attendance or academic achievement in terms of gender. However, fewer girls pursue tertiary education than boys and the gender disparity is even greater in graduate and doctoral programs than in undergraduate studies, so fewer women than men are entering the STEM workforce in Korea. National assessments do suggest a developing "gap" in achievement between students in different social class levels and between students living in different regions of the country. Researchers attribute differences in educational advancement between boys and girls to historical gender inequities and differences in achievement are attributed to economic disparities in different regions of the country (e.g., rural versus urban/suburban areas) and between the social classes. Economically disadvantaged families (especially those who tend to live in rural areas) cannot afford private tutoring fees (or access tutors), so these students are not as competitive on the annual national college entrance exam. These gender and class inequalities have their roots in sociohistorical, political traditions, which have helped to shape Korea's education system over the last 500 years. In addition to these issues, Korea is facing new challenges with regards to educating an increasingly culturally, ethnically, and linguistically diverse student population resulting from the development of new immigration policies seeking to create an international workforce.

In Singapore, there are disparities in educational outcomes between students of differing demographic characteristics, such as ethnicity and socioeconomic status. The achievement gap in Singapore is defined largely in terms of ethnicity, especially the ethnic Malay minority's persistent educational gaps vis-à-vis the ethnic Chinese majority and socioeconomic class. However, official data are often scant, especially in the case of socioeconomic gaps. The little data that are available for STEM achievement are based on ethnicity and highlight Malay students falling behind in mathematics and science at the primary level and in mathematics at the secondary level. This is despite the existence over the past three decades of various state-supported Malay community initiatives such as private tutoring schemes to boost overall Malay educational achievement. No evidence is provided about the effectiveness of these initiatives in reducing the achievement gaps.

The achievement gap in England is primarily defined in terms of socioeconomic status. There is a considerable and persistent gap in England in the rates of participation in higher education between those from higher and lower socioeconomic groups.

The gap is often expressed as the difference between those who are eligible for free school meals and the rest of the student population. There are also gender and ethnicity achievement gaps but the current political emphasis in England is on closing the socioeconomic status achievement gap. The STEM attainment gap during compulsory education appears to be driven by similar factors to the general attainment gap; however, there seems to have been less progress. More concerning is the gap in the proportions who continue studying STEM subjects in postcompulsory education, particularly between males and females. There has been more progress in closing the overall socioeconomic status achievement gap in the period between 1997 and 2010, using a diverse range of strategies. Many of these have run counter to the general thrust of increased market competition to drive school improvement.

In the Turkish context, achievement gaps refer to the differences of students' mathematics or science achievement depending on educational factors (e.g., school types or students' socioeconomic backgrounds), especially in the national context. These achievement gaps or differences can be observed on individual, group, school, and/or regional levels. In general, Turkey has a large achievement gap contributed to four major challenges which are all connected: quality differences in school types, competitive nationwide examinations, standardized and teacher-centered science and mathematics teaching from elementary school through college, and the effects of socioeconomic background differences on science and mathematics. There is a need to explore relationship patterns between these challenges.

In Turkey, the number of high school types is very high. While elementary school types are at expected levels (public and private schools), there are more than 20 types of secondary schools. In addition to this school-type variability, there are big STEM gaps, particularly in mathematics and science achievement gaps, between these schools. This challenge seems to be the biggest factor widening mathematics and science achievement gap in Turkey. As a solution to narrow differences between high schools, the Ministry of National Education (MONE) has started to decrease the number of school types at high school levels. In 2010, 350 general high schools were converted into Anatolian high schools, and by the end of 2013, all general high schools are going to be Anatolian high schools. In the near future, MONE is planning to convert Anatolian teacher high schools into Anatolian high schools. In the long run, the aim of MONE is to collect similar high schools under one umbrella and to narrow science and mathematics achievement gaps between high schools.

In Australia the achievement gap is identified in relation to socioeconomic status, Indigeneity and geographical location with students in rural and remote schools generally achieving significantly lower results than their peers attending city schools. Importantly, these three components interact with rural locations having a higher population of Indigenous students and populations with lower SES compared to many affluent suburbs in cities. These achievement gaps have been considered in government policy for educational planning in the past; however, access to international data sets like the Program for International Student Assessment (PISA) have provided the hard evidence around the extent of this achievement gap. While Australian students compare favorably to most other Western countries regarding their scientific literacy, significant gaps emerge in relation to Indigenous and low socioeconomic status (SES) students. For example, PISA 2009 highlighted that Indigenous students achieved a mean score that was 81 points below the Australian mean score while students from low socioeconomic backgrounds attained a mean score 96 points below the Australian mean. Unfortunately, these gaps are substantively equating to 2–2.5 years of schooling with an equivalent gap identifiable for PISA 2009 mathematics.

The achievement gap in China is geographical, economical, and political.

China's achievement gap is influenced by economic factors. In China, achievement has been narrowly defined as the *Gaokaao* scores. The unequal educational opportunities are marked between urban and rural areas, between the Eastern and the Central/Western regions, and between more and less prosperous provincial areas. In contrast with the United States, where race/ethnicity is the primary concern for the achievement gap, ethnicity is a much smaller factor in China. Instead, both the general achievement gap and STEM gap in China are influenced by economic favors. However, more important, both gaps are the product of social, political, and historical factors such as the Hukou requirement, the quota system, and policies of school choices.

South Africa's focus is on apartheid that created racial discrimination/segregation and fiscal inequality. At the height of the apartheid era, public spending on white children was around five times the amount for Africans. South Africa has participated in seven cross-country comparative studies and the results were that South Africa performed poorly compared to many of its more impoverished neighbors, and very poorly in relation to developing countries in other parts of the world. Poorer children receive schooling inferior to that of their more affluent peers. There are continuing large disparities in the outcomes produced by different kinds of schools linked to past racial affiliation.

A concern of Canada's public education is that of achievement differences in national, provincial/territorial, or group performance. In Canada there are interprovincial, gender, and indigenous status gaps. The critical issue for Canada has been the engagement and performance of Indigenous students (male and female). Canada has used international, national, and provincial (BC) test, participation, and graduation data to identify gaps across nations, provinces, schools, gender, and ethnicity (Indigenous/nonindigenous). The nation and province differences are apparent but not super interesting. The gender differences of the past have closed to where females perform as well or better than males, except their participation in mathematical sciences in the postsecondary level remains less than males. This is critical for the STEM pipeline issues. Attention is given to language arts, science, and mathematics. Technology and engineering are not a central part of the school curriculum in most provinces. To assist in closing the achievement gap, postsecondary institutions in BC participate in a program that provides scientists, engineers, technologists, and mathematicians as speakers for schools. The University of Victoria and other universities have offered informal or extracurricular in summer camps, after school, and Saturday programs on STEM to encourage and interest girls and boys in these disciplines and future careers. Similar events and internships have been offered during the Pacific Centre for Research in Youth, Science Teaching, and Learning (CRYSTAL) Project and other projects to allow young Indigenous people to learn their IKW and transition to WMS. These projects have real potential, but they are small in number and a rather recent arrival in STEM education platform. Clearly, the New Framework for Science Education K-12 (NRC, 2012) that recognizes science and engineering practices will provide a justification for more in- and out-of-school opportunities like these for Indigenous and nonindigenous girls and boys.

An inequality gap exists in the great cultural and socioeconomically diversity of the Mexican population, characterized by the large differences among those more or less marginalized, with a very high percentage of population in poverty and a high percentage in extreme poverty. There is concern on the achievement of students in rural, urban of high marginalization, and urban of low marginalization. By addressing this concern, it is hopeful that significant improvement toward closing the gap will occur.

In Brazil, the achievement gap focuses on the prevalence of socioeconomic differentials between Black and White Brazilians. There is the persistence of gaps in the quality of education provided to Blacks and Whites. In Brazil, test scores in the southeast top those in the northeast. In the United States, Blacks graduate from high school at far lower rates than their White and Asian peers. Around the world, in countries rich and poor, some groups succeed educationally—attending school, earning high grades and test scores, and completing college degrees—while others struggle, for a complex mix of historical, cultural, and economic reasons.

In the United States, there are two achievement gaps in its education systems. The first of these—well-documented, widely discussed, and the focus of education reform efforts for the past decades or longer—is the gap between the quality of schooling that most middle-class children (wealthy) get in America and the quality of schooling available for most minority and poor children—and the consequent disparity results. The second one is the global achievement gap—the gap between what even our best suburban, urban, and rural public schools are teaching and testing versus what *all* students will need to succeed as learners, workers, and citizens in today's global knowledge economy. There is also a large gap in STEM education in the United States compared to many countries. Achievement in the United States since its founding has been concentrated in just a few places, which has created a gap that correlates with economic and educational disparities observed today.

Concluding Remarks

Across the globe, both rich and poor nations struggle with stubborn educational and social-mobility gaps that divide members of disadvantaged groups from their more privileged peers. From country to country, the size of the gap between advantaged and disadvantaged groups varies greatly. The problem is not uniform. The size of the gaps, the severity of the deprivation, and the identity of the disadvantaged vary from culture to culture. A very important concern is the level of education and inequitable distribution of support to schools in low income and minority communities. Economics is a critical determinant to access. To improve student achievement and opportunities demand access and equity.

The near universality of the educational gap masks profound global diversity. Each society defines its disadvantaged groups differently—by geography, gender, race, ethnicity, class, religion, or language. Groups that succeed in one country may stumble in another. Immigrants do well in Canada, but struggle in Europe.

Data from the various countries suggest several conclusions. First, they confirm that socioeconomic status is a strong and consistent determinant of academic achievement in all countries and contribute to the achievement gap. The issues of equality and equity are important concerns in education reform and become salient political concerns. In practice, however, in many countries, the children who most need an extra educational boost are the least likely to get it. Lower-quality schooling appears to help perpetuate inequality rather than combating it.

The time is ripe for a concerted effort to enhancethe achievement of all of our students. By focusing our attention on closing the achievement gap, with immediate attention to STEM, we will be able to give local, State, and Federal educational agencies a call for action that is substantive, timely, and sufficiently targeted that it is reasonable to anticipate progress.

Although at different stages of development, many countries are implementing initiatives in an effort to close the gap. A few of them will be mentioned here. However, other initiatives are discussed in more details in the various chapters of this book. Korean society has a well-organized, highly developed system in place for educating their citizenry, and they have a long and powerful history on which they can reflect to apply lessons learned from the past to help continually reshape their future. In addition, Korean researchers can capitalize on the opportunities they have to learn from other countries that have been addressing achievement-related problems resulting from differences in gender, class, and race. The challenge for today's educational reformers is to not only maintain these gains but also continue to expand equitable opportunities for educational advancement. Finding ways to prevent the achievement gap from widening and to implement innovative polices that expand opportunities for all students to pursue STEM careers will be important areas for research in Korea over the next two decades.

Currently, researchers and policy makers in Korea are challenged with the task of developing innovative policies, research initiatives, and changes in science teacher practices and teacher education to address inequities in achievement and opportunity, including creating free after-school science programs designed to promote interest and positive attitudes in science for boys and girls; implementing free, high-quality distance learning programs for those students who are living in rural areas or who are economically disadvantaged and cannot afford private tutoring; and developing policies, research programs, and teacher education coursework to educate teachers about multicultural education strategies.

In an effort to close the achievement gap and increase STEM achievement in Singapore, various state-supported Malay community initiatives, such as private tutoring schemes to boost overall Malay educational achievement, are provided. Many studies including national and international examination results (e.g., PISA) show that Turkish elementary and middle school students' socioeconomic backgrounds were related to their mathematics and science achievement. To reduce the effect of socioeconomic background on student achievement, high-quality preschool education has been provided for children whose parents volunteer to take this education. Accordingly, MONE has recently accelerated the spread of preschool education across the country in order to resolve the problem of large differences in socioeconomic backgrounds. The number of preschools is continually increasing, and parallel with this development, parents' interest in preschool education is also greatly increasing.

In Australia, different strategies and projects have been implemented to target the achievement of Indigenous students with few demonstrating substantive improvements. While this is still a focus, considerable effort currently is targeted at addressing the inequity around socioeconomic status. Importantly, governments are concentrating on primary, secondary, and tertiary education, which initiated the project described in the Australian chapter within this volume.

Many initiatives are being considered in the United States in an effort to improve the quality of STEM education in schools and to make mathematics and science accessible to all students. These initiatives have gained wide distribution and have been or are being implemented by a wide range of U.S. schools, universities, industries, and science organizations. These comprehensive initiatives are the *No Child Left behind Act* (NCLB), *America COMPETES Act*, and *Race to the Top*.

Countries highlighted in this book have shown that access and equity are compelling factors in closing the achievement gap. Providing all students (rich and poor, male and female, Black, Hispanic, White, and other ethnic groups) with well-prepared and qualified teachers, adequate funding and resources, rigorous mathematics and science curriculum, opportunities with high expectations, will go a long way to promoting excellence and in closing the achievement gap.

Various countries have provided information on reform initiatives, policy implementation for closing the gap. The book strives to gather the best available evidence on the need for closing the achievement gap. The issues of access, equality, and equity in education reform become salient political concerns. The authors have shared interest in education. We hope that the story presented, and the findings and analysis in each book chapter and the overall issues of the achievement gap will benefit not only the further development of each region but also other international communities.

Education is the key to developing the intellectual capacity of our children. Nothing is more vital to our country's future than ensuring that all students receive a quality education. Gains in student achievement can most likely be realized wherever along the development continuum the effort is made. The success of education in this century and the century to come will depend on the extent to which we educate all of our children and the achievement gap is closed so that No Child is Left Behind.

We live in an era in the history of nations when there is a greater need than ever for coordinated political action and responsibility. It is my hope that this book will serve to elevate an international dialogue on the critical issues associated with the achievement gap and provide concrete examples to foster a solution. Perhaps most importantly, our country will keep the goal of closing the achievement gap and raising the achievement performance of all the children in STEM at the forefront of their attention. In this way, we would be working together to solve a problem of global significance.

As reported by the Education Commission of the States (1990) in the *Education Agenda 1990,* to serve the needs and aspiration of all Americans, and to fulfill the promise of American democracy, the United States education system must display and encourage inclusiveness for all racial, ethnic, and cultural groups. I believe this could be applied to all of the countries presented in this book. Curricular and teacher quality and equality and equity need to be considered. Research suggests that when these factors are applied to practice, they can improve learning.

We have much to be proud of in our education system, but we ought always to be seeking to address our weaknesses and to improve our performance. Overall, it is understood that this is not a panacea, but it is the belief that it will go a long way toward the pursuit of excellence and in closing the achievement gap.

The Contributors

Motoko Akiba is Associate Professor in the Department of Educational Leadership and Policy Studies at Florida State University. Her research expertise is in teacher policy, teacher learning, and comparative and international education. Dr. Akiba's publications include "Improving Teacher Quality: The U.S. Teaching Force in Global Context" (Teachers College Press, 2009), "Teacher Salary and National Achievement: A Cross-National Analysis of 30 Countries" (*International Journal of Educational Research,* 2012), and "Professional Learning Activities in Context: A Statewide Survey of Middle School Mathematics Teachers" (Education Policy Analysis Archives, 2012). She is currently working on a National Science Foundation (NSF)-funded project titled, "Work Contexts, Teacher Learning Opportunities, and Mathematics Achievement of Middle School Students."

Jake Anders studied Philosophy, Politics and Economics at New College, University of Oxford, where he earned an academic scholarship. He is now a Ph.D. researcher at the Institute of Education, University of London, working on issues surrounding access to Higher Education. His article, "The Link Between Household Income, University Applications and University Attendance," was published in a recent special issue of the journal, *Fiscal Studies*, on social mobility. With his supervisors, Lorraine Dearden and John Micklewright, he is also working as part of a Nuffield Foundation-funded project, "Higher Education Funding and Access: Exploring Common Beliefs," based at the Institute of Education and the Institute for Fiscal Studies.

John O. Anderson (Ph.D., 1978, University of Alberta) is Professor and Chair of the Department of Educational Psychology and Leadership Studies, Faculty of Education, University of Victoria, British Columbia, Canada. John taught secondary school sciences in Manitoba and Victoria (Australia). His main area of research interest is educational measurement, with a current focus on large-scale student assessment at the provincial, national, and international levels. John led the secondary data analysis project in the Pacific CRYSTAL project and conducted research and analysis on provincial (British Columbia), national, and international student assessment programs. The results from these endeavors served as the basis for the coauthored chapter on achievement gaps in Canada.

Seung-Urn Choe is Professor in Earth Science Education at Seoul National University in Seoul, Republic of Korea. Professor Choe is currently serving as Chair for the Earth Science Education Department. His research focuses on student's developing and practicing models co-constructively to learn science in the context of inquiry in the science classroom. As an astronomer, he is also interested in developing learning modules based on astronomy and mathematical knowledge for use with all students, including science-gifted students. Choe is Director of the Gwanak Institute of Gifted Education at Seoul National University, where he manages research projects for gifted education. Choe is also Chief Editor of the *Journal of the Korean Society of the Gifted*. He recently published a Korea language book entitled "Understanding Astrophysics Using Excel."

Edmund W. Gordon is the John M. Musser Professor of Psychology, Emeritus at Yale University; Richard March Hoe Professor Emeritus of Psychology and Education at Teachers College, Columbia University; and Director Emeritus of the Institute for Urban and Minority Education (IUME) at Teachers College. Dr. Gordon also serves as Chair of the Gordon Commission on the Future of Assessment in Education.

Professor Gordon's career spans professional practice, scholarly life as a minister, clinical and counseling psychologist, research scientist, author, editor, and professor. He has held appointments at Howard, Yeshiva, Columbia, Yale, and City University of New York. Additionally, Dr. Gordon has served as visiting professor at City College of New York and Harvard University. From July 2000 to August 2001, he was Vice President for Academic Affairs and Interim Dean of Faculty at Teachers College, Columbia University. Dr. Gordon has been recognized as a preeminent member of his discipline. He is an elected Fellow of the American Psychological Association, American Society of Psychological Science, and American Association for Orthopsychiatry, and he is Fellow and Life Member of the American Association for the Advancement of Science. In 1968, he was elected to be a member of the National Academy of Education. The "Edmund W. Gordon Chair for Policy Evaluation and Research" was created by the Educational Testing Service to recognize his contributions to developments in education including Head Start, compensatory education, school desegregation, and supplementary education. In 2005, Columbia University named its Harlem campus the Edmund W. Gordon Campus. Dr. Gordon has been named one of America's most prolific and thoughtful scholars. He has authored over 200 articles and 18 books. He has been married to Susan G. Gordon, M.D., since 1948. They have raised four children, whom they claim as their most important achievements.

Linda Darling-Hammond is Charles E. Ducommun Professor of Education at Stanford University, where she is Co-Director of the Stanford Center for Opportunity Policy in Education. She launched the Stanford Educational Leadership Institute and the School Redesign Network. She has also served as faculty sponsor for the Stanford Teacher Education Program. She is former president of the American Educational Research Association and Member of the National Academy of Education. Her research, teaching and policy work focus on issues of school restructuring, teacher quality, and educational equity. From 1994 to 2001, she served as executive director of the National Commission on Teaching and America's Future, whose 1996 report, "What Matters Most: Teaching for America's Future," led to sweeping policy changes affecting teaching and teacher education. In 2006, this report was named one of the most influential reports affecting the US education, and Darling-Hammond was named one of the nation's ten most influential people affecting educational policy over the last decade. Darling-Hammond is the author of over 400 publications, including "The Flat World and Education: How America's Commitment to Equity Will Determine Our Future" (2010) and "Powerful Teacher Education" (2006). She holds a BA magna cum laude from Yale University and an Ed.D. (Urban Education) from Temple University. She began her career as a school teacher. She has served as Director and Senior Social Scientist for the RAND Corporation's Education and Program and William F. Russell Professor of Education and Co-Director, National Center for Restructuring Education, Schools and Teaching at Teachers College, Columbia University. She is a member of the boards of directors of the National Council for Educating Black Children, Alliance for Excellent Education, National Commission on Teaching and America's Future, and Center for Teaching Quality. She was education adviser to Barack Obama during the 2008 election campaign, and she led his education policy transition team.

Chan-Jong Kim is Professor in Earth Science Education at Seoul National University (SNU) in the Republic of Korea. Prior to joining SNU, he had worked for Korean National Board of Educational Evaluation, and Chongju National University of Education as professor. Kim received his doctoral degree from the University of Texas at Austin in 1989. He served as Chairperson of International Geoscience Education Organization (IGEO) from 2006 to 2010 and as Chairperson of Advisory Committee for International Earth Science Olympiad (IESO) from 2004 to 2010. He has been trying to understand science learning in formal and informal settings with sociocultural perspectives. His current research focuses on co-construction of scientific models in science classrooms and understanding and improving scaffolding in science museums for visitors' better learning. He is trying to improve science learning and teaching in Korean science classrooms by introducing co-constructing scientific models among students and teachers considering Korean classroom culture, including Confucianism. He is also studying the characteristics of various media in science museums, including panels, hand-held devices, worksheets and docents in terms of scaffolding.

Youngsun Kwak is a fellow researcher at Korea Institute for Curriculum and Evaluation (KICE) in Seoul, Republic of Korea. Her current research focuses on developing science teacher education that promotes teacher professionalism and instructional consulting based on curriculum revision in Korea. She is particularly interested in exploring ways to improve science teaching and science curriculum in the competency-based curriculum context. She recently published qualitative research in science education. Kwak taught secondary school general and Earth science in the Seoul public high school district, during which time she also taught science for the gifted children in Korea. After 7 years of teaching in Korea, she went to the USA to get a better perspective in science teaching and teacher education. In 2001, Kwak completed her doctoral research through Ohio State University while serving as a research associate in the Eisenhower National Clearinghouse (ENC). Her dissertation research is titled "Profile Change in Pre-service Science Teacher's Epistemological and Ontological Beliefs about Constructivist Learning: Implications for Science Teaching and Learning." For this study, she investigated preservice teachers' understandings of the ontology and epistemology underlying constructivist notions of learning. Kwak researches science teachers' pedagogical content knowledge through teaching consultations based on videotaped science lessons.

Guodong Liang completed his Ph.D. at the University of Missouri, with an emphasis on policy studies, in 2011. He has worked as a postdoctoral researcher in the Department of Educational Leadership and Policy Analysis at the University of Missouri. His research areas include comparative and international education, teacher policy, and social justice, and his publications include "Performance-Related Pay: District and Teacher Characteristics (*Journal of School Leadership*, 2011)." He is a former Barbara Jackson Scholar and David L. Clark Scholar.

Ricardo A. Madeira is Assistant Professor of Economics at the University of Sao Paulo (USP). Madeira holds a Ph.D. in Economics from Boston University, an M.A. in Economics from Fundação Getúlio Vargas de São Paulo (FGV-SP) and a B.A. in Economics from USP. His research agenda is concentrated on development microeconomics and evaluation of public policies. Before joining USP, Madeira was a consultant for the World Bank. In the last few years, Madeira has been in charge of evaluations of policies undertaken by Brazilian NGOs and the Inter-American Development Bank (IDB).

Sonya N. Martin is Assistant Professor in Science Education at Seoul National University in Seoul, Republic of Korea. Prior to moving to Korea, Martin was a tenured faculty member at Drexel University in Philadelphia, where she was principal investigator of a National Science Foundation (NSF)-funded (HRD 1036637) study examining the intersections of gender, ethnicity, and language learning in the context of middle school science instruction. In G-SPELL (Gender and Science Proficiency for English Language Learners), she focused on identifying science teacher practices that promoted language learning in the context of science inquiry with English Language Learners. She has interest in exploring ways to improve collaborative teaching between content and ESL teachers to promote beneficial science teaching practices for all students. In addition, she became interested in the science education experiences of the students in the study who had recently immigrated to the Philadelphia from Asian countries. To learn more about science education in Asia, Sonya accepted an international faculty position at Seoul National University and moved to Korea in 2011, where she is learning Korean and is engaging in research with colleagues in Korea and in Asia. She serves as an editorial board member for several journals, including *Journal of Research in Science Teaching*, *Research in Science Education*, and *Cultural Studies of Science Education*, and she recently co-edited "Re-visioning Science Education from Feminist Perspectives: Challenges, Choices and Careers."

Armando Sánchez Martínez is a Mexican education consultant with master's Degrees in Physical Chemistry (UNAM), Education (UAEMorelos) and Teachers' Training for all education levels, and he completed a doctoral study at the Education Institute, London University. He worked from 1993 to 2004 in the Federal Public Education Ministry (SEP) in curriculum design and training programs, coordinating the national textbooks and other educational materials of science for basic education and the curricular field for secondary education reform. He also participated in international meetings. Since 2005, he has been working in Editorial Santillana, and he has been the High School Manager for textbooks since 2008. His articles in education magazines include "Demógraphie scolaire et réforme de l'enseignement y Curriculum scientifique et innovation" (*Revue internationale d'éducation*), "Educando para educar" (*División de Estudios de Posgrado de la BECENE de S.L.P.*). His co-authored books include "PISA en el aula" (INEE, 2008), "¿Cómo promover el interés por la cultura científica?" (UNESCO, 2005), "La enseñanza de las ciencias en la escuela secundaria como parte de la educación básica" (SEP, 2003), and "¿Qué educación secundaria para el siglo XXI?" (UNESCO, 2002).

Todd M. Milford (Ph.D., 2009, University of Victoria) is a Lecturer in the Art, Law, and Education Group at Griffith University, Mt. Gravatt, Queensland, Australia. Todd taught elementary and secondary science and special education (British Columbia), as well as in the online environment. Since 2005, his postsecondary teaching has been primarily in the areas of science education, mathematics education, and classroom assessment. Todd's research continues to be varied; however, the constant theme is the use of data and data analysis to help teachers and students in the classroom. He was part of the Natural Sciences and Engineering Research Council (NSERC)-funded Pacific CRYSTAL project and was awarded the Andy Farquharson Award for Excellence in Graduate Student Teaching (UVic).

Johan Muller holds a Ph.D. in Education from the University of Cape Town. He taught at Rhodes University, the University of Limpopo, and the University of the Witwatersrand, where he became the first head of the Education Policy Unit. In 1990, he was appointed Chair of Curriculum [WORDING?] at the University of Cape Town. He is currently Visiting Professor at the University of London's Institute of Education. He has authored, co-authored, and edited several books and has published numerous papers and book chapters.

Brian W. Neill (Ph.D. candidate, University of Victoria, British Columbia, Canada). Brian's doctoral research focuses on the development of science and technology experiences for indigenous students. Brian was a science teaching consultant in China on a Canadian International Development Agency project (Strengthening Basic Education in Western China) with Tibetan, Uighur, Kazak, and Hui indigenous peoples. He has taught middle school science and mathematics in China (Beijing) and Canada (Ontario, Alberta, and British Columbia) for over 35 years, with half of this time as Department Head. He has developed media, distance education, and textual materials for elementary and senior secondary science courses. Brian was a research scientist in ecological parasitism prior to beginning his teaching career; he wrote a seminal research paper entitled "Spermatogenesis in the Hologonic Testis of the Trichuroid Nematode, *Capillaria hepatica*" in *Journal of Ultrastructure Researc* (1973).

Debra Panizzon is Associate Professor Biology Education at Monash University in Melbourne, Australia, with research interests in cognition, student acquisition of scientific concepts, rural and regional education, and assessment. Debra was Deputy Director for the Flinders Centre for Science Education in the 21st Century at Flinders University in South Australia. Here, much of Debra's work became focused around STEM policy, working as a conduit between stakeholders and school communities to overcome emerging STEM-related issues in the state. Much of her writing at this time reports on projects implemented by the South Australian government to enhance student and teacher engagement with science and mathematics. This work built upon earlier experiences as Deputy Director for the National Centre of Science, Information and Communication Technology, and Mathematics Education for Rural and Regional Australia (SiMERR) at the University of New England. While working with preservice secondary teachers, Debra was involved in collaborative research projects involving secondary science and mathematics teachers and their students. Debra has participated at international and national conferences and currently reviews for two international science journals.

Leslee Francis Pelton (Ph.D., 1989, Brigham Young University) is Associate Professor and Chair of the Department of Curriculum and Instruction, Faculty of Education, University of Victoria, British Columbia, Canada. Leslee taught middle and secondary school mathematics in Alberta and Utah; postsecondary mathematics at BYU, where she coordinated pre-calculus programs focusing on student assessment; and mathematics education courses for preservice teachers at UVic. Leslee focuses on developing and assessing students' problem-solving skills through hands-on activities and investigations integrating mathematics and science, as well as communication of their understanding of mathematical concepts, outreach education, and STEM activities both within the curriculum and as extracurricular programs. She has engaged in evaluation and development of mathematics curricula and frameworks through provincial task forces and the Western and Northern Canadian Protocol for Education research project. In collaboration with Tim Pelton, she has designed, deployed, and evaluated iOS apps to support mathematics and science investigations and concept consolidation.

Tim Pelton (Ph.D., 2002, Brigham Young University) is an Associate Professor in the Department of Curriculum and Instruction, Faculty of Education, University of Victoria, British Columbia, Canada. Tim taught middle and secondary school mathematics (British Columbia) and currently teaches postsecondary mathematics pedagogy and educational technology courses for preservice and in-service teachers. Tim focuses on helping children make sense of mathematics and on examining the potential of new technologies to support mathematics and science learning. He has engaged in research in provincial foundations and problem gambling assessments, concept inventory development, outreach education, and technology applications. He has designed, developed, and validated iOS applications that explore mathematics and science concepts and provide individualized experiences for students to achieve mastery. Tim incorporated various technology-based enrichment activities (robotics, comics, clickers, geotrekking) into grades 3–11 mathematics classes as Co-Principal Investigator (with Dr. Leslee Francis Pelton) of the project, Making Mathematics and Science Meaningful: Investigating the Effects of Enrichment Activities on Students, funded by NSERC through the Pacific CRYSTAL initiative and by the Constructivist Education Resources Network (CER-Net) through the Faculty of Education.

Marcos A. Rangel is Associate Professor of Economics and Public Policy at the University of Sao Paulo (USP) and Research Affiliate at the Population Research Center (NORC/University of Chicago) and the Abdul Latif Jameel Poverty Action Lab (MIT). His research agenda is concentrated on development microeconomics and economic demography. Before joining USP, Rangel held an Assistant Professor position at the University of Chicago's Harris School of Public Policy Studies (2004–2009). Rangel holds a Ph.D. in Economics from the University of California at Los Angeles (UCLA), and B.A. and M.A. degrees in Economics from PUC-Rio/ Brazil. Rangel was awarded the Article of the Year prize by the Royal Economic Society (UK) in 2007.

Jason TAN is Associate Professor of Policy and Leadership Studies at the National Institute of Education, Singapore. He obtained his doctoral degree in comparative education at the State University of New York at Buffalo. His research interests include educational policy and educational reform. Among his publications are "Going to School in East Asia" (co-edited with Gerard Postiglione) and "Education in Singapore: Taking Stock, Looking Forward."

Nick Taylor holds a master's degree in Geology and a Ph.D. in Mathematics Education from the University of the Witwatersrand. He taught math and science at the high school level for 10 years, followed by a period as subject advisor in mathematics in Soweto, 1984–1988. He conducted policy research at the Education Policy Unit at the University of the Witwatersrand before being appointed Executive Director of the nonprofit organization Joint Education Trust Education Services in 1993. In 2012, he was invited to head the National Education Evaluation and Development Unit by the Minister of Basic Education. He has co-authored three books on school improvement and is currently Visiting Researcher at the University of the Witwatersrand.

Mustafa Sami Topcu is Associate Professor of Science Education in the Department of Elementary Science Education at Muğla Sıtkı Koçman University, Turkey. He received a B.S. in science education from the 19 Mayıs University, Turkey, a M.Sc. in science education from the 9 Eylul University, Turkey, and a Ph.D. in science education from the Middle East Technical University, Turkey, in 2008. He also studied as a research scholar at University of Florida in the USA, where he completed part of a doctoral dissertation in 2007. He worked as an elementary school teacher of science in Izmir, Turkey. After teaching science in elementary school, he worked as a research assistant in the Middle East Technical University. As a Muğla Sıtkı Koçman University faculty member, Topcu teaches courses for an undergraduate science education program and for masters' and a doctorate elementary science education programs. His research interests are teachers' epistemological beliefs and educational practices, students' achievement gaps in science, argumentation, and socioscientific issues.

Geoff Whitty was educated at the University of Cambridge and the Institute of Education, University of London, UK. He taught in primary and secondary schools before working at Bath University, the University of Wisconsin-Madison, King's College London, Bristol Polytechnic and Goldsmiths College, University of London. He joined the Institute of Education, University of London as the Karl Mannheim Professor of Sociology of Education in 1992 and served as its Director between 2000 and 2010. He is Professor in the School of Management at the University of Bath, UK. His publications include "Making Sense of Education Policy" (Sage, 2002) and "Education and the Middle Class" (Open University Press, 2003). He is past president of the British Educational Research Association. In the Queen's Birthday Honors 2011, he was awarded the CBE for services to teacher education.

Larry D. Yore (Ph.D. 1973, University of Minnesota) is a University Distinguished Professor Emeritus in the Department of Curriculum and Instruction, Faculty of Education, University of Victoria, British Columbia, Canada. Larry taught secondary and elementary school science and served as a K–12 science coordinator, secondary science department head, and instructor and supervisor of student teachers (Minnesota).

In more than four decades of postsecondary teaching and research, he has engaged in developing provincial science curricula, national science frameworks, and national K–12 assessment projects in North America. His research focuses on the roles of language (reading, writing, representing, and metacognition) in science and science education and the ways language arts affect scientific inquiry. Larry received the 2005 Association for Science Teacher Education's Science Teacher Educator of the Year Award and the 2012 National Association for Research in Science Teaching's Distinguished Contributions to Science Education through Research Award.

Gaoming Zhang is Assistant Professor in the Department of Teacher Education in the School of Education at the University of Indianapolis. Her research interests include comparative education, technology integration, and language learning. Her work has appeared in the *Journal of Early Childhood Teacher Education, the Asia Pacific Journal of Education, EDUCAUSE Review, On the Horizon,* and the *International Encyclopedia of Education.*

Yong Zhao is Presidential Chair and Associate Dean for Global Education, College of Education at the University of Oregon. He is a fellow of the International Academy for Education. His research interests include educational policy, computer gaming and education, diffusion of innovations, teacher adoption of technology, computer-assisted language learning, and globalization and education. Zhao has published over 20 books and 100 articles. His most recent books include "World Class Leaders: Educating Creative and Entrepreneurial Students," "Catching Up or Leading the Way: American Education in the Age of Globalization" and the "Handbook of Asian Education."

CPSIA information can be obtained at www.ICGtesting.com
Printed in the USA
LVOW10*2129060314

376391LV00008B/31/P